GROUP THEORY

The Application to Quantum Mechanics

PAUL H. E. MEIJER
The Catholic University of America, Washington, D. C.

EDMOND BAUER
Laboratoire de Chimie Physique, Paris

DOVER PUBLICATIONS, INC.
Mineola, New York

Bibliographical Note

This Dover edition, first published in 2004, is a corrected republication of the second (1965) printing of the work first published by North-Holland Publishing Company, Amsterdam, in 1962. Dr. Meijer has written a new preface specially for this edition.

International Standard Book Number: 0-486-43798-1

Manufactured in the United States of America
Dover Publications, Inc., 31 East 2nd Street, Mineola, N.Y. 11501

PREFACE TO THE DOVER EDITION

I am very happy that Dover is republishing this book, because it gives me the opportunity to mention a few things about the life of my coauthor Edmond Bauer (1880–1963). Born in Paris, he worked at the Ecole Supérieure de Physique et Chimie Industrielles de la Ville de Paris (ESPCI) under Langevin in 1905. For his biography see ref (1). Travelling through Europe, he met and became a friend of Paul and Tatiana Ehrenfest (2) and was inspired by them. When appointed in 1919 in Strasbourg he became the first to teach quantum theory in France (3). He attracted all of two students. The students in Strasbourg were discouraged by his colleagues from following his classes since he taught such revolutionary things as the quantum theory. However, he participated in a crucial discussion on the question of whether the revolutionary assumption by Planck was the only way to explain the observations.

Bauer had a tumultuous life: serving in the French army in the First World War, he was shot through the lung, and he had to flee and hide in World War II. Despite all this he remained a very mild mannered scholar. His experiences in World War I are described in the obituary by K. Darrow (4). At his grave Francis Perrin said: "He did not pontificate, and spoke to the young students as an equal." He was also an accomplished pianist.

In the Oral History in the Niels Bohr Library, Bauer stated (5) that "Before World War I physics was quite old fashioned . . ." and added "I always said that I have a great gratitude to my teachers because they taught me so badly that I learned to work by myself." His complete scientific biography can also be found in a publication of 1935 (6).

Bauer's book on Group Theory resulted from a request by the members of the Institut Henri Poincaré that he give a series of lectures in 1932. Yvette Cauchois (7) mentioned in her eulogy in 1963 that the 1962 English edition of his Group Theory book of 1933 showed in a striking way how up-to-date it still was. I hope that this will also be the case in 2004!

PAUL MEIJER

Washington, D.C.
May 25, 2004

References

1) *Dictionary of Scientific Biography*, ed. C. C. Gillispie. Vol. 1, pp. 519–520.

2) M. J. Klein, *Paul Ehrenfest*. Amsterdam: North-Holland Publishing Company, 1970, p. 50

3) Maria C. Bustamante, "Rayonnement et quanta en France 1900–1914." *Physis, Revista Internazionale de Storia della Scienza,* Vol. 39 (2002) nuova ser. Fasc. 1, pp. 63–107.

4) K. Darrow, *Physics Today,* June 1964.

5) AIP, Oral History Project, Tape No. 42, transcript p. 3.

6) *Notices sur les travaux scientifiques de E. Bauer.* Paris: Imprimerie Hemmerle Petit, 1935.

7) Y. Cauchois, in: M. Letort, G. Champetier, J. Wyart et al., "Hommage à Edmond Bauer." *Journ. de chimie physique*, **61** (1964), pp. 963–967.

PREFACE

Seldom has an application of so-called pure mathematics to mathematical physics had more appeal than the use of group theory to quantum mechanics. Almost every student in this subject, after going through the necessary theorems, felt the satisfaction of overlooking a broad field, mastering it in its complete generality, as an award to his efforts.

In recent years the availability of tables of coefficients has increased the applicability of many ideas introduced one or two decades ago and the number of papers applying the results of representation theory has been steadily increasing.

The application of group theory to problems in Physics can be classified in two types. As an example of the first type we mention the considerations based on the symmetry of a crystal used to reduce the 6 by 6 stress-strain-matrix (the generalization of Hooke's law). An example of the same type is the heat conduction tensor in a crystal. Instead of nine components the number of different elements is reduced by the symmetry of the crystal and still further reduced by the Onsager relations (which are based on microscopic time reversal). Still another example of the first type is the set of piezo-electric constants.

The second type of problems are those eigenvalue problems where the differential equation or the boundary are of such a geometric nature that certain rotations or translations leave the problem unchanged. In this case it may happen that the eigenvalue connected with the solution of the problem is degenerated; that is, more than one eigenfunction belongs to the same eigenvalue.

The central problem of the book is the study of this second type through the transformation properties of these eigenfunctions. In the first cases the application of "group theory" is hardly more than the application of symmetry considerations. In the second case the application of group theory, or actually the application of representation theory, is a much more essential matter. The general idea of the transformation induced in function space by a rotation (or translation) in configuration space as explained in Chapter 4 is not only useful in quantum mechanics but in any other eigenvalue problem such as vibrating systems (molecules or lattices) or waveguides as well.

The crucial point in the developing of representation theory is Schur's Lemma. The proof has been illustrated with a symbolic diagram and in the subsequent sections the theory is developed on the basis of this lemma.

Great value is attached to represent the ideas in a geometrical fashion: For instance the similarity transformation is described as a rotation in multidimensional function space and the reduction is described in terms of mutual orthogonal spaces.

Although there are many books written on group theory as well as on the connection between physics and group theory, the number of books of introductory nature are only very few. The general references are chosen with emphasis on clarity and readability and are mainly mentioned for further study in this field.

The book is based on a French monograph entitled "Introduction à la théorie des groupes et ses applications en physique quantique" which appeared in 1933 in the Annales de l'Institut Henri Poincaré. Chapters 1 through 5 are a translation with addition of the monograph. The subjects treated in the additional chapters deal mainly with developments since then. The basic ideas of the application of group theory to quantum mechanics have not considerably changed and hence a fairly literal translation of the material of the first edition is still valuable today.

I would like to quote from the introduction to the first edition the following statement: "The pages that follow are simple introduction, having as main goal to familiarize the reader with some new concepts and to permit him to read the original papers of Weyl and Wigner with less difficulty."

I would like to thank Mrs. Peretti, Drs. Morrison and Barry and Mrs. Mielczarek for their cooperation.

CONTENTS

CHAPTER 3. GROUP THEORY

Chapter 5. Rotations in 3-dimensional space: group $\mathscr{D}_3$

Chapter 6. Continuation of the theory of the rotation group

Chapter 7. space groups

Chapter 8. finite groups

CHAPTER 1

VECTOR SPACES — UNITARY GEOMETRY

1. *n*-dimensional Vector or Affine Spaces

1.1. *n*-DIMENSIONAL VECTOR SPACES

They are obtained by generalizing the properties of ordinary space which is an abstraction made from the concept of measuring lengths. Let us remember that in affine geometry the vectors $\boldsymbol{x}$, $\boldsymbol{y}$, . . . are defined only with regard to the operations of addition $\boldsymbol{x}+\boldsymbol{y}$ (commutative and associative), subtraction $\boldsymbol{x}-\boldsymbol{y}$, and multiplication by a complex number (distributive and associative). The n vectors, $\boldsymbol{e}_1, \boldsymbol{e}_2, \ldots, \boldsymbol{e}_n$, are linearly independent if one cannot satisfy the equation

$$x_1\boldsymbol{e}_1+x_2\boldsymbol{e}_2+\ \ldots\ x_n\boldsymbol{e}_n = 0,$$

unless *all the numbers* x_n *are equal to zero.*

An n-dimensional vector space is defined by n linearly independent *basis vectors.* It consists of the set of vectors, obtained by taking all possible linear combinations from these n basis vectors

$$\boldsymbol{x} = \sum_{i=1}^{n} x_i\boldsymbol{e}_i. \tag{1.1}$$

In ordinary space, $n = 3$, the $\boldsymbol{e}_i$ are *unit vectors* laid out in three arbitrary directions, the coordinate axes, and x_i, the components of the vector $\boldsymbol{x}$, are real. In quantum theory the quantities considered are in general complex; thus one is led to take complex x_i.

1.2. TRANSFORMATION OF COORDINATES

If $\mathfrak{R}_n$ is a vector space spanned by n basis vectors $\boldsymbol{e}_i$, let us select n other independent vectors $\boldsymbol{e}'_k$. From (1.1) the $\boldsymbol{e}'_k$ will be linear combinations of the $\boldsymbol{e}_i$ (as all vectors of $\mathfrak{R}_n$ are). Thus we will have

$$\boldsymbol{e}'_k = \sum_i \boldsymbol{e}_i s_{ik}, \tag{1.2}$$

where $\mathrm{S} = (s_{ik})$ is the *transformation matrix* with n rows and columns.

A fixed vector $\boldsymbol{x}$ will have the coordinates x_i and x_i' in these two systems. Equations (1.1) and (1.2) give:

$$\boldsymbol{x} = \sum_i x_i \boldsymbol{e}_i = \sum_k x_k' \boldsymbol{e}_k' = \sum_{ik} x_k' \boldsymbol{e}_i s_{ik},$$

from which

$$x_i = \sum_k s_{ik} x_k', \tag{1.3}$$

or

$$x_k' = \sum_i s_{ki}^{-1} x_i. \tag{1.3a}$$

Notice the difference between (1.3a), the transformation formula for the components of the vector, and (1.2), the transformation of the unit vectors. One transformation is the *transposed inverse* of the other.

The notion of a vector can be generalized to tensors, quantities that have n^k components instead of n components. k is the rank or order of the tensor. By definition these components transform like the products of vector components. For instance a tensor of the second rank has n^2 components that have the transformation property

$$x_{ij} = \sum_k \sum_l s_{ik} s_{jl} x_{kl}'. \tag{1.3b}$$

The coefficients may have a symmetry relation $x_{ij} = x_{ji}$ among them. Such a *symmetric tensor of the second rank* has only $\frac{1}{2}n(n+1)$ independent components. We shall not make use of this concept till Chapter 5, § 2.2.

1.3. LINEAR MAPPING OF THE SPACE $\mathfrak{R}_n$ ON ITSELF (LINEAR TRANSFORMATION)

To each vector $\boldsymbol{x}$ of $\mathfrak{R}_n$ let there correspond a vector of the same space $\boldsymbol{y} = \mathsf{A}\boldsymbol{x}$ defined in the same coordinate system by its components:

$$y_i = \sum_k a_{ik} x_k; \qquad \mathsf{A}\boldsymbol{x} = \sum_{i,k} \boldsymbol{e}_i a_{ik} x_k, \tag{1.4}$$

$\mathsf{A} = (a_{ik})$ is the *mapping matrix* or projection matrix.
Its order or degree is n (the number of rows or columns). There exists a matrix of the order n that makes the vector $\boldsymbol{x}$ correspond with the vector $\boldsymbol{x}$ itself that is it leaves the space $\mathfrak{R}_n$ unchanged. According to (1.4) all the elements are zero except those of the main diagonal and these are equal to one. This is the *unit matrix*:

$$\mathsf{I} = (\delta_{ik}), \qquad \delta_{ik} = 0\ (i \neq k) \qquad \delta_{ii} = 1.$$

One can map $\mathfrak{R}_n$ on another space $\mathfrak{R}_m$ where m may be different from n and in this case, the matrix A is rectangular.

The mapping, A, of $\mathfrak{R}_n$ on itself transforms a vector which is originally identified with one of the basis vectors $\boldsymbol{e}_k$ into $\boldsymbol{\eta}_k = \mathsf{A}\boldsymbol{e}_k$ and since $\boldsymbol{e}_k$ has the components δ_{ik} [1] one finds in accordance with (1.1),

$$\boldsymbol{\eta}_k = \mathsf{A}\boldsymbol{e}_k = \sum_{il} \boldsymbol{e}_i a_{il} \delta_{lk} = \sum_i \boldsymbol{e}_i a_{ik}. \tag{1.4a}$$

Note the identity in form of the equations (2.1) and (1.4a). A change of coordinates is a type of mapping limited to the basis vectors only.

The mappings are transformation operations on the n-dimensional space. Their symbols A are *operators* acting on the vectors of this space in order to transform them into other vectors. These operators are analytically expressed by matrices of the degree n. The mathematical problem presented by quantum mechanics is to generalize the properties of these operators to the case where n is infinite.

1.4. COMPOSITION OF MAPPINGS. MULTIPLICATION OF MATRICES

Let us make two successive mappings of $\mathfrak{R}$ on itself, first A then B:

$$\boldsymbol{z} = \mathsf{B}\boldsymbol{y} = \mathsf{BA}\boldsymbol{x} = \mathsf{C}\boldsymbol{x}.$$

The successive operations are carried out in the proper order by reading from the right to the left in this equation. This rule will always be applied in the calculus of operators. Equation (1.4) gives us

$$z_i = \sum_l b_{il} y_l = \sum_l b_{il}(\sum_k a_{lk} x_k) = \sum_k c_{ik} x_k,$$

that is

$$\mathsf{C} = \mathsf{BA} = (c_{ik}) = (\sum_l b_{il} a_{lk}). \tag{1.5}$$

The matrix C is obtained by the usual multiplication rule of determinants. The same result is also found when the mappings are carried out from one space to another provided that the number of dimensions lend themselves to it, as we shall see below.

It is convenient to represent a vector $\boldsymbol{x}$ by a matrix X, where all the components are arranged in a single column (the first one for example) and where

[1] One has obviously

$$\boldsymbol{e}_i = \sum_k \delta_{ik} \boldsymbol{e}_k = \sum_k \boldsymbol{e}_k \delta_{ki}.$$

all other places are filled with zeros. Without difficulty one sees, from (1.4) and (1.5) that the formula $y = \mathsf{A}x$ becomes

$$\mathsf{Y} = \mathsf{AX}. \tag{1.4b}$$

A still simpler scheme is to consider vectors as a special case of rectangular (i.e. non-square) matrices. We recall that in the case of a product of two rectangular matrices A and B the width of A has to match the height of B (see Fig. 1.1).

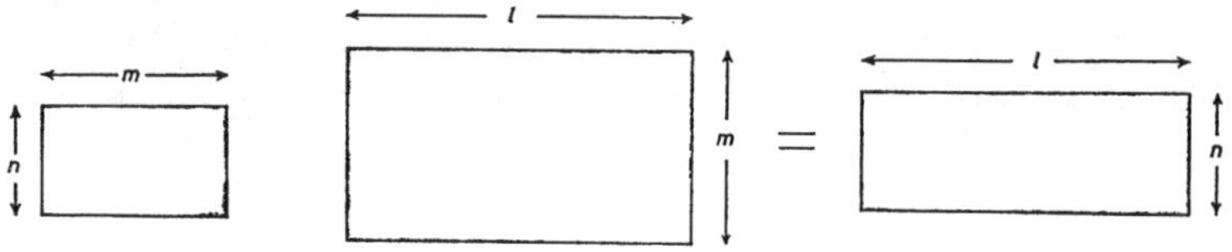

Fig. 1.1.

As in equation (1.4b) A stands in front of X. This implies that X has to be a one-column matrix.

All our operations are thus reduced to the multiplication of matrices. In particular, the change of coordinates (1.3) becomes

$$\mathsf{X} = \mathsf{SX}'. \tag{1.3a}$$

1.5. INVERSE MATRIX

In order to insure that the mapping A will be reversible the determinant $|\mathsf{A}| = |a_{ik}|$ has to be different from zero; the equations (1.4) are then solvable for x and one has

$$x_i = \sum_k a_{ik}^{-1} y_k \quad \text{or} \quad \mathsf{X} = \mathsf{A}^{-1}\mathsf{Y} = \mathsf{A}^{-1}\mathsf{AX}.$$

So the inverse matrix A^{-1} of A is defined by

$$\mathsf{AA}^{-1} = \mathsf{A}^{-1}\mathsf{A} = \mathbb{I} = (\delta_{ik}), \tag{1.6}$$

that is

$$\sum_l a_{il}^{-1} a_{lk} = \sum_l a_{il} a_{lk}^{-1} = \delta_{ik}. \tag{1.6a}$$

1.6. TRANSFORMATION OF A MAPPING OR PROJECTION MATRIX BY A CHANGE OF COORDINATES (SIMILARITY TRANSFORMATION)

Let e_i be the initial system of basis vectors in which we will carry out the transformation $x \to y = \mathsf{A}x$. Now let us change the basis and let e_i' be the new basis vectors and x_i', y_i' the new set of components of the vectors $\boldsymbol{x}$ and $\boldsymbol{y}$. The mapping or projection A, that is to say the correspondence between the vectors $\boldsymbol{x}$ and $\boldsymbol{y}$, is expressed in the second coordinate system by a matrix A′ such that

$$\boldsymbol{y}' = \mathsf{A}'\boldsymbol{x}'. \tag{1.4c}$$

The transformation that converts A into A′ is called a *similarity transformation* or canonical transformation.

The form of A′ can be calculated without trouble. Let us indicate the change of coordinates $\boldsymbol{e} \to \boldsymbol{e}'$ by S and the inverse $\boldsymbol{e}' \to \boldsymbol{e}$ by S^{-1}, then we have (compare § 1.2) $\boldsymbol{x} = \mathsf{S}\boldsymbol{x}'$ and $\boldsymbol{x}' = \mathsf{S}^{-1}\boldsymbol{x}$, that is $x_i' = \sum_k s_{ik}^{-1} x_k$. The mapping A′ can be done in two different ways, either directly in the system $\boldsymbol{e}'$, or indirectly by going from the axes $\boldsymbol{e}'$ to the axes $\boldsymbol{e}$ using S^{-1} and then carrying out the mapping A in the system $\boldsymbol{e}$ and finally going back to the system $\boldsymbol{e}'$ by means of S. These two procedures are equivalent. The second is the series of transformations

$$\boldsymbol{y}' = \mathsf{S}^{-1}\boldsymbol{y},\ \boldsymbol{y} = \mathsf{A}\boldsymbol{x}, \quad \boldsymbol{x} = \mathsf{S}\boldsymbol{x}',$$

combining these three gives

$$\boldsymbol{y}' = \mathsf{S}^{-1}\mathsf{A}\mathsf{S}\boldsymbol{x}' = \mathsf{A}'\boldsymbol{x}',$$

that is

$$\mathsf{A}' = \mathsf{S}^{-1}\mathsf{A}\mathsf{S} \quad \text{and} \quad \mathsf{A} = \mathsf{S}\mathsf{A}'\mathsf{S}^{-1}. \tag{1.7}$$

Two transformations A and A′ are said to be equivalent if they are related by an equation of the type (1.7), where S is an arbitrary transformation, having an inverse. They go over into each other by a change of axes.

As a simple example let us consider the operation in which every vector of a three-dimensional vector space $\mathfrak{R}_3$ is rotated through an angle $\vartheta = 45°$ around a given axis $\boldsymbol{a}$ coinciding with $\boldsymbol{e}_3$. This operation can be expressed by $\boldsymbol{y} = \mathsf{A}\boldsymbol{x}$ where the matrix A is given by (compare Fig. 1.2a):

$$\mathsf{A} = \begin{pmatrix} \varepsilon & \varepsilon & 0 \\ -\varepsilon & \varepsilon & 0 \\ 0 & 0 & 1 \end{pmatrix} \quad (\varepsilon = \tfrac{1}{2}\sqrt{2}).$$

If we now perform a mapping of $\mathfrak{R}_3$ on itself with the matrix:

$$\mathsf{S} = \begin{pmatrix} 1 & 0 & 0 \\ 0 & 0 & -1 \\ 0 & 1 & 0 \end{pmatrix},$$

we obtain the new basis vectors e'_1, e'_2 and e'_3 indicated in Fig. 1.2b. The 45°-operation described above is now given by $y' = \mathsf{A}'x'$ where

$$\mathsf{A}' = \begin{pmatrix} \varepsilon & 0 & -\varepsilon \\ 0 & 1 & 0 \\ \varepsilon & 0 & \varepsilon \end{pmatrix},$$

and one easily verifies the relationship (1.7).

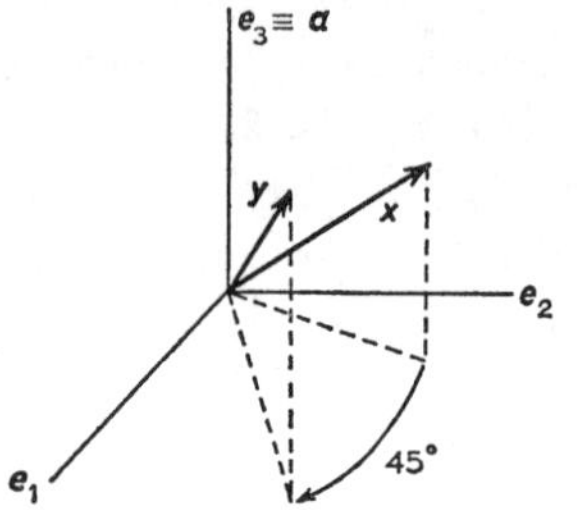

Fig. 1.2a

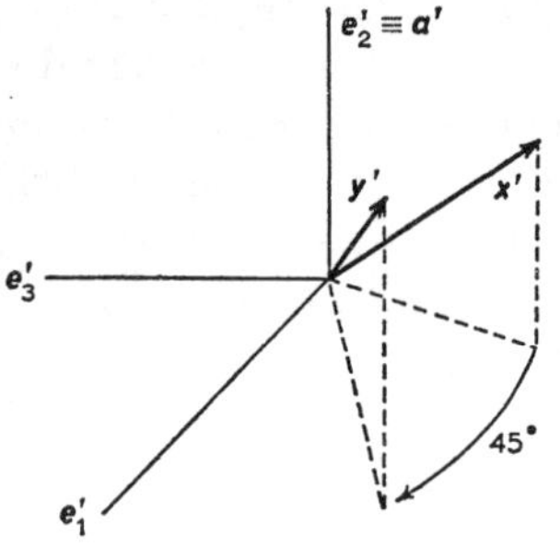

Fig. 1.2b

Fig. 1.2. The operation "rotate every vector of the 3-dimensional space $\mathfrak{R}_3$ through an angle $\vartheta = 45°$ around a given axis a" performed in two different systems of reference.

2. Euclidean and Unitary Spaces

2.1. UNITARY METRIC

The preceding definitions are sufficient to determine the affine properties of a vector space. In order to complete the analogy with ordinary usual space, it is necessary to attribute additional properties to it (the so-called *metric* properties) and for this it is sufficient to give an invariant and positive definite form which is a function of the components of an arbitrary vector and which defines *the square of the length* or *the norm* of that vector,

$$x^2 = \sum_{ik} g_{ik} x_i x_k \text{ [1]}.$$

[1] A metric in which g_{ik} is non-diagonal is not very often used, at least in the case of constant coefficients. A non-diagonal metric geometrically interpreted means that we have an oblique coordinate system. In this case the projections perpendicular and parallel to the other coordinate axes are the co- and the contravariant components. If the vectors are functions (compare § 1.4), a non-diagonal metric means a set of non-orthogonal wave functions, which are used in some exceptional cases, for instance the Orthogonalized Plane Wave method in solid state physics.

In general g_{ik} may be functions of the coordinates x_i. Here already in the diagonal case

An affine space with a metric in a real domain is *Euclidean* if one can reduce (g_{ik}) to a unit matrix by a change of axes S:

$$\mathsf{S}^{-1}(g_{ik})\mathsf{S} = (\delta_{ik}) \quad \text{i.e.} \quad x^2 = \boldsymbol{x} \cdot \boldsymbol{x} = \sum_{i=1}^{n} x_i^2 .$$

In quantum mechanics one deals in general with complex quantities, but the norm (or fundamental metric invariant) of an arbitrary vector has necessarily to be real in order to have a meaning in physics. The space is *unitary* because this invariant can be brought into the following form by a change of axes S:

$$x^2 = (\boldsymbol{x} \cdot \boldsymbol{x}) = \sum_{i=1}^{n} x_i^* x_i , \tag{1.8}$$

where x_i^* *is the complex conjugate of* x_i.
The coordinate system is thus unitary. In such a system one can even define a *scalar product* of two vectors $\boldsymbol{y}$ and $\boldsymbol{x}$:

$$(\boldsymbol{x} \cdot \boldsymbol{y}) = \sum x_i^* y_i = (\boldsymbol{y} \cdot \boldsymbol{x})^* . \tag{1.8a}$$

These equations show that the axis of a unitary coordinate system are *orthogonal unit vectors*; their norm is equal to one and the scalar products of any pair are zero because the components e_{il} and e_{kl} of $\boldsymbol{e}_i$ and $\boldsymbol{e}_k$ are δ_{il} and δ_{kl} respectively. Thus we have:

$$\boldsymbol{e}_i \cdot \boldsymbol{e}_i = 1; \quad \boldsymbol{e}_i \cdot \boldsymbol{e}_k = 0. \tag{1.8b}$$

One verifies at the same time that the component x_i is equal to the scalar product of the vector $\boldsymbol{x}$ and the unit vector $\boldsymbol{e}_i$:

$$x_i = (\boldsymbol{e}_i \cdot \boldsymbol{x}), \tag{1.8c}$$

because the components of $\boldsymbol{e}_i$ are real.

In § 1.4 we mentioned that vectors may be considered as special (rectangular) matrices. In order to obtain invariants, i. e. one-by-one dimensional matrices, like the scalar product (1.8) and (1.8a) one matrix has to be of the one-column type, the other matrix of the one-row type. If we write equation

---

one must distinguish between co- and contravariant components (MORSE and FESHBACH [1953] pp. 30 and 44). A simple example of a variable orthogonal metric is a polar coordinate system. The most general form, where g_{ik} are functions and non-diagonal, is used in the general theory of relativity. (Compare e.g. H. WEYL [1922] sections 2, 3, 4 and 11.)

We would like to remind the reader that there are linear spaces used in mathematics in which no metric is introduced, for instance, the group-algebra mentioned in Chapter 3, § 10.3.

(1.8) in matrix notation, the left-hand side of the product becomes the transpose of the complex conjugate of the right-hand side.

2.2. UNITARY TRANSFORMATION

A unitary transformation is a change in coordinate system, it is a transformation from one unitary system $\boldsymbol{e}_i$ to another $\boldsymbol{e}'_i$. If x_i and x'_i are the components of an arbitrary vector $\boldsymbol{x}$ in the two systems, one must have

$$\boldsymbol{x} \cdot \boldsymbol{x} = \sum_i x_i^* x_i = \sum_i x_i'^* x_i',$$

hence this requires that the coefficients in (1.3) obey the following restriction,

$$\sum_i x_i^* x_i = \sum_{i,k,l} (s_{ik} x'_k)^* s_{il} x'_l = \sum_{k,l} x_k'^* x'_l (\sum_i s_{ik}^* s_{il}),$$

that is to say

$$\sum_i s_{ik}^* s_{il} = \delta_{kl},$$

or using (1.6a),

$$s_{ik}^* = s_{ki}^{-1}.$$

The transpose of the complex conjugate of S, is called the *adjoint matrix* S, that is

$$s_{ik}^\dagger = s_{ki}^*.$$

Using this definition the condition which unitary transformations have to satisfy becomes

$$\mathsf{S}^\dagger = \mathsf{S}^{-1}, \quad \text{i.e.} \quad \mathsf{S}^\dagger \mathsf{S} = \mathrm{I}. \tag{1.9}$$

The unitary transformations, considered as projections of the space $\mathfrak{R}_n$ onto itself, are those which leave the scalar products of vectors invariant and which consequently conserve their orthogonal relations. This definition leads again to (1.9).

2.3. BILINEAR FORMS AND HERMITIAN MATRICES

Let us consider the scalar product,

$$\boldsymbol{x} \cdot \mathsf{A}\boldsymbol{y} = \sum_{i,k} x_i^* a_{ik} y_k. \tag{1.10}$$

This is a *bilinear form* of the variables x_i and y_k.[1]

[1] From its very definition this form is invariant, that is to say it does not change under a *unitary* coordinate transformation. Hence if one considers the x_i as covariant components and the x_i^* as contravariant components, the matrix A is a *mixed tensor of the second order* in unitary space.

It is *Hermitian* if the projection A can operate just as well on the first as on the second vector, that is if,

$$\boldsymbol{x} \cdot \mathsf{A}\boldsymbol{y} = \mathsf{A}\boldsymbol{x} \cdot \boldsymbol{y}, \tag{1.11}$$

thus

$$\sum_{i,\,k} x_i^* a_{ik} y_k = \sum_{k,\,i} a_{ki}^* x_i^* y_k,$$

and finally

$$a_{ik} = a_{ki}^*; \qquad \mathsf{A} = \mathsf{A}^\dagger. \tag{1.11a}$$

One can verify without any trouble that the *quadratic form* $\mathsf{A}\boldsymbol{x} \cdot \boldsymbol{x}$ is real if and only if A is Hermitian. The expression $\boldsymbol{x} \cdot \boldsymbol{x}$ is always real by virtue of its definition (1.8) and obviously $\boldsymbol{x} \cdot \boldsymbol{y}$ is Hermitian.

2.4. GENERALIZATION OF UNITARY SPACE INTO BRA- AND KET SPACES

To establish a scalar product of quantities that have complex components we could follow a different line of thought. Instead of using *one* space, the unitary space, and the attribution of a certain metric to this space, we will proceed as follows.

Take two independent spaces $\mathfrak{R}$ and $\overline{\mathfrak{R}}$ of the same dimensionality. To every vector $\boldsymbol{x}$ in the first space corresponds one and only one vector $\bar{\boldsymbol{x}}$ in the second space. This one-to-one correspondence should be chosen such that to every transformation $\boldsymbol{x} \to \boldsymbol{y} = \mathsf{A}\boldsymbol{x}$, there corresponds a transformation $\bar{\boldsymbol{x}} \to \bar{\boldsymbol{y}} = \mathsf{A}\bar{\boldsymbol{x}}$, in the $\overline{\mathfrak{R}}$-space, and also that there is a one-to-one correspondence between the transformations A and $\bar{\mathsf{A}}$. Finally instead of defining a metric we postulate that in order to obtain a number one has to take the components of $\boldsymbol{x}$ and multiply each one with the corresponding component of $\bar{\boldsymbol{y}}$. This is possible since i-th component x_i is the coefficient of the unit vector $\boldsymbol{e}_i$ and this unit vector is related to a definite unit vector $\bar{\boldsymbol{e}}_i$ in $\overline{\mathfrak{R}}$. As a result a component (which can be a complex number) of vector $\boldsymbol{x}$ corresponds to a definite component of $\bar{\boldsymbol{y}}$ and one can perform the sum over all these products $\sum x_i \bar{y}_i$. If we take this scalar-product type of sum of a vector and its own adjoint in $\overline{\mathfrak{R}}$-space, we call it the "length" of the vector.

Instead of indicating the unit vector of a certain baisis by $\boldsymbol{e}_i$ or $\bar{\boldsymbol{e}}_i$ we use, following DIRAC [1958], only the indices. A unit vector in space $\mathfrak{R}$ is designated by $|i\rangle$ a ket-unit vector, and in $\overline{\mathfrak{R}}$ by $\langle i|$ a bra-unit vector. These strange names stem from the fact that the length of a vector is always the product of a bra and a ket and only these bra-ket's are ordinary numbers (or in case of application to quantum mechanics; only these are observable quantities).

The notation of a length is as follows. A vector in $\mathfrak{R}$ or ket space is

$$| \rangle = \sum_i x_i |i\rangle;$$

the adjoint vector in $\overline{\mathfrak{R}}$ or bra space is

$$\langle | = \sum_i \bar{x}_i \langle i|,$$

the result is the following bracket:

$$\langle | \rangle = (\sum_i \langle i|\bar{x}_i)(\sum_j x_j|j\rangle) = \sum_i \bar{x}_i x_i$$

since the brackets $\langle i|i\rangle$ are all equal to one as $|i\rangle$ was a *unit* ket and $\langle i|$ was a *unit* bra and we assume for simplicity that the scalar product of two different bra's and ket's is zero

$$\langle i|j\rangle = 0, \quad i \neq j.$$

An operator that establishes a projection in ket space can be fully characterized by the mappings of all the unit vectors,

$$\mathsf{A}|i\rangle = \sum_j a_{ij}|j\rangle,$$

and if we are interested in the scalar product of an arbitrary bra $\langle j|$ and the *projected ket* $\mathsf{A}|i\rangle$ we obtain a set of numbers.

$$\langle j|\mathsf{A}|i\rangle = a_{ij}\langle j|j\rangle = a_{ij}$$

which are representative for this particular operation A. Similarly if we study the same operation in bra space the transform of $\langle j|$ will be

$$\langle j|\bar{\mathsf{A}} = \bar{a}_{ji}\langle i|$$

(we write $\bar{\mathsf{A}}$ to the left in bra space for reasons of symmetry in the final result, one has to remember, however, that the action of this operator is from right to left). The scalar product with an arbitrary vector in ket space gives again a set of numbers that characterize the operation completely

$$\langle j|\bar{\mathsf{A}}|i\rangle = \bar{a}_{ji}.$$

We can calculate in a similar way as § 1.6 the matrix elements of A with respect to a new set of basis kets $|\nu\rangle$ and bras $\langle\mu|$.

$$\mathsf{A}' = \mathsf{S}^{-1}\mathsf{A}\,\mathsf{S} \rightarrow \langle\mu|\mathsf{A}'|\nu\rangle = \langle\mu|i\rangle\,\langle i|\mathsf{A}|j\rangle\,\langle j|\nu\rangle.$$

The same index in the "opposite" positions means a sum over that particular index and is called a contraction. The relation between the coefficients of transformation in bra space and the coefficients of transformation in ket space is

$$\langle \mu | i \rangle = \overline{\langle i | \mu \rangle}.$$

Finally we want to mention a quantity which consists of a ket and a bra: $|i\rangle\langle\mu|$. This is not a number, but an operator. Its matrix elements between the bra $\langle j|$ and the ket $|\nu\rangle$ are

$$\langle j | i \rangle \langle \mu | \nu \rangle \equiv \delta_{ij} \delta_{\mu\nu}.$$

It is sometimes called a *projection operator* since this special operator picks out a particular ket, i.e. $|\mu\rangle$, and projects it upon one and only one other ket: $|i\rangle$. This quantity is comparable to the direction cosines in an ordinary change of axes. One could call it the dyadic product of a ket and a bra.

All this is so far only a different language and a different notation. Nothing new is added that could not have been described by a unitary space with a certain metric, if we had considered the bra vectors as the complex conjugate of the ket vectors. It is, however, possible to generalize the preceding picture quite easily in cases where the bra vectors have a different relation to the ket vectors, and this is used in relativistic quantum mechanics.

3. Reduction to Main Axes

We will admit, without specifying its proof, the following theorem which is the generalization of theorems on the equations in S and the reduction of quadratic forms to a sum of squares.

3.1. DIAGONALIZABILITY

In unitary space every Hermitian form may be written in the form:

$$\sum \alpha_i x_i^* x_i,$$

by a convenient choice of orthogonal coordinate axes; α_i *being real numbers.* Or, that is: *Every Hermitian matrix* A *can be brought in diagonal form by a unitary transformation:*

$$\mathsf{A}' = \mathsf{S}^{-1}\mathsf{A}\mathsf{S} = \begin{pmatrix} \alpha_1 & 0 & \dots & 0 \\ 0 & \alpha_2 & \dots & 0 \\ \cdot & \cdot & \cdot & \cdot \\ \cdot & \cdot & \cdot & \cdot \\ 0 & 0 & \dots & \alpha_n \end{pmatrix}; \qquad \mathsf{S}^\dagger \mathsf{S} = \mathbb{I}. \tag{1.13}$$

A similar statement holds if the matrix is unitary.

We would like to end this section with a short note on the diagonalization of matrices in general. The question "Which class of matrices can be fully diagonalized?", is usually not treated in physics, hence the mathematical techniques in physics sometimes create the misunderstanding that all matrices can be diagonalized.

A *normal matrix* N is a matrix that commutes with its Hermitian conjugate:

$$\mathsf{N}\mathsf{N}^{\dagger} - \mathsf{N}^{\dagger}\mathsf{N} = 0. \tag{1.12}$$

An arbitrary matrix can always be brought into a form in which half the off-diagonal elements, i.e. at one side of the diagonal only, are zero [1].

If we have a normal matrix, the relation

$$\begin{pmatrix} \lambda_1 & N_{12} \cdots \\ 0 & \lambda_2 & \cdot \\ \vdots & \vdots & \cdot \end{pmatrix}\begin{pmatrix} \lambda_1^* & 0 & \cdots \\ N_{12}^* & \lambda_2^* & \cdots \\ \cdots & \cdots & \cdots \end{pmatrix} = \begin{pmatrix} \lambda_1^* & 0 & \cdots \\ N_{12}^* & \lambda_2^* & \cdots \\ \cdots & \cdots & \cdots \end{pmatrix}\begin{pmatrix} \lambda_1 & N_{12} \cdots \\ 0 & \lambda_2 & \cdots \\ \cdots & \cdots & \cdots \end{pmatrix}$$

gives a set of relations:

$$\lambda_1^*\lambda_1 + |N_{12}|^2 + |N_{13}|^2 + \ldots = \lambda_1\lambda_1^*$$

hence all $N_{ij} (i \neq j)$ are zero.
Conversely if a matrix is diagonal (1.12) holds and since this equation is invariant it holds if the matrix is not diagonal.

If we diagonalize the normal matrix its eigenvalues are complex. If they are all real the matrix is Hermitian and vice versa. In the unitary case the eigenvalues have absolute value 1.

3.2. DETERMINATION OF DIAGONAL ELEMENTS

The problem of determining the diagonal elements arises in geometry, in mechanics, and in the physics of crystals. Consider for instance an anisotropic dielectric. The distribution of dielectric constants in different directions in space is represented by a quadratic form: the ellipsoid of specific dielectric constants. To bring this ellipsoid on its axis means to find the "principal directions" of the crystal, where the vector of the dielectric displacement $\boldsymbol{D}$, is parallel to the vector of the electric field strength $\boldsymbol{E}$. One knows that there are in general three directions [2] and that they are orthogonal. Hence

[1] It takes more effort to show that one can simplify an arbitrary matrix to a form consisting of a set of diagonal elements and off-diagonal elements $a_{i+1,i} = 1$, the Jordan normal form.

[2] Except the case in which the ellipsoid of dielectric constants is an ellipsoid of revolution. Then the two main values of the dielectric constant are equal (a degeneracy). Thus only one main direction is determined; it is normal to a plane in which the two other axes can be chosen arbitrarily.

there are three rectangular axes for which one has

$$\boldsymbol{D}_i = \varepsilon_i \boldsymbol{E}_i, \qquad (i = 1, 2, 3)$$

the ε_i are the main dielectric constants of the medium. Similarly in the general case in which we are actually interested, we have to search for a unitary system of axes $\boldsymbol{e}_i$ which brings the matrix A′ (the transformed of A) into the diagonal form (1.13). This means solving the following problem: find n directions $\boldsymbol{e}'_1, \boldsymbol{e}'_2, \ldots, \boldsymbol{e}'_n$ such that every vector $\boldsymbol{x}$ parallel to one of the directions $\boldsymbol{e}'_i$ will be transformed by A into a vector $\boldsymbol{y}$ which is parallel to $\boldsymbol{x}$:

$$\boldsymbol{y} = \mathsf{A}\boldsymbol{x} = \alpha \boldsymbol{x}, \tag{1.14}$$

where α is a constant.

Writing the "components" of this vector equation, we obtain n linear homogeneous equations of the form

$$a_{i1}x_1 + \ldots (a_{ii} - \alpha)x_i + \ldots a_{in}x_n = 0 \qquad (i = 1, 2, \ldots, n), \tag{1.14a}$$

which are only compatible with each other if the undetermined constant α is a root of the *secular equation*[1]

$$\operatorname{Det} |\mathsf{A} - \lambda \mathsf{I}| = \begin{vmatrix} (a_{11}-\lambda) & a_{12} & \ldots & a_{1n} \\ a_{21} & (a_{22}-\lambda) & \ldots & a_{2n} \\ \cdot & \cdot & \cdot & \cdot \\ a_{n1} & \cdot & \ldots & (a_{nn}-\lambda) \end{vmatrix} = 0. \tag{1.15}$$

This equation has in general n roots $\lambda = \alpha_1, \ldots \alpha_n$ (which may be distinct or not) to which correspond the n directions of the axes given by equation (1.14a). We know only the directions of the vectors $\boldsymbol{x}$ but not their magnitude because (1.14a) determines only the ratios between the components. The roots α_i of (1.15) are the eigenvalues, propervalues, or characteristic constants of the matrix A. They are real in the Hermitian case; in the unitary case their absolute value is 1. The $\boldsymbol{e}'_i$ are the eigenvectors or principal directions[2]. When (1.15) represents a multiple root of order $p\,(\alpha_1 = \alpha_2 = \ldots$

[1] This name originates from a problem in astronomy, where a similar equation determines the perturbation over long time intervals.

[2] The proof of these theorems can be made by choosing a root α_1 of (1.15), determining the corresponding eigenvector $\mathbf{e}_1$, and completing the system with $(n-1)$ vectors which have to be orthogonal to the first one in order to form a unitary system of axes. As a result of the symmetry properties of the Hermitian and unitary matrices the coefficients $a_{12} \ldots a_{1n}, a_{21} \ldots a_{n1}$, are all zero and the matrix A takes the form:

$$\begin{pmatrix} \alpha_1 & 0 & \ldots & 0 \\ 0 & a_{22} & \ldots & a_{2n} \\ \cdots & \cdots & \cdots & \cdots \\ 0 & a_{n2} & \ldots & a_{nn} \end{pmatrix}.$$

$\alpha_p = \alpha$), the vectors which have the property $y = \mathsf{A}x = \alpha x$ form a *subspace of p dimensions* in which the direction of the axes is not determined. (See the second note on page 12).

3.3. JOINT DIAGONALIZATION OF A SET OF MATRICES

In order that all the matrices of an Hermitian or unitary system can be reduced at the same time to their principal axes, it is necessary and sufficient that they all commute with each other.

First we will show that this condition is necessary: let A and B be two matrices. By a transformation of axes S we reduce them simultaneously to diagonal form A′ and B′. Then obviously they commute with each other:

$$\mathsf{A}'\mathsf{B}' = \mathsf{B}'\mathsf{A}'. \text{ Hence } \mathsf{S}^{-1}\mathsf{A}\mathsf{S}\mathsf{S}^{-1}\mathsf{B}\mathsf{S} = \mathsf{S}^{-1}\mathsf{A}\mathsf{B}\mathsf{S} = \mathsf{S}^{-1}\mathsf{B}\mathsf{A}\mathsf{S} \text{ i.e. } \mathsf{A}\mathsf{B} = \mathsf{B}\mathsf{A}.$$

Now we will show that the condition is sufficient: let us suppose that AB = BA and let us make a transformation of coordinates such that B is diagonal:

$$\mathsf{B} = (\beta_i).$$

We have

$$(\mathsf{AB})_{ik} = (\mathsf{BA})_{ik} \quad \text{or} \quad a_{ik}(\beta_i - \beta_k) = 0.$$

If

$$\beta_i \neq \beta_k, \quad a_{ik} = 0.$$

So we find that A is a *step-wise matrix*, of which all the terms are zero, except those which are situated on the main diagonal, or those inside certain squares that share the diagonal; they correspond with the case in which $\beta_i = \beta_k$. These squares are each related to a subspace $\mathfrak{R}$ in which the principal directions of the matrix B are undetermined (the matrix possesses a circular, spherical, or hyperspherical symmetry). One may finally choose these undetermined axes in such a way that A will be completely diagonal.

3.4. INVARIANCE OF A SECULAR EQUATION

A transformation of coordinates does not change the form of the secular equation (1.15). Let us carry out the transformation of coordinates S:

$$\mathsf{A} \rightarrow \mathsf{A}' = \mathsf{S}^{-1}\mathsf{A}\mathsf{S}; \qquad \mathbf{I} \rightarrow \mathsf{S}^{-1}\mathbf{I}\mathsf{S} = \mathbf{I}.$$

The rule of multiplication of determinants gives us

$$\begin{aligned} |\mathsf{S}^{-1}\mathsf{A}\mathsf{S} - \lambda\mathbf{I}| &= |\mathsf{S}^{-1}| \cdot |\mathsf{A} - \lambda\mathbf{I}| \cdot |\mathsf{S}| \\ &= |\mathsf{S}^{-1}| \cdot |\mathsf{S}| \cdot |\mathsf{A} - \lambda\mathbf{I}| = |\mathsf{A} - \lambda\mathbf{I}|. \end{aligned} \tag{1.16}$$

3.5. TRACE OR SPUR OF A MATRIX

Let $\mathsf{A} = (a_{ik})$ be a matrix, the trace is then the sum of all the diagonal elements

$$\mathrm{Tr}\,\mathsf{A} = \sum_{i=1}^{n} a_{ii}. \tag{1.17}$$

To prove that the trace is invariant we shall write equation (1.15) in the form:

$$(-\lambda)^n + (-\lambda)^{n-1}(a_{11}+a_{22}+\ldots a_{nn}) + \ldots = (-\lambda)^n + (-\lambda)^{n-1}\,\mathrm{Tr}\,\mathsf{A} + \ldots.$$

Because this equation keeps its form under a transformation of coordinates S, all the coefficients are invariant, in particular the second:

$$\mathrm{Tr}\,\mathsf{A} = \text{invariant}. \tag{1.17a}$$

4. Function Space. Complete Sets of Orthogonal Functions

Every function of continuous variables and particularly the wave function in Schrödinger mechanics can be represented by vectors in *a function space* in which the *number of dimensions is infinite*. The operators acting on the wave functions will transform these functions into other functions producing a transformation of this space into itself. On the other hand quantum mechanics can also be formulated with the help of certain relations among matrices. The matrices in this so-called *matrix mechanics* are matrices with an infinite number of rows and columns. They can again be considered as operators that transform a space with an infinite number of dimensions into itself. This analogy, brought forward by Hilbert, made it possible for Schrödinger and later for Dirac to show the equivalence between wave mechanics and matrix mechanics.

There exists between the two spaces just mentioned a difference which appears to be essential: the matrices operate on a space in which the number of dimensions is denumerably infinite; on the contrary the number of dimensions in function space is of the order of a continuum. We will see, in a moment that this difference is more apparent than real (see § 4.3).

4.1. FUNCTION SPACE

$\psi(x)$ defines a function of one variable x. The simplest case is that of a discontinuous function where the value is given only for a finite number of values of the variable $x : x_1, x_2, \ldots, x_n$, i.e. the domain of the variable x consists of a set of discrete points. The n corresponding values of the function $\psi_1, \psi_2 \ldots \psi_n$ ($\psi_i = \psi(x_i)$) can be considered as components of a vector in a space of n dimensions. To each different function $\varphi(x)$ defined for the same values of x there corresponds a distinct vector which is the geometrical representation of that particular function.

Normally the functions in which we are interested depend on continuous variables. So we have to go over to the limit in which the number n of dimensions of the function space becomes infinite. In such a case one may consider the value ξ of the variable x as an index and say that $\psi(\xi)$ is the component of a vector ψ along an axis characterized by the index ξ. For functions of several variables, such as $\psi(x, y, z)$ the set x, y, z constitutes one single index and the function space shows the same properties as those with a single variable.

In the next section we will reason mainly by analogy and be satisfied by indicating now and then the mathematical difficulties. We will always assume that conditions of convergence are realized.

4.2. SCALAR PRODUCT; NORM

In n-dimensional space, the scalar product of two vectors $\boldsymbol{x}$ and $\boldsymbol{y}$ has been defined by

$$\boldsymbol{x} \cdot \boldsymbol{y} = \sum_{k=1}^{n} x_k^* y_k. \tag{1.8a}$$

By analogy in the case of two functions of a discontinuous variable, which are defined for the values $x_1, x_2, \ldots, x_n$ of the variable, we will put:

$$(\psi \cdot \varphi) = \sum_{k=1}^{n} \psi^*(x_k)\varphi(x_k).$$

If the functions $\varphi(x)$ and $\psi(x)$ are defined in a continuous domain D of the variable x the only possibly generalization is

$$(\psi \cdot \varphi) = \int_{\mathrm{D}} \psi^*(x)\varphi(x)\mathrm{d}x. \tag{1.18}$$

This integral is the *scalar product of two functions* $\psi(x)$ and $\varphi(x)$. The wave function in quantum mechanics depends on the coordinates of the particles $x_1, y_1, z_1, x_2, \ldots, z_p$, the domain D is the finite or infinite *configuration space* in which we will indicate the element by $\mathrm{d}\tau = \mathrm{d}x_1 \cdot \mathrm{d}y_1 \cdot \mathrm{d}z_1 \cdot \mathrm{d}x_2 \ldots \mathrm{d}z_p$ and we have

$$(\psi \cdot \varphi) = \int_{\mathrm{D}} \psi^* \varphi \, \mathrm{d}\tau. \tag{1.18a}$$

In particular the *norm* of the function ψ, which corresponds to the square of the length of the representing vector can be written

$$(\psi \cdot \psi) = \int_{\mathrm{D}} \psi^* \psi \, \mathrm{d}\tau. \tag{1.19}$$

We will assume that all the functions which we will consider make the integrals converge, whatever the domain D is, even if it is infinite. These are called *square integrable functions.*

4.3. FOURIER SERIES. COMPLETE SETS OF ORTHOGONAL FUNCTIONS

Every function $\psi(x)$ satisfying certain broad conditions about continuity inside a domain contained between the values $x = \pi$ and $x = -\pi$ can be developed into a convergent trigonometric series, a series of *fundamental basis functions* $\sin nx$ and $\cos nx$ ($n = 0, 1 \ldots \infty$). If x is an angle, this domain is a circle with unit radius. This is the well-known theorem by Fourier. It will be convenient to use complex variables. Then the basis functions are the exponentials $e^{inx} = \exp(inx)$ where n takes on all integers between $-\infty$ and $+\infty$. The possibility of these power-series developments is related to two essential properties of the fundamental functions:

1. In the domain $-\pi, +\pi$ the basis functions multiplied by a convenient "normalization factor", which happens to be equal to $(1/\sqrt{2\pi})$, *are orthogonal and normal* or orthonormal which means that they satisfy the well-known and easy-to-check equations

$$(\varphi_n \cdot \varphi_m) = \frac{1}{2\pi}\int_{-\pi}^{+\pi} e^{-inx}e^{imx}\,dx = \delta_{mn};$$

$$\varphi_n = (2\pi)^{-\frac{1}{2}}e^{inx}; \qquad \varphi_m = (2\pi)^{-\frac{1}{2}}e^{imx}.$$

2. These functions form in this domain a *complete or closed system.* We will give this proposition the following meaning: in the domain considered we can develop every function that interests us into a convergent series of basis functions[1] for instance continuous functions which have a sufficient number of derivatives.

We know that the Fourier development can be extended without trouble

[1] For reasons of rigor, mathematicians give in general a more precise definition of a complete system: a system of fundamental functions $\psi(x)$ is complete if one can find for each continuous function $\psi(x)$ coefficients β_k such that

$$\lim_{n\to\infty}\int_D |\psi(x) - \sum_k \beta_k \psi_k(x)|^2 dx = 0, \qquad (1.21a)$$

$|\ |^2$ means the absolute square. So we deal with an unlimited approximation *in the mean* on the domain D or *convergence in the mean.* It is not necessary that $\sum_k \beta_k \psi_k$ is convergent. The formulas (1.22) and (1.23) can result just as well from this definition as from our restricted definition. All the calculations, including those in perturbation theory in Chapter 2, can be performed by making use of the rigorous definition (1.21a). We have preferred to sacrifice generality in order to simplify the arguments.

to an arbitrary number of variables. We know also that the limits of the domain of validity of the trigonometric series can be modified by a change of variables. We will return to this point in a moment.

Spherical functions, Hermite functions and Laguerre functions which one encounters in quantum theory, possess the same properties. They form, each in their domain (i.e. -1, $+1$ for the polynomials of Legendre, the surface of a unit sphere for the spherical harmonics, $-\infty$, $+\infty$ for those of Hermite and 0, $+\infty$ for those of Laguerre) a complete set of orthogonal functions. There are many other sets of functions which possess these properties. Therefore we will consider in a general way an infinite sequence of functions $\psi_1 \psi_2 \ldots$ of a certain number of variables and we will say that they constitute a complete system of orthogonal and normal functions inside a certain domain D, if the following two conditions are fulfilled. First we demand that

$$(\psi_i \cdot \psi_k) = \int \psi_i^* \psi_k \mathrm{d}\tau = \delta_{ik}, \tag{1.20}$$

the integrals being extended over the domain D. Second that all the continuous functions which occur in the applications in physics can be expressed in this domain by a series development of the form

$$\psi = \sum_k \beta_k \psi_k, \tag{1.21}$$

k being an index which is allowed to take all integer values between 0 and ∞ or between $-\infty$ and $+\infty$ (as in the case of a series of exponentials). If one compares (1.21) with (1.1) and (1.20) with (1.8b) one sees that these equations can be expressed in the following geometrical language:

A complete system of orthogonal functions establishes a unitary system of coordinates which spans completely the function space. The β_k or the *Fourier coefficients* are the *components of the vector* ψ in this system. These unitary axes establish, by their very definition, a denumerable set. *Therefore, they permit reduction of the properties of the functional space to those of a space with an infinite denumerable number of dimensions.*

If we form the scalar product of ψ_i and ψ and take (1.20) into account, we obtain

$$(\psi_i \cdot \psi) = \sum_k \int \psi_i^* \beta_k \psi_k \mathrm{d}\tau = \beta_i,. \tag{1.22}$$

a well-known formula from Fourier which allows us to calculate the coefficients β_i. It expresses that β_i is the "orthogonal projection" of ψ on the axis ψ_i. We obtain in the same way the fundamental formula of Parseval

$$(\psi \cdot \psi) = \sum_{ik} \int \beta_i \psi_i \beta_k \psi_k \mathrm{d}\tau = \sum_i \beta_i \beta_i . \qquad (1.23)$$

If one can show that this relation is correct for an arbitrary continuous function, one is sure that the set of ψ_i is complete. The choice of orthogonal functions that span the function space depends essentially on the problem that is studied. There is an infinite number of possible choices, whether the domain D is finite or infinite. The different complete sets can be derived from each other by a change of coordinates or unitary transformation (compare § 2.2).

5. Operators

5.1. TRANSFORMATION OF THE FUNCTION SPACE INTO ITSELF BY A LINEAR OPERATOR

An *operator* is a symbol which establishes a correspondence between any vector ψ in function space and any other vector. For example, the operator x acting on $\psi(x)$ establishes a correspondence between $\psi(x)$ and a function $\varphi(x) = x\psi(x)$; in the same way the operator $\mathrm{d}/\mathrm{d}x$ acting on $\psi(x)$ gives us

$$\psi'(x) = (\mathrm{d}/\mathrm{d}x)\psi(x).$$

The operators that interest us in quantum mechanics are *linear operators.* Any linear operator A satisfies the following three conditions:

1. If α is a number: $\mathrm{A}(\alpha\psi) = \alpha \mathrm{A}\psi$.
2. $\mathrm{A}(\psi+\varphi) = \mathrm{A}\psi+\mathrm{A}\varphi$.
3. Both the functions ψ and $\mathrm{A}\psi$ are normalizable .This implies that $\mathrm{A}\psi$ is defined in the same domain as ψ.

By definition, and by analogy with ordinary space, a mapping or transformation of the function space on itself is a correspondence established among its vectors by a linear operation.

Let us span the function space by a complete set of orthogonal axes $\psi_1, \psi_2, \ldots$. Let ψ be one of its vectors with the components $\beta_1, \beta_2, \ldots$, then the development in a series of orthogonal basis functions can be written as

$$\psi = \sum_k \beta_k \psi_k . \qquad (1.21)$$

The projection A makes this vector correspond to a new vector

$$\psi' = \mathrm{A}\psi,$$

from which the components β'_i, in the same coordinate system can be expressed as a linear function (a series with constant coefficients) of the coefficients β_i, assuming of course that A was linear,

$$\beta'_i = \sum_k a_{ik}\beta_k . \tag{1.24}$$

We write:

$$\psi' = \mathrm{A}\sum_k \beta_k\psi_k = \sum_k \beta_k(\mathrm{A}\psi_k),$$

where $\mathrm{A}\psi_k$, the function ψ_k transformed by the operator A, can be developed itself in a series of fundamental functions $\psi_1, \psi_2, \ldots$ with Fourier coefficients a_{ik}:

$$\mathrm{A}\psi_k = \sum_i \psi_i a_{ik} . \tag{1.25}$$

From which we find that:

$$\psi' = \sum_{i,k} (a_{ik}\beta_k)\psi_i = \sum_i \beta'_i\psi_i .$$

This implies (1.24).

The equations (1.25) are infinite in number when one deals with continuous functions ($k = 1, 2 \ldots$). They are important because they define the matrix $(a_{ik}) = \mathsf{A}$ which represents the linear operator A in the coordinate system ψ_i. The matrices which we have introduced this way to associate with every linear operator and which we will call representation matrices depend essentially on the fundamental basis functions. Their components can be obtained, according to (1.25), by forming the scalar products

$$(\psi_i \cdot \mathrm{A}\psi_k) = \int \psi_i^*\mathrm{A}\psi_k \mathrm{d}\tau = \sum_j \int \psi_i^*\psi_j a_{jk}\mathrm{d}\tau = a_{ik}; \tag{1.26}$$

(1.25) and (1.26) are constantly used in quantum theory.

The equation (1.25) may still be considered from another point of view: as a "transformation of axes" (compare (1.4a) and (1.2)). Particularly if we span the function space by another complete set of orthogonal functions φ, we will have by developing each φ_k in a series of functions ψ_i:

$$\varphi_k = \sum_i \psi_i u_{ik} . \tag{1.27}$$

Hence if we introduce a new set of orthogonal functions the linear operators will undergo a similarity transformation (see (1.7)). Since the transformation matrix $\mathsf{U} = (u_{ik})$ conserves the orthogonality of the basis functions it is a *unitary matrix.* One can verify that this satisfies (1.9).

The arguments of § 1.6 are still valid under the condition that we give the infinite matrices, with which we are dealing here, convenient *convergence* properties. We will obtain for the expression of the transformation A in the coordinate system φ_k,

$$\mathsf{A}' = \mathsf{U}^{-1}\mathsf{A}\mathsf{U}. \tag{1.27a}$$

5.2. BILINEAR FORMS. HERMITIAN OPERATORS

In n-dimensional space the scalar product $(\boldsymbol{x} \cdot \mathrm{A}\boldsymbol{y})$ is a bilinear form of the components of the two vectors $\boldsymbol{x}$ and $\boldsymbol{y}$ (compare (1.10)). We will establish the same product $(\psi \cdot \mathrm{A}\varphi)$ supposing that our functions are developed according to (1.21):

$$\psi = \sum_k \beta_k \psi_k; \qquad \varphi = \sum_j \gamma_j \psi_j.$$

Taking into account (1.25) we obtain

$$(\psi \cdot \mathrm{A}\varphi) = \int \psi^* \mathrm{A}\varphi \, \mathrm{d}\tau = \sum_{i,k,l} \beta_i^* \psi_i^* \psi_l a_{lk} \gamma_k \mathrm{d}\tau = \sum_{i,k} \beta_i^* a_{ik} \gamma_k, \tag{1.28}$$

which is an infinite bilinear form of Fourier coefficients of the components of the two functions ψ and φ in the basis system ψ_i. Since one defines an *Hermitian* operator in an n-dimensional space by the condition

$$(\psi \cdot \mathrm{A}\varphi) = (\mathrm{A}\psi \cdot \varphi) = \sum_{k,i,l} \int \psi_k^* a_{ki}^* \beta_i^* \gamma_l \psi_l \mathrm{d}\tau = \sum_{i,k} \gamma_k a_{ki}^* \beta_i^*, \tag{1.29}$$

from which, according to (1.28)

$$a_{ki}^* = a_{ik}, \tag{1.29a}$$

thus, the corresponding matrix is Hermitian too.

5.3. REDUCTION OF AN HERMITIAN OPERATOR TO ITS MAIN AXES

Let us reason again by analogy with the n-dimensional space: we want to find a system of basis vectors $\psi_1, \psi_2 \ldots$ to span the function space such that each vector $\mathrm{A}\psi$ is "parallel" to ψ, or more precisely that the function $\mathrm{A}\psi_i$, the transformation of ψ_i by the operator A, be equal to ψ_i multiplied by a constant:

$$\mathrm{A}\psi_i = \alpha_i \psi_i. \tag{1.30}$$

The functions found in this way are the *eigenfunctions* or *eigenvectors* of the operator A; the numerical values α_i its *eigenvalues.*

The solution of this problem for a space with a finite number of dimensions is relatively simple at least in principle (§ 3) and can be done by purely algebraic methods. In function space, on the contrary, complications and considerable difficulties appear and rigorous discussion of the equation (1.30)[1] has not been completed by the mathematicians.

The problem of finding α_i is algebraic, differential, or integral according to whether the operator A is algebraic, differential, or integral. The Schrödinger equation (2.15) is a particular case: The Hamilton operator to which it refers is a differential operator. Discussion of physical problems presented by quantum mechanics has lead to a better understanding of equation (1.30) since the work of Hilbert.

Depending on the nature of the operator A and on the size of the domain D, which restricts the variables on which the function ψ depends, we are dealing with two different cases.

1. The easiest case arises when the square summable solutions of equation (1.30) form a denumerable set, corresponding to a discontinuous and denumerable set of eigenvalues $\alpha : \alpha_1, \alpha_2, \ldots$. There may correspond several eigenfunctions ψ_k to certain constants α_i, but this multiplicity is always supposed to be finite. This is expressed by saying that the *spectrum of eigenvalues of the eigenfunctions is discontinuous or discrete.*

This case generalizes in a most direct way the results obtained in n-dimensional space. It occurs most often if the domain D is finite (as e.g. the case of free particles confined to a box), but this condition appears to be neither necessary nor sufficient. It happens for instance in the quantum theory of a harmonic oscillator that the levels form a discontinuous spectrum although the size of the domain is infinite.

It can be shown in the case of discrete eigenvalues that the eigenfunctions ψ_i of the operator A form a complete set of orthogonal functions which span the function space completely[2]. In order to prove the orthogonality of two eigenfunctions ψ_i and ψ_k corresponding to two different eigenvalues of an *Hermitian* operator A, we have

$$\mathrm{A}\psi_i = \alpha_i \psi_i; \quad \mathrm{A}\psi_k = \alpha_k \psi_k; \quad \alpha_i \neq \alpha_k.$$

Multiplying the first equation by ψ_k and the second by ψ_i and substracting,

[1] One finds a basic discussion in J. v. NEUMANN [1932].

[2] See, for example, HILBERT and COURANT [1930], Chap. 5 and 6 or P. M. MORSE and H. FESHBACH [1953] p. 727 and p. 728.

and using

$$(A\psi_i \cdot \psi_k) = (\psi_i \cdot A\psi_k),$$

we have

$$(\alpha_i - \alpha_k)(\psi_i \cdot \psi_k) \doteq 0 \quad \text{i.e.} \quad (\psi_i \cdot \psi_k) = 0.$$

If several eigenfunctions, for example ψ_1, ψ_2, ψ_3, belong to the same eigenvalue α we have *degeneracy*: every linear combination of these three functions is again an eigenfunction belonging to the value α, because the three equations

$$A\psi_1 = \alpha\psi_1; \qquad A\psi_2 = \alpha\psi_2 \quad \text{and} \quad A\psi_3 = \alpha\psi_3,$$

after multiplication with three arbitrary constants β_i and summation will give

$$\sum_{i=1}^{3} \beta_i A\psi_i = A \sum_i \beta_i \psi_i = \alpha \sum_i \beta_i \psi_i.$$

Hence one can always choose three orthogonal linear combinations among them.

In order to make use of the analogy that exists between the function space and a n-dimensional space to the largest extent, we will consider an arbitrary function ψ having components $\beta_1, \beta_2, \ldots$ in a coordinate system of orthogonal functions, as in equation (1.21). The scalar product $(\psi \cdot A\psi)$ can be written as a quadratic form in β_i according to (1.28):

$$(\psi \cdot A\psi) = \sum_{i,k} \beta_i^* a_{ik} \beta_k.$$

But if the basis functions are eigenfunctions of the operator A, (1.25) can be replaced by (1.30) and we obtain

$$(\psi \cdot A\psi) = \sum_{i,k} \int \beta_i^* \psi_i^* \beta_k \alpha_k \psi_k \mathrm{d}\tau = \sum_i \alpha_i \beta_i^* \beta_i. \tag{1.31}$$

Thus the quadratic form $(\psi \cdot A\psi)$ is reduced to a "sum of squares".

2. It happens very often and particularly if the domain D is infinite that in addition to the discontinuous spectrum we have a continuous spectrum of eigenvalues, and it may even happen that the discontinuous spectrum is not there at all. The theory of atoms with a central force field furnishes an example: the spectrum of energy levels is discontinuous up to a certain limit (the ionization potential) and becomes continuous above this. Since the essential character of the orthogonal functions is that they form a denumer-

able set, obviously, one does not have the right to apply without modification the language of linear vector spaces[1].

The development in Fourier series (1.21) is then replaced at least partially (compare (1.32) below) by integrals. These integrals can be considered as the limits of series of orthogonal functions. This permits us at least to a certain extent to give a sense to the expression of orthogonal functions in the continuous case. It will be sufficient to give two examples. Let us consider first functions of a single variable x defined in an arbitrary finite or infinite domain D and the operator "multiplication by x": (1.30) has now the form

$$x\psi(x) = \alpha\psi(x),$$

an equation which has to hold everywhere in the domain D. It is obviously impossible to construct an analytical function which satisfies this condition, but we will imagine a function ψ_α which is everywhere zero except at the point $x = \alpha$ and we would consider this as an eigenfunction of the preceding equation. In order that this has a sense one has, moreover, to demand that the integral over the product of this function and an arbitrary function $f(x)$ is not zero. This makes it necessary that the ψ_α becomes infinite for $x = \alpha$. We will take this integral equal to $f(\alpha)$:

$$\int f(x)\psi_\alpha(x)\mathrm{d}x = f(\alpha).$$

If we permit the existence of such functions we observe that the spectrum of eigenvalues and eigenfunctions of the operator x is continuous in the domain D as the value of α is arbitrary. Our function $\psi_\alpha(x)$ coincides with the Dirac delta function $\delta(x-\alpha)$. It is possible to construct analytical expressions that have this function as their limit[2].

We have obviously

$$\int_{\mathrm{D}} \psi_\alpha(x)\psi_\beta(x)\mathrm{d}x = 0, \qquad \alpha \neq \beta$$

---

[1] The label i in (1.20) has to be replaced by a parameter E that varies continuously

$$\int \psi_i^*(x)\psi_k(x)\mathrm{d}\tau \to \int \psi^*(x, E)\psi(x, E')\mathrm{d}\tau$$

and we will see below that the Kronecker δ_{ik} must be replaced by a new kind of function of the variable $(E-E')$.

[2] For explicit representations see HEITLER, [1957], section 8.

This function was accepted reluctantly by mathematicians because of its non-rigorous definition. It was known, however, that the results obtained via the δ-function could always be repeated without its use. Later a French mathematician, L. Schwartz, incorporated the underlying ideas into mathematics by defining a generalized notion of the idea of a function (LIGHTHILL [1958]).

since one of the functions becomes zero when the other one is different from zero. So we may qualify the symbols $\psi_\alpha(x) = \delta(x-\alpha)$ to a certain extent as orthogonal and normal functions depending on a *continuous index* α. The "norm", however, is infinite as we can see from:

$$\delta(x-x_1)\delta(x-x_2) = \delta\left(x-\frac{x_1+x_2}{2}\right)\delta(x_1-x_2)$$

if now $x_1 = x_2$ the "norm" is $\delta(0)$.

The second example will be given with Fourier integrals. It will be useful in the next chapter. We know that to extend the domain of validity of the Fourier series to an arbitrary interval $-a$, $+a$ (a is real) it is sufficient to make a change of the variable: $t = \pi x/a$. In this way we obtain for functions ψ defined in this domain, having the periodicity $2a$ and satisfying certain continuity conditions, the well-known developments:

$$\psi(x) = (2a)^{-\frac{1}{2}} \sum_{n=-\infty}^{+\infty} c_n \exp(in\pi x/a),$$

$$c_n = (2a)^{-\frac{1}{2}} \int_{-a}^{+a} \psi(x) \exp(-in\pi x/a)\mathrm{d}x,$$

(n is an integer).

The orthogonal basis functions are the exponentials

$$(1/\sqrt{2a}) \exp(in\pi x/a);$$

they form a complete set. These are the eigenfunctions of the operator $-\mathrm{i}(\mathrm{d}/\mathrm{d}x)$, because we have

$$-\mathrm{i}(\mathrm{d}/\mathrm{d}x) \exp(in\pi x/a) = (n\pi/a) \exp(in\pi x/a),$$

n has to be an integer in order that the functions are periodic over a distance $2a$. The normalizing factor $1/\sqrt{2a}$ decreases when the domain D extends. If we let a go to infinity it is convenient to introduce:

$$\nu = n\pi/a; \qquad \Delta\nu = \pi/a;$$

$$\beta_\nu = c_n\sqrt{a/\pi}; \qquad \beta_\nu \Delta\nu = c_n\sqrt{\pi/a}.$$

We obtain

$$\psi(x) = (2\pi)^{-\frac{1}{2}} \sum_{\nu=-\infty}^{+\infty} \beta_\nu \Delta\nu \exp(\mathrm{i}\nu x), \qquad \beta_\nu = (2\pi)^{-\frac{1}{2}} \int_{-a}^{+a} \psi(x) \exp(-\mathrm{i}\nu x)\mathrm{d}x,$$

and if $a \to \infty$

$$\psi(x) = (2\pi)^{-\frac{1}{2}} \int_{-\infty}^{+\infty} \beta_\nu \exp(i\nu x) d\nu, \quad \beta_\nu = (2\pi)^{-\frac{1}{2}} \int_{-\infty}^{+\infty} \psi(x) \exp(-i\nu x) dx. \tag{1.32}$$

These are the formulas of Fourier where the index ν is a continuous variable. The series development is replaced by an integral.

The formulas similar to (1.23) and (1.31) are

$$(\psi \cdot \psi) = \int_{-\infty}^{+\infty} \psi^* \psi \, dx = \int_{-\infty}^{+\infty} \beta_\nu^* \beta_\nu d\nu,$$

and

$$(\psi \cdot -i(d/dx)\psi) = -i \int \psi^*(d\psi/dx) dx = \int_{-\infty}^{+\infty} \nu \beta_\nu \beta_\nu d\nu,$$

as one can easily verify.

The preceding argument is not a proof but the passing to the limit ∞ can easily be justified[1]. As a result we see that the eigenvalues ν of the operator $-i(d/dx)$ form a continuous spectrum in the interval $-\infty$, $+\infty$. The corresponding eigenfunctions:

$$\lim_{\Delta\nu \to 0} \sqrt{\Delta\nu/2\pi} \exp(i\nu x)$$

have a normalizing factor that goes to zero if the domain D goes to infinity i.e. in the limit their "amplitude" is zero.

We have seen in these two examples the difficulties that will occur if one extends the idea of eigenfunctions to the case of continuous spectra. It is therefore better in physical problems to avoid if possible the continuous spectrum by a limitation of the domain D. The theory of black body radiation is a well-known example of this procedure: following Rayleigh and Jeans one supposes the radiation enclosed in a rectangular box with perfectly reflecting walls. The amplitude ψ of the waves is zero at the boundaries and therefore may be developed in a Fourier series and it is not necessary to use an integral expression.

A similar procedure is used in solid state physics where the Born-von Kármán periodic boundary conditions enable us to use a Fourier series also.

In the general case in which we have a juxtaposition of a continuous spectrum and a discontinuous spectrum of eigenfunctions the last, although

[1] See the standard books on calculus (e.g. WEBSTER, [1955] p. 153.)

consisting of an infinite number of functions, does not form a complete set. The combination of the two spectra, however, does form a complete system. The development of an arbitrary function[1] consists then of a sum of a series corresponding to the discontinuous spectrum, and of an integral providing the continuous spectrum.

$$\psi = \sum_k \beta_k \psi_k + \int \beta_\lambda \psi_\lambda \mathrm{d}\lambda \tag{1.33}$$

a formula that includes as a particular case the series and the integrals of Fourier.

[1] At least the functions that occur in physics.

Comparative table of the properties of function space and n-dimensional vector space

Orthogonal basis vectors	$\boldsymbol{e}_1\boldsymbol{e}_2 \ldots \boldsymbol{e}_n.$		$\psi_1\psi_2 \ldots \psi_n \ldots$	
Components of a vector	$\begin{cases} \boldsymbol{x} = \sum_{i=1}^{n} x_i \boldsymbol{e}_i \\ \boldsymbol{y} = \sum_i y_i \boldsymbol{e}_i \end{cases}$	(1.1)	$\left.\begin{array}{l} \psi = \sum_i \beta_i \psi_i, \\ \varphi = \sum_i \gamma_i \psi_i, \end{array}\right\}$	(1.21)
Length or norm of a vector	$x^2 = (\boldsymbol{x} \cdot \boldsymbol{x}) = \sum_i x_i^* x_i$	(1.8)	$(\psi \cdot \psi) = \int \psi^* \psi \mathrm{d}\tau = \sum_i \beta_i^* \beta_i,$	(1.23)
Scalar product	$(\boldsymbol{x} \cdot \boldsymbol{y}) = \sum_i x_i^* y_i$	(1.8a)	$(\psi \cdot \varphi) = \int \psi^* \varphi \mathrm{d}\tau = \sum_i \beta_i^* \gamma_i,$	(1.18)
Expression for one component	$x_i = (\boldsymbol{e}_i \cdot \boldsymbol{x})$	(1.8c)	$\beta_i = \int \psi_i^* \psi \mathrm{d}\tau = (\psi_i \cdot \psi).$	(1.22)
Linear operators (mappings and coordinate transformations)	$\boldsymbol{y} = \mathrm{A}\boldsymbol{x} = \mathrm{A}\left(\sum_i x_i \boldsymbol{e}_i\right) = \sum \boldsymbol{e}_i a_{ik} x_k$		$\psi' = \mathrm{A}\psi = \mathrm{A}\left(\sum_i \beta_i \psi_i\right) = \sum_{i,k} \psi_i a_{ik} \beta_k,$	
	$y_i = \sum_k a_{ik} x_k$	(1.4)	$\beta_i' = \sum_k a_{ik} \beta_k,$	(1.24)
	$\eta_k = \mathrm{A}\boldsymbol{e}_k = \sum_i \boldsymbol{e}_i a_{ik}$	(1.4a)	$\psi_k' = \mathrm{A}\psi_k = \sum_i \psi_i a_{ik}.$	(1.25)
Bilinear forms	$(\boldsymbol{x} \cdot \mathrm{A}\boldsymbol{y}) = \sum_{i,k} x_i^* a_{ik} y_k,$	(1.10)	$(\psi \cdot \mathrm{A}\varphi) = \int \psi^* \mathrm{A}\varphi \mathrm{d}\tau = \sum_{ik} \beta_i^* a_{ik} \gamma_k.$	(1.28)
Hermitian forms	$(\boldsymbol{x} \cdot \mathrm{A}\boldsymbol{y}) = (\mathrm{A}\boldsymbol{x} \cdot \boldsymbol{y}),$	(1.11)	$(\psi \cdot \mathrm{A}\varphi) = (\mathrm{A}\psi \cdot \varphi),$	(1.29)
	$a_{ki}^* = a_{ik}$	(1.11a)	$a_{ki}^* = a_{ik},$	(1.29a)
Reduction to main axes	$\mathrm{A}\boldsymbol{x}_i = \alpha_i \boldsymbol{x}_i$	(1.14)	$\mathrm{A}\psi_i = \alpha_i \psi_i$	(1.30)
Eigenvalues Eigenvectors	$\boldsymbol{x} \cdot \mathrm{A}\boldsymbol{x} = \sum_i \alpha_i x_i^* x_i$	(1.12)	$(\psi \cdot \mathrm{A}\psi) = \sum_i \alpha_i \beta_i^* \beta_i$	(1.31)

CHAPTER 2

THE PRINCIPLES OF QUANTUM MECHANICS

1. Waves

1.1. CLASSICAL WAVES

In classical physics wave motion in a continuum is described by the three-dimensional wave equation,

$$\nabla^2\psi - \frac{1}{c^2}\frac{\partial^2\psi}{\partial t^2} = 0. \tag{2.1}$$

A possible elementary solution of this equation is given by

$$\psi(x, y, z, t) = A \exp\left\{-2\pi \mathrm{i}\nu\left(t - \frac{\alpha x + \beta y + \gamma z}{c}\right)\right\} \tag{2.2}$$

where A is the amplitude, ν is the frequency, c the velocity of propagation, and α, β and γ the direction cosines of the normal to the plane wavefront. The wavelength $\lambda = c/\nu$ which we prefer to replace by the wave vector $\boldsymbol{k}$, which has the components

$$k_x = 2\pi\alpha\nu/c; \qquad k_y = 2\pi\beta\nu/c; \qquad k_z = 2\pi\gamma\nu/c. \tag{2.3}$$

The magnitude $\boldsymbol{k}$ is 2π times the number of waves per cm; the direction of $\boldsymbol{k}$ is the direction of the wavefront.

Each complete solution of the wave equation (in a box or in infinite space) can be obtained by superposition of elementary solutions. The function that satisfies (2.1),

$$\psi(x, y, z, t) = \sum_{\boldsymbol{k}=-\infty}^{\infty} A_{\boldsymbol{k}} \exp \mathrm{i}\{\boldsymbol{k}\cdot\boldsymbol{r} - 2\pi\nu_{\boldsymbol{k}} t\} \tag{2.2a}$$

represents an arbitrary disturbance which may be electromagnetic or elastic. The right-hand-side of equation (2.2a) represents a Fourier series or, in case of an infinite space, a Fourier integral and either side of the equation describes at a given moment t a spatial distribution of waves. If $\psi(x, y, z, t)$

is substantially different from zero over a localized region the function is usually called a *wave packet*.

If we take together all waves of the same frequency, that is of the same absolute k-value, but of different directions (2.2a) becomes

$$\psi = \sum_{k=0}^{\infty} a_k(x, y, z) \exp\{-2\pi i \nu_k t\}. \tag{2.4}$$

The coefficients of (2.2a) and (2.4) are generally complex because the different spectral components usually have different phases. The relative phases will determine whether we will have constructive or destructive interference, one of the most striking properties of waves.

The product $a_k^* a_k(x, y, z)$ is the *intensity* of a particular spectral component ν_k of the wave at a certain point x, y, z. On the other hand, if we look for the average value $\langle\psi^*\psi\rangle$ of $\psi^*\psi$ over a time T long compared to the periods $1/\nu_k$, we have:

$$\langle\psi^*\psi\rangle = \lim_{T\to\infty} \frac{1}{T} \sum_{rs} \int_0^T a_r^* a_s e^{2\pi i(\nu_r-\nu_s)t} dt = \sum_r a_r^* a_r(x, y, z) \tag{2.5}$$

or, *the average intensity* $\langle\psi^*\psi\rangle$ *at a point* (x, y, z) *is the sum over all the intensities of the different spectral components.*

The expression (2.5) is obtained by integrating separately the terms in which $\nu_r \neq \nu_s$ and those in which $\nu_r = \nu_s$. The first set of integrals are oscillating functions of T with a constant amplitude. After multiplication with T^{-1} the product goes to zero for $T \to \infty$. The second integral is proportional to T hence it is unnecessary to take the limit of the product.

1.2. QUANTUM MECHANICAL WAVES

Early in the development of quantum mechanics a postulate was introduced that the energy was proportional to a frequency

$$E = h\nu \tag{2.6}$$

or

$$E = \hbar\omega \tag{2.7}$$

where $\hbar = h/2\pi$.

It was successively realized that the nature of the wave was not always electromagnetic and an additional postulate

$$p = \hbar k \tag{2.8}$$

which was originally suggested by electromagnetism and relativity, was proposed for matter waves.

The result is that a particle is associated with a quantum wave. In particular a free particle is represented by:

$$\psi = A \exp\left\{\frac{2\pi i}{h}(\boldsymbol{p}\cdot\boldsymbol{r} - Et)\right\}. \tag{2.2b}$$

The complementarity principle, that is the statement that electrons can have either a wave aspect or a corpuscular aspect but not both, can be clarified very beautifully by considering wave packets. Consider the following two cases.

a) If we have a single wave (2.2b), that is one frequency only, the wave is a "monochromatic" wave, the momentum is exactly determined by (2.8) and the position is completely washed out;

b) However, we do have superpositions like (2.2a) and the possibility exists to choose the spatial distribution of

$$\psi^*\psi = \sum_{k,l} a_k^* a_l e^{2\pi i(\nu_k - \nu_l)t}$$

such that $\psi^*\psi$ is only different from zero in a certain region of space, say at $t = 0$. If we now interpret the intensity as proportional to the probability to locate a particle, we obtain in this case a reasonable determination for the position of the particle. This occurred at the expense of a well-determined momentum since we must use not one but many different k-values in our wave packet.

It is interesting to notice that the question as to *which* of the two pictures is adequate is completely determined by the experiment. This is a special example of a general pattern in quantum mechanics viz. that the choice of the wave function is dictated by the experiment. This choice is called the *state* of the system.

If one tries to construct wave packets with the smallest possible spread in velocity and position, one is led to a Gaussian packet which has the property that the product of the root mean squares of the momentum and of the position obeys Heisenberg's uncertainty principle. Hence (2.2b) or a superposition of similar waves gives a complete and consistent description of the free particle.

1.3. THE FREE PARTICLE

Which equation has (2.2b) as a solution? At first one would be tempted to quote (2.1) but it turns out that by introducing a slightly different form one can give the formalism an analogy to classical mechanics such that one can generalize to a non-free particle, i.e. bound by some potential field, in a natural way.

The function (2.2b) is a solution of an equation of the first order in time, (but with an imaginary coefficient), i.e.:

$$\frac{\hbar^2}{2m}\nabla^2\psi = \frac{\hbar}{\mathrm{i}}\frac{\partial\psi}{\partial t}, \tag{2.9}$$

and by comparing this with the classical energy momentum relation

$$\frac{p^2}{2m} = E$$

we find the following correspondence:

$$H \leftrightarrow -\frac{\hbar}{\mathrm{i}}\frac{\partial}{\partial t} \tag{2.10}$$

$$\boldsymbol{p} \leftrightarrow \frac{\hbar}{\mathrm{i}}\boldsymbol{\nabla}. \tag{2.10a}$$

Hence by introducing (2.10) and (2.10a) and by proclaiming that: *A physical quantity or observable is an operator applied to a wave function,* we can drop the postulates (2.7) and (2.8). The reason is that the operators (2.10) and (2.10a) can be related to certain *eigenvalues*, and upon substitution of the *eigenfunction* (2.2b) in their respective eigenvalue equations

$$\mathrm{H}_{\mathrm{op}}\psi = E\psi; \tag{2.11}$$

$$\mathbf{p}_{\mathrm{op}}\psi = \boldsymbol{p}\psi \tag{2.11a}$$

one obtains (2.7) and (2.8). Since the operators are Hermitian, the eigenvalues are always real (comp. Chapter 1, § 5). There are of course circumstances in which the *state* is not pure (monochromatic), i.e. that a distribution of eigenvalues is found and thus the physical observable is not sharply determined. The interpretation of such a case will be discussed in § 5 of this chapter.

2. The Schrödinger Equation

In this section we discuss the Schrödinger equation, i.e. the general wave equation for conservative mechanical systems, and also we will introduce the many-particle equation.

2.1. SCHRÖDINGER EQUATION

A conservative system can be described by a Hamiltonian. According to (2.10) we have

$$\frac{\hbar}{i}\frac{\partial\psi}{\partial t}+\mathrm{H}\psi = 0. \tag{2.13}$$

The Hamiltonian can usually be separated into a kinetic and a potential energy:

$$\mathrm{H} = \frac{1}{2m}\mathbf{p}^2+V, \tag{2.14}$$

hence we have the following equation, named after Schrödinger

$$\frac{\hbar}{i}\frac{\partial\psi}{\partial t}-\frac{\hbar^2}{2m}\nabla^2\psi+V(x, y, z, t)\psi = 0. \tag{2.13a}$$

If V does not depend on t we obtain in general an infinite set of functions $\psi_k(x, y, z)$ such that the operation H is reduced to multiplication with a real constant, i.e.

$$\mathrm{H}\psi_k = E_k\psi_k. \tag{2.15}$$

If we are able to determine E_k we find by integrating (2.13a) that ψ_k is a periodic function of the time t. Hence we call this the *stationary* state of the system:

$$\psi_k(x, y, z, t) = \psi_k(x, y, z)\mathrm{e}^{-\mathrm{i}E_k t/\hbar}. \tag{2.16}$$

In this state the energy H has a well-defined value E_k. The spatial function $\psi(x, y, z)$ is a solution of the time independent Schrödinger equation:

$$\nabla^2\psi+\frac{2m}{\hbar^2}(E-V)\psi = 0, \tag{2.17}$$

which can be obtained by separation of variables from (2.13a), provided, of course, the potential energy does not depend on t.

After introducing the potential energy the eigenfunctions of H are no longer simultaneous eigenfunctions of p_{op}. This has a very good physical reason because if we would ask for the value of the momentum at a fixed E we would find that it is no longer a constant. Hence if we apply p_{op} to the eigenfunction we do not expect to find a time independent distinct eigenvalue.

2.2. THE n-PARTICLE PROBLEM

It is important to realize that one has to add a new concept in order to handle two or more particles. V is now of the form $V(x_1, y_1, z_1, x_2, y_2, z_2,$

$\ldots, t)$ and ∇^2 is generalized to

$$\nabla_1^2+\nabla_2^2\ldots = \frac{\partial^2}{\partial x_1^2}+\ldots\frac{\partial^2}{\partial x_2^2}\ldots. \tag{2.13b}$$

We have instead of (2.13a)

$$\frac{\hbar}{i}\frac{\partial\psi}{\partial t}-\sum_i\frac{\hbar^2}{2m_i}\nabla_i^2\psi+V(x_1\ldots t)\psi = 0$$

and the time independent equation (2.17) becomes:

$$\sum_i\frac{\hbar^2}{2m_i}\nabla_i^2\psi+(E-V)\psi = 0. \tag{2.17a}$$

An n-particle system in quantum mechanics is described in a $3n$-dimensional space: *the configuration space.*
Since the configuration space is infinite and since the potential goes to zero at large distances[1], the eigenvalue spectrum consists of a continuous and a discontinuous part[2].

The continuous eigenvalues correspond to functions which are everywhere bounded, but which are not square summable since they do not go to zero at infinitity. In the case of a central Coulomb force they correspond to hyperbolic orbits. The discrete eigenvalues form a denumerable set finite or infinite. In the last case they approach the zero energy value rapidly when they have their largest amplitude at places farther and farther away from the attractive center.

3. Angular Momentum

3.1. OPERATOR

The only physical quantities that we have studied so far from a quantum point of view are the energy and the momentum of a particle. They appeared as operators acting on the wave function.

$$\mathrm{H} = -\frac{\hbar}{i}\frac{\partial}{\partial t}, \qquad \mathrm{p}_x = \frac{\hbar}{i}\frac{\partial}{\partial x}\ldots$$

It is easy to define in the same way the components of the angular momentum. Let us consider a particle of coordinates x, y, z and momentum p_x, p_y, p_z. Classically the components of its angular momentum with respect to the origin are $L_x = yp_z-zp_y$; $L_y = zp_x-xp_z$; $L_z = xp_y-yp_x$.

[1] Provided of course the (piece-wise continuous) potential function has regions in which it is sufficiently negative, otherwise the discrete eigenvalues may not occur at all.
[2] This is not the case for harmonic oscillators (compare Chapter 1, § 5.3)).

To translate them into quantum language we replace p_y and p_z by the corresponding operators and obtain

$$L_x = \frac{\hbar}{i}\left(y\frac{\partial}{\partial z} - z\frac{\partial}{\partial y}\right), \qquad \text{etc.} \tag{2.18}$$

3.2. OPERATORS AND GROUPS

Now we can connect the operators and the physical quantities which they represent with some simple groups or rather with the infinitesimal operations or *infinitesimal transformations* which generate them.

First let us consider a small, so-called virtual, translation δ_x that displaces without deformation the spatial distribution of the function ψ. After this translation there corresponds at the point x, y, z a value of ψ which belonged originally to the point $x-\delta x, y, z$. Then we have $\delta\psi = -(\partial\psi/\partial x)\delta x$. The component p_x of the momentum is the differential operator of this infinitesimal transformation multiplied by $i\hbar$. When the system contains f particles and undergoes a total virtual translation:

$\delta x = \delta x_1 = \delta x_2 = \ldots = \delta x_f$ the corresponding variation of ψ *in the configuration space* can be written as

$$\delta\psi = -\sum_{i=1}^{f}\frac{\partial\psi}{\partial x_i}\,\delta x$$

and the projection of the total momentum P_x on the x axis is represented by the differential operator of this transformation multiplied by $i\hbar$:

$$P_x = +\frac{\hbar}{i}\sum\frac{\partial}{\partial x_i} = -\frac{\hbar}{i}\frac{\delta}{\delta x}.$$

Let us come back to the case of one particle. If we perform a virtual rotation $\delta\theta_x$ of the distribution of the ψ around the Ox axis: $\delta x = 0$, $\delta y = -z\delta\theta_x$, $\delta z = y\delta\theta_x$ we obtain:

$$\delta\psi = -\frac{\partial\psi}{\partial x}\delta x - \frac{\partial\psi}{\partial y}\delta y - \frac{\partial\psi}{\partial z}\delta z = \left(z\frac{\partial}{\partial y} - y\frac{\partial}{\partial z}\right)\psi\delta\theta_x.$$

If we compare this formula with (2.18) we see that the component L_x of the angular momentum is represented by the differential operator of its right-hand side multiplied by $i\hbar$.

We have

$$L_x\psi = -\frac{\hbar}{i}\frac{\delta\psi}{\delta\theta_x}$$

or symbolically

$$\mathrm{L}_x = -\frac{\hbar}{\mathrm{i}}\frac{\delta}{\delta\theta_x}. \tag{2.19}$$

This definition is general. It can be extended to a system containing any number of particles provided one calculates first the modifications of ψ in the configuration space. We obtain immediately

$$\mathrm{L}_x = \frac{\hbar}{\mathrm{i}}\sum_i \left(y_i \frac{\partial}{\partial z_i} - z_i \frac{\partial}{\partial y_i}\right). \tag{2.18a}$$

Hence there exists in quantum theory a correspondence between the momentum and the operator which generates the translation group, as well as between the angular momentum and the operator which generates the rotation group. The equation (2.10)

$$\mathrm{H} = -\frac{\hbar}{\mathrm{i}}\frac{\partial}{\partial t}$$

shows that the energy is the operator which generates real translations in time. This remark about the connection between a group and certain operators will be generalized and treated in further detail in Chapter 4, § 1.2.

3.3. COMMUTATION RELATIONS

According to (2.18a) we have:

$$(\mathrm{L}_x\mathrm{L}_y - \mathrm{L}_y\mathrm{L}_x)\psi$$

$$= -\hbar^2 \sum \left[\left(y\frac{\partial}{\partial z} - z\frac{\partial}{\partial y}\right)\left(z\frac{\partial}{\partial x} - x\frac{\partial}{\partial z}\right) - \left(z\frac{\partial}{\partial x} - x\frac{\partial}{\partial z}\right)\left(y\frac{\partial}{\partial z} - z\frac{\partial}{\partial y}\right)\right]\psi$$

$$= -\hbar^2 \sum \left(y\frac{\partial}{\partial x} - x\frac{\partial}{\partial y}\right)\psi,$$

from which it follows that:

$$\begin{aligned} \mathrm{L}_x\mathrm{L}_y - \mathrm{L}_y\mathrm{L}_x &= -\frac{\hbar}{\mathrm{i}}\mathrm{L}_z, \\ \mathrm{L}_y\mathrm{L}_z - \mathrm{L}_z\mathrm{L}_y &= -\frac{\hbar}{\mathrm{i}}\mathrm{L}_x, \\ \mathrm{L}_z\mathrm{L}_x - \mathrm{L}_x\mathrm{L}_z &= -\frac{\hbar}{i}\mathrm{L}_y. \end{aligned} \tag{2 20}$$

These are the fundamental commutation relations for the operators which represent the components of the angular momentum. One can prove by a similar calculation the commutation relations of Heisenberg.

$$\mathrm{p}_x x - x\mathrm{p}_x = \frac{\hbar}{\mathrm{i}}, \qquad \mathrm{p}_x y - y\mathrm{p}_x = 0. \tag{2.20a}$$

The operators discussed above are corresponding to the most common invariants in classical mechanics: energy conservation, conservation of linear momentum and conservation of angular momentum. If a certain number of these quantities happen to be conserved in a problem, for instance L_x, L_y and L_z the three components of angular momentum, then we can also say that any polynomial in L_x, L_y and L_z:

$$P = P(\mathrm{L}_x, \mathrm{L}_y, \mathrm{L}_z)$$

is a quantity that is conserved. In the classical case such generalizations are not very interesting since they can be reduced again to the fact that the components of angular momentum are separately conserved. In quantum mechanics they are very helpful as we will see in Chapter 6.

4. The Postulates of Quantum Mechanics

We have now reached a point where we can generalize the results which we have obtained and thus make a table that lists the fundamental postulates of quantum mechanics. We shall use for this a geometrical language, i.e. the theory of function space which we have discussed in Chapter 1, § 4.

Let us make two preliminary remarks:

1. The operators A to which a physical meaning has been given, either differential or multiplicative, are all linear and Hermitian (Chapter 1, §§ 5.1 and 5.2). This is obvious for real multiplicative operators. For the case of momentum and energy which are differential operators we can easily verify equation (1.29) by partial integration provided that ψ and φ vanish at the boundaries of the domain D [1].

We shall see later why a physical quantity can only be represented by an Hermitian operator.

2. The Schrödinger equation, (2.11) or (2.17), and more generally equation (1.30) for the definition of the eigenfunctions of an operator are linear and homogeneous. Thus the wave functions are defined except for a multi-

[1] To verify eq. (1.29) for the Hamiltonian H, one can use (2.14) taking equation (2.10a) into account.

plicative constant; hence they are not really vectors of the function space but directions or *rays*. If we wish $\psi^* \psi d\tau$ to represent a probability we must suppose that the functions ψ are *normalized*, i.e.

$$\int \psi^* \psi \, d\tau = 1.$$

With this condition ψ is completely determined except for a multiplicative constant which has *the absolute value* 1 and which can be considered as *a phase factor*.

In a certain sense quantum mechanics provides a dictionary of relations between the classical representations, which are easily seized by the imagination but do not fit exactly the facts, and an abstract formalism which allows us to predict exactly all that is predictable — at least for our present knowledge. The quantum formalism can be expressed in two different dialects: one is connected with the wave picture and the other (which is absolutely equivalent to the first) with function space and matrices. This is the reason why we shall present the first four postulates of quantum physics in the form of a tri-lingual dictionary.

I State of a system	Wave function ψ	Ray ψ of the function space
II Physical or observable quantity	Hermitian operator A acting on ψ	Hermitian matrix mapping the function space onto itself
III Observable values of this quantity	Eigenvalues α_i of this operator or characteristic constants of the equation $A\psi = \alpha\psi$	Eigenvalues of this matrix, i.e. diagonal elements α_i of this matrix, which has been put in a diagonal form by a unitary transformation of the coordinates
IV State of the system where A has a definite value α_i	Eigenfunction of A corresponding to the eigenvalue α_i	Ray of the function space which is multiplied by α_i when the operator A acts on it

V. Moreover we shall allow, following DIRAC [1958], that every Hermitian operator has a physical meaning[1].

[1] One can find in the book by WEYL [1950] (Chapter IV D, section 14) a remarkable argument that justifies this postulate or at least connects it to a very general and natural irreducibility postulate.

It follows from the postulates III and IV that if a system is in the state ψ_i, a measurement of the quantity A gives with certainty the value α_i.

VI. But if ψ is not an eigenfunction of the operator A we do not know with certainty the value of the physical quantity which this operator represents. We can expect to find the different possible values $\alpha_i, \alpha_k \ldots$ with different probabilities. In order to make our postulates more precise we have only to generalize what has been said in § 1 of this chapter: ψ can be expanded according to (1.33) in which we shall neglect the continuous spectrum for simplicity. Then the state appears as the superposition of the pure states ψ_k. Each of them is weighted by a coefficient β_k whose modulus is $\leqq 1$ and each corresponds to a given value α_k of the physical quantity A.

We shall assume that a measurement made on the system in the state ψ can yield any of those values; the probability that this value is α_k being $\beta_k^* \beta_k$.

This last postulate foreseen by Einstein, enunciated precisely by Born, and developed by Dirac can be considered as the keystone of quantum mechanics. It is the link between experiments and wave theory, giving to the latter the character of an essentially statistical theory.

If we repeat the same experiment a great number of times with identical systems all in the same state ψ we shall find for the quantity A either the value α_i, or α_k, or α_l and the relative frequency of these different results will be in the limit equal to the corresponding probabilities. The average value of A in the state ψ is:

$$\begin{aligned} \langle \mathrm{A} \rangle &= \sum_i \alpha_i \beta_i^* \beta_i = \sum_{ik} \int \alpha_i \beta_k^* \psi_k^* \beta_i \psi_i \mathrm{d}\tau \\ &= \int \psi^* \mathrm{A} \psi \, \mathrm{d}\tau = (\psi \cdot \mathrm{A}\psi), \end{aligned} \tag{2.21}$$

an important formula due to Dirac.

Let us suppose that the function space is spanned by the eigenfunctions $\varphi_1, \varphi_2, \ldots$ of another operator H. In this system of axes the quantity A is represented by a non-diagonal matrix (a_{ik}) defined by:

$$\mathrm{A}\varphi_k = \sum_i \varphi_i a_{ik}. \tag{1.25}$$

The state ψ can be considered as a superposition of states φ_k

$$\psi = \sum_k \gamma_k \varphi_k.$$

We have

$$\mathrm{A}\psi = \mathrm{A}(\sum_k \gamma_k \varphi_k) = \sum_{ik} \varphi_i a_{ik} \gamma_k,$$

and according to (2.21)

$$\langle A\rangle = \int \psi^* A\psi \, d\tau = \sum_{l,i,k} \int \gamma_i^* \varphi_i^* \varphi_i a_{ik} \gamma_k d\tau = \sum_{ik} a_{ik} \gamma_i^* \gamma_k . \tag{2.22}$$

$\langle A\rangle$ is then a quadratic form of the Fourier coefficients γ_i. This form which represents the average of a number of experimental results *is necessarily real: it must be Hermitian,* since the condition that $a_{ik}\gamma_i^*\gamma_k + a_{ki}\gamma_k^*\gamma_i$ is real implies $a_{ik} = a_{ki}^*$.

Hence we see the reason why a physical quantity must be represented only by an Hermitian operator.

Let us suppose that the state we consider is an eigenstate of the operator H, for example $\psi = \varphi_n$, $\gamma_n = 1$, $\gamma_k = 0$, $k \neq n$. From (2.22) one obtains:

$$\langle A\rangle = a_{nn} . \tag{2.23}$$

If the operator A *is represented by a matrix* a_{ik} *in the basis* $\varphi_1, \varphi_2, \ldots,$ *its expectation value for the state* φ_n *is the element* a_{nn} *of the main diagonal.*

In the foregoing sections the time t has not been explicitly considered since the wave functions which have been used depended only on the coordinates. Now we shall study the change of the states and the physical observables in time. This will allow us to elaborate the meaning of the equations (2.22) and (2.23).

5. Time Dependence of a State and of a Physical Observable

5.1. GENERAL THEORY

The easiest way to tackle this subject seems to be the following. Let $\psi(x_1, y_1, z_1, x_2, \ldots, t)$ be a wave function which will be written more concisely as $\psi(x, t)$; its variations in time follow from "the propagation equation" (2.13) or (2.13a). As in classical non-relativistic mechanics the time t is considered as a parameter and one deals with the function space $\psi(x)$, which are functions of space coordinates only. To span the function space one chooses *fixed axes* which are a complete set of orthogonal functions, for example the eigenfunctions $\psi_k(x)$ of the Hamiltonian operator H, i.e. the solutions of the Schrödinger equation (2.17a) which we shall call, following Dirac, *the Schrödinger axes.*

In the expansion of an arbitrary wave function $\psi(x, t)$

$$\psi(x, t) = \sum_k \gamma_k(t)\psi_k(x), \tag{2.24}$$

the coefficients γ vary with the time. A convenient picture is the following:

the vector $\psi(x, t)$ of the function space whose length (1.19) is equal to unity rotates around the origin obeying equation (2.13a). The problem is to determine the expression $\gamma_k(t)$ of its components along the fixed axes $\psi_k(x)$. We have

$$\mathrm{H}\psi = -\frac{\hbar}{\mathrm{i}}\frac{\partial\psi}{\partial t} = -\frac{\hbar}{\mathrm{i}}\sum_k \psi_k \frac{\mathrm{d}\gamma_k}{\mathrm{d}t}$$

and since the γ_k are ordinary numbers and the ψ_k the eigenfunctions of H we have

$$\mathrm{H}\psi = \sum_k \gamma_k \mathrm{H}\psi_k = \sum_k \gamma_k E_k \psi_k .$$

Identifying term by term one obtains

$$-\frac{\hbar}{\mathrm{i}}\frac{\mathrm{d}\gamma_k}{\mathrm{d}t} = E_k\gamma_k .$$

Finally by integration we obtain

$$\gamma_k = c_k \exp\left(-\frac{\mathrm{i}E_k}{\hbar}t\right) \tag{2.25}$$

and

$$\psi(x, t) = \sum_k c_k\psi_k(x) \exp\left(-\frac{\mathrm{i}E_k}{\hbar}t\right) \tag{2.25a}$$

the c_k being constants which may be complex.

The γ_k's are always harmonic functions of time with frequency $\nu_k = 2\pi E_k/\hbar$. In a limited sense equation (2.25a) looks like a Fourier expansion. Particularly if ψ is a steady state wave function, this expansion reduces only to one term ($c_k = 1$, $c_l = 0$ for $l \neq k$) and we find again (2.16). A steady state is an harmonic oscillation along a Schrödinger axis, this oscillation does not give rise to any radiation and therefore cannot be detected by experiment; since the modulus of the phase factor $\exp(-2\pi \mathrm{i}\nu_k t)$ remains equal to unity, the system remains on the same *ray* of the Hilbert space. If the state is not steady, i.e. if there is more than one non-zero term in (2.25a), the motion in Hilbert space will be some complicated rotation.

Let us suppose that the Hilbert space is spanned by another complete set of orthogonal functions φ_j; the transition from the Schrödinger system φ is obtained by using a unitary transformation $\mathsf{U} = (u_{jk})$ *not depending on time.* The inverse transformation corresponds to the operator U^{-1}

$$\varphi_j(x) = \sum_k \psi_k(x)u_{kj}; \qquad \psi_k(x) = \sum_j \varphi_j(x)u_{jk}^{-1}.$$

By expanding $\psi(x, t)$ in a series of the functions $\varphi_j(x)$ we have:

$$\psi(x, t) = \sum_j \eta_j(t)\varphi_j(x) = \sum_k c_k \psi_k(x) \exp(-2\pi i \nu_k t)$$

from which it follows:

$$\eta_j(t) = \sum_k u_{jk}^{-1} c_k \exp(-2\pi i \nu_k t). \tag{2.25b}$$

In the basis φ, each of the components $\eta_j(t)$ of the wave function appears as a Fourier expansion.

5.2. HEISENBERG REPRESENTATION

In expression (2.25a) it is often convenient to attach the periodic factor to the functions ψ_k instead of the coefficients c_k, i.e. to consider that the function space is spanned by variable axes which will be called *Heisenberg axes.*

$$\psi_k(x, t) = \psi_k(x) \exp(-2\pi i \nu_k t).$$

Then the state $\psi(x, t)$ appears as a superposition of steady states $\psi_k(x, t)$ where coefficients c_k are independent of time.

This change of axes has certain consequences with respect to the representation of the physical observables. This transformation corresponds to a change in our point of view which makes it possible to indicate a more precise connection between wave theory and matrix mechanics and even classical mechanics.

We have used a Dirac matrix as an abstract representation of a physical observable A. The elements a_{jk} of this matrix are certain constants defined, for example in the Schrödinger coordinate system by the equations

$$A\psi_k(x) = \sum_j \psi_j(x) a_{jk}. \tag{1.25}$$

We have been led to this representation because the possible values of a physical observable (eigenvalues of the corresponding operator) are constants independent of the state of the observed system. But physically the measurement of the quantity A is done for a real material system, the state of which varies according to the equations (2.24), (2.25) or (2.25a). The probabilities of the different eigenvalues α_i of A and its expectation value $\langle A \rangle$ are functions of time.

For example (2.22) and (2.25) give for the Schrödinger axes

$$\langle A \rangle = \sum_{jk} \gamma_j^* \gamma_k a_{jk} = \sum_{jk} c_j^* c_k a_{jk} \exp\left(i \frac{E_j - E_k}{\hbar} t\right). \tag{2.26}$$

Hence we shall not define the matrix which represents an operator A by its action on the invariable wave functions $\psi_k(x)$, but by its action on the complete stationary wave functions. Equation (1.25) becomes

$$\mathrm{A}\psi_k(x, t) = \mathrm{A}\psi_k(x) \exp\left(-\mathrm{i}\frac{E_k}{\hbar}t\right) = \sum_j \psi_j(x) a_{jk} \exp\left(-\mathrm{i}\frac{E_k}{\hbar}t\right)$$

$$= \sum_j \psi_j(x, t) a_{jk} \exp\left(\mathrm{i}\frac{E_j - E_k}{\hbar}t\right).$$

Thus we get *the Heisenberg matrices*

$$\mathsf{A} = (A_{jk}) = (a_{jk} \exp(\mathrm{i}\omega_{jk}t)); \qquad \omega_{jk} = \frac{E_j - E_k}{\hbar} \tag{2.27}$$

which differ from the Dirac matrices only by periodic factors of modulus unity. One can write

$$\langle \mathrm{A} \rangle = \sum_{jk} c_j^* c_k A_{jk}. \tag{2.26a}$$

On the other hand, (2.27) gives

$$\frac{\hbar}{\mathrm{i}} \frac{\mathrm{d}A_{jk}}{\mathrm{d}t} = (E_j - E_k) A_{jk}$$

or using the matrix notation, E_j being an element of the diagonal matrix E,

$$\frac{\hbar}{\mathrm{i}} \frac{\mathrm{d}\mathsf{A}}{\mathrm{d}t} = \mathsf{EA} - \mathsf{AE}. \tag{2.27a}$$

The transition to the arbitrary system of orthogonal functions φ_j is made by using the unitary transformation U. We have according to (1.27a)

$$\mathsf{E} \to \mathsf{H} = (H_{jk}) = \mathsf{U}^{-1}\mathsf{EU};$$

$$\mathsf{A} \to \mathsf{Q} = (q_{jk}) = \mathsf{U}^{-1}\mathsf{AU} = \left(\sum_{lm} u_{jl}^{-1} A_{lm} u_{mk}\right).$$

The coefficients q_{jk} are complicated functions of the time since the sum $\sum_{lm}$ is a series expansion which involves all the frequencies ω_{jk}; equation (2.27a) becomes:

$$\frac{\hbar}{\mathrm{i}} \frac{\mathrm{d}\mathsf{Q}}{\mathrm{d}t} = \frac{\hbar}{\mathrm{i}} \mathsf{U}^{-1} \frac{\mathrm{d}\mathsf{A}}{\mathrm{d}t} \mathsf{U} = \mathsf{U}^{-1}(\mathsf{EA} - \mathsf{AE})\mathsf{U}$$

$$= \mathsf{U}^{-1}\mathsf{EUU}^{-1}\mathsf{AU} - \mathsf{U}^{-1}\mathsf{AUU}^{-1}\mathsf{EU}$$

or

$$\frac{\hbar}{i}\frac{dQ}{dt} = HQ - QH \tag{2.28}$$

the well-known equation of Born, Heisenberg, and Jordan[1]. We shall rarely use this equation. The foregoing discussion is merely a change in notation, or a change of axes in function space. Considered as *rays* of the Hilbert space, the Heisenberg axes are not essentially different from the Schrödinger axes. The formulae (2.22) and (2.26a) differ only in the way the time factors are grouped on the right-hand side. For us, the first one is more directly connected with the principles. The time variations of $\langle A \rangle$ are apparently connected with the changes in the state of the system and with the coefficients $\gamma_k(t)$ of the expansion of $\psi(x, t)$.

In order to understand better the meaning of the expressions "expectation value $\langle A \rangle$ of a physical observable A "and "probability $|\gamma_i|^2$ of a possible value α_i", we must state more precisely what is understood in quantum theory by "measurement of a physical observable in a given system".

To perform an experiment under well-defined conditions the observer has first to determine, by the way he sets things up, the state of the system he is studying. At the initial instant this state is represented by a wave function $\psi(x, 0)$ which changes according to the Schrödinger equation (2.13) and becomes $\psi(x, t)$ at the time t of the measurement. We do not suppose that $\psi(x, t)$ is an eigenfunction of the operator A. At this time the observer changes *suddenly* the state $\psi(x, t)$ of the system and transforms this state into an eigenfunction of the operator A by the very performance of the measurement. The observed value is α_i. He performs again the experiment and starts always from the same initial state, acting always at the same instant t. This intervention brings according to quantum theory an unavoidable element of chance despite the fact that it is always the same: it gives sometimes the state ψ_i, sometimes ψ_k.... But after a great number of experiments the frequency of the different possible results allows us to attribute a definite probability to each of them, and to calculate their average (expectation value) at the time t. These probabilities and these expectation values are what the theory can predict "a priori" using the expansion of $\psi(x, t)$ in a series of eigenfunctions ψ_i.

[1] This equation which is sometimes called the equation of motion of matrix mechanics leads to a direct connection between quantum and classical mechanics. Classical equations such as Newton's law, the virial theorem, and the Lorentz force can be mimicked by using matrix notation. (Compare SCHIFF [1949] Chapter VI or DIRAC [1958] Chapter VI.)

In order that an experimental intervention of the observer can be considered as a *sudden* intervention it is necessary that the rate of spontaneous change of the system be small with respect to the rate at which the external perturbation is established.

A very simple case is when one considers the eigenstates ψ_n of the Hamiltonian operator H: all the coefficients in the expansion (2.25a) are zero except c_n the modulus of which is equal to unity. The equation (2.26) shows that $\langle A \rangle = a_{nn}$ is independent of the time t and can be considered *as a time average* of the quantity A for a system in the state ψ_n. Let us clarify the meaning of this statement: If a series of measurements of an *arbitrary* quantity A is performed for *different* systems of the same kind and in the same state ψ_n their average does not depend on the instant at which each measurement is done. *This is the reason why the states corresponding to the level E_n are said to be stationary states.*

In the old quantum theory the meaning of the expression "time average" was rather intuitive. It is not possible to indicate such a thing for a quantity represented by a non-diagonal matrix.

One must note that the stationary states of the atomic systems with which the quantum theory is principally concerned are only approximately stationary, provided "the radiation damping" is neglected (compare the next section). As the frequency increases this damping becomes more and more important, it accelerates more and more the rate of the spontaneous evolution of the systems, it decreases their "mean life" which can be less than the duration of an observation.

6. Transition Probabilities and Radiation Theory

The original basis of the present quantum theory is the Bohr postulate, i.e. an atom radiates when it changes from an energy level E_m to a lower energy level E_n; during this transition it emits a quantum of electromagnetic energy

$$\hbar\omega = W = E_m - E_n. \tag{2.29}$$

To calculate the relative intensity of the different spectral lines one must calculate the probabilities P_{mn} of the different possible transitions. A first attempt based on the correspondence principle was made by Kramers. Later, Schrödinger using a semi-classical wave picture established the exact formulae. Finally by utilizing the theory of Jeans which had been improved by Lorentz and Debye, Dirac succeeded in deriving the transition probabilities from the general principles of quantum mechanics by rigorous reasoning which made the proof of postulate (2.29) possible. We shall be satisfied only with a short and rough justification using some considerations of the correspondence principle.

In classical theory radiation is produced by a periodic change of the electric distribution. Let μ_x be the electric moment of an atom. Let us suppose that this moment is parallel to the x axis and harmonic with the frequency ν

$$\mu_x(t) = \sum_k e_k x_k = \mu \cos 2\pi\nu t.$$

According to the Maxwell-Lorentz theory, the light radiated by this atom is linearly polarized; the electric vector being parallel to Ox; the energy radiated per unit time is: (if μ is measured in e.s.u.)

$$\frac{\mathrm{d}W}{\mathrm{d}t} = \frac{2}{3c^3}(2\pi\nu)^4\mu^2 = \frac{4\pi\nu}{3c^3}(2\pi\nu)^3\mu^2 \tag{2.30}$$

where c is the velocity of light[1].

According to the correspondence principle the emission of light is determined by the electric moment in quantum theory. Thus when the atom is in a stationary state ψ_m, one must decompose this moment into its different components related to the different possible transitions, i.e. one must form the matrix (μ_{mn}). This is very simple: μ, as any physical observable, is an operator acting on the wave function which represents the state of the atom, i.e. on the ψ_k and according to (1.25) we have

$$\mu\psi_n = \sum_k \psi_k \mu_{kn}. \tag{2.31}$$

The ψ_k being eigenfunctions of the energy H are usually not eigenfunctions of the electric moment operator and hence the eigenvalue of μ is not determined for the level E_n. In other words the matrix μ_{kn} has off-diagonal elements. In this way the possibility for the emission of several lines from a given state is explained in quantum theory. Equation (2.31) is the quantum analogue of the expansion of the electric moment in a Fourier series. When the function ψ_n and the general form of the operator μ are known, one uses (1.26) to obtain the elements of the matrix (μ_{mn}) and one gets:

$$(\psi_m \cdot \mu\psi_n) = \int \psi_m^* \sum_k \psi_k \mu_{kn} \mathrm{d}\tau = \mu_{mn}. \tag{2.32}$$

When m and n are both large and when their difference is small, the formula

[1] Quadrupole emission is neglected.

(2.30) must be valid. This use of the correspondence principle leads to the following result, which can be obtained rigorously using time-dependent perturbation theory (the so-called "Golden-rule"). The energy radiated per second is equal to the average number of the transitions, i.e. to the probability of transition P_{mn} multiplied by the value of the emitted quantum. Then one gets

$$P_{mn} = \left(\frac{1}{h\nu}\frac{\mathrm{d}W}{\mathrm{d}t}\right) \quad \text{quantally translated as} \quad P_{mn} = \frac{4\pi}{3}\frac{(2\pi\nu)^3}{hc^3}\,|\mu_{mn}|^2. \tag{2.33}$$

It was later shown by Dirac, that a similar formula is valid for spontaneous emission, i.e. when the atom is not subject to an external radiation field.

The preceding result is the basis for the selection rules: if

$$\mu^x_{mn} = \mu^y_{mn} = \mu^z_{mn} = 0$$

the corresponding line does not exist.
If $\mu^x_{mn} = \mu^y_{mn} = 0$ but $\mu^z_{mn} \neq 0$, the line is linearly polarized in the Oz direction.

7. Perturbation Theory

7.1. FORMULATION OF THE PROBLEM

The number of problems in quantum mechanics that can be solved explicitly is rather limited, a situation similar to classical mechanics, or statistical mechanics. To study more complex questions one must often be satisfied with successive approximations starting from the cases for which solutions are known.

This method is called the perturbation method: the energy levels E_1, E_2, and *the complete set* of eigenfunctions $\psi_1, \psi_2, \ldots$ of a given Hamiltonian are supposed to be known. They correspond to the Schrödinger equation

$$\mathrm{H}\psi_i = E_i\psi_i. \tag{2.34}$$

The aim is to calculate the energy levels E' and the eigenfunctions ψ' of the system disturbed by a perturbing energy which is added to the Hamiltonian H and which is generally supposed to be expanded in a power series of the parameter λ. Taking into account the terms of only first order, one gets

$$\mathrm{H}' = \mathrm{H} + \lambda \mathrm{W}$$

and (2.34) becomes

$$(\mathrm{H} + \lambda\mathrm{W})\psi' = E'\psi'. \tag{2.35}$$

One can proceed with this approximation to any order. The formal calculations remain simple, but the practical difficulties become quickly inextricable. Particularly it often happens that the series which are thus obtained are divergent. We will discuss here neither this kind of difficulty nor the case in which the spectrum of eigenvalues is partially continuous.

7.2. NON-DEGENERATE PROBLEMS

Let us suppose that all the levels E_i of the unperturbed equation (2.34) are single. The eigenvalues E_i' and eigenfunctions ψ_i' of the perturbed problem (2.35) are slightly different from the energy levels E_i and the eigenfunctions ψ_i. Therefore we can put

$$E_i' = E_i + \lambda w_i \tag{2.36}$$

$$\psi_i' = \psi_i + \lambda u_i \tag{2.37}$$

and we expand u_i as a series of the orthogonal functions ψ_i.

$$u_i = \sum_l c_{il} \psi_l . \tag{2.38}$$

Using those three expressions in (2.35), taking into account (2.34), and neglecting λ^2, we get

$$\mathrm{W}\psi_i + \sum_l c_{il} E_l \psi_l = w_i \psi_i + \sum_l c_{il} E_i \psi_l \tag{2.35a}$$

where W appears as an operator acting on the ψ_i. Then we can use the expansion (1.25) and write:

$$\mathrm{W}\psi_i = \sum_l \psi_l w_{li} . \tag{2.39}$$

[The matrix (w_{ik}) represents the perturbing function in the coordinate system ψ_i.] Finally (2.35a) and (2.39) yield

$$\sum_l \psi_l [w_{li} - w_i \delta_{il} - c_{il}(E_i - E_l)] = 0 \tag{2.35b}$$

$$\delta_{il} = \begin{cases} 0 & (i \neq l) \\ 1 & (i = l) \end{cases}$$

and as the basis vectors ψ_l are orthogonal, the brackets are all zero.

Then, (a) for $i \neq l$

$$c_{il} = \frac{w_{li}}{E_i - E_l} \tag{2.40}$$

(b) for $i = l$

$$w_i = w_{ii}, \tag{2.41}$$

and one can suppose $c_{ii} = 0$ (in such a way that the norm of ψ_i' is 1).

The problem is completely solved in the first order and we have to calculate the matrix (w_{ik}) which according to (2.39) is given by the expressions

$$\int \psi_i^* \mathrm{W} \psi_i \mathrm{d}\tau = w_{li} \tag{2.42}$$

$$\int \psi_i^* \mathrm{W} \psi_i \mathrm{d}\tau = w_i . \tag{2.43}$$

The preceding calculation is valid when $\lambda w_{li}/(E_i - E_l)$ can be considered as small as can be seen from (2.40). When this condition is not fulfilled, i.e. when the two levels E_i and E_l are such that $E_i - E_l$ *is of the same order of magnitude as the perturbation*, the series expansion whose first term is given by (2.37) becomes poorly convergent or even divergent. This is *a quasi degeneracy* and the problem can be treated almost as if there was a true degeneracy (compare § 7.4).

7.3. DEGENERACY

Let us suppose that E is an α-times degenerate eigenvalue of (2.34); this means that the equation

$$\mathrm{A}\psi_i = E\psi_i \tag{2.34a}$$

allows α different eigenfunctions $\psi_i (i = 1, 2, \ldots \alpha)$ which may or may not be orthogonal. These eigenfunctions are determined only up to an arbitrary unitary transformation. One can build α independent orthogonal linear combinations with these eigenfunctions which also obey (2.34a). As previously, we assume that we know the complete set of eigenfunctions ψ of the operator H which contains the α functions ψ_i.

Generally the perturbation splits the energy level E into α different levels, close to the unperturbed level

$$E_i' = E + \lambda w_i . \tag{2.36a}$$

To each level E_i' there corresponds a wave function ψ_i' which does not necessarily become one of the functions ψ_i when λ goes to zero, but instead might become a linear combination of the ψ_i's.

Since we do not know beforehand which linear combination we will need to make a proper start if λ is becoming gradually different from zero, let us begin with a set of α arbitrary linear combinations of the α-wave functions

$$\psi_l' = \sum_{k=1}^{\alpha} \gamma_{lk} \psi_k , \qquad l = 1 \ldots \alpha. \tag{2.44}$$

We will find eventually which combinations we really do want from the form in which the perturbation takes place. The situation is similar to the determination of the main axes of a circle, a problem which gets a meaning as soon as we "squeeze" the circle slightly into an ellipse. This can be formulated mathematically as follows. Suppose we try to solve the degenerate problem in the same way as before, then the coefficient determined by (2.40) would become infinite for those indices i and l for which $E_i = E_l$. The reason is that in deriving (2.40) from equation (2.35b) we would have divided by zero, which is not allowed. We can circumvent this difficulty by making $w_{li} = 0$ for $l \neq i$; $i, l = 1 \dots \alpha$. This condition means diagonalization of the so-called secular matrix. The eigenvectors determine the choice of our linear combinations (2.44), i.e. the coefficients γ_{ik}.

We obtain in first order by taking into account (2.38),

$$\psi_i' = \sum_{k=1}^{\alpha} \gamma_{ik}\psi_k + \lambda u_i = \sum_{k=1}^{\alpha} \gamma_{ik}\psi_k + \lambda \sum_{l} c_{il}\psi_l \tag{2.44a}$$

$$i, k = 1, 2, \dots \alpha$$
$$l = \alpha+1, \alpha+2 \dots$$

where the zeroth order coefficients γ_{ik}, and the first order coefficients c_{il} are the unknowns.

The Schrödinger equation (2.35) becomes (after 2.36a)

$$(\mathrm{H}+\lambda \mathrm{W})\psi_i' = (E+\lambda w_i)\psi_i'$$

or taking into account (2.44), (2.34a) and dividing by λ

$$\sum_{l=\alpha+1}^{\infty} c_{il}(E-E_l)\psi_l = \sum_{k=1}^{\alpha} \gamma_{ik}(\mathrm{W}\psi_k - w_i\psi_k). \tag{2.45}$$

As in the first case let us expand $\mathrm{W}\psi_k$ in a series of functions ψ_l, but writing separately the α first eigenfunctions ψ_j, equation (2.39) becomes

$$\mathrm{W}\psi_k = \sum_{j=1}^{\alpha} \psi_j w_{jk} + \sum_{l=\alpha+1}^{\infty} \psi_l v_{lk} \tag{2.39a}$$

where (w_{ik}) is a square matrix with α rows and columns and the elements v_{lk} are the other elements of the infinite matrix W. Equation (2.45) becomes

$$\sum_{l=\alpha+1}^{\infty} c_{il}(E-E_l)\psi_l = \sum_{k,\, j=1}^{\alpha} (\gamma_{ik} w_{jk} - \gamma_{ij} w_i)\psi_j + \sum_{kl} \gamma_{ik} v_{lk}\psi_l. \tag{2.45a}$$

But we can complete all our partial matrices (γ_{ik}), (w_{jk}), (v_{lk}), (c_{il}) with zeros in order to change them into matrices which are defined for all the values of the indices from 1 to ∞. This allows us to put (2.45a) into the following condensed form

$$\sum_{l=1}^{\infty} \psi_l[c_{il}(E-E_l) - \sum_k \gamma_{ik} w_{lk} + \gamma_{il} w_i - \sum_k \gamma_{ik} v_{lk}] = 0. \qquad (2.45b)$$

The brackets are zero since the ψ_l are orthogonal. The equations which are obtained this way can be classified in two groups:

(i) $l \leqq \alpha$, $E_l = E$, $v_{lk} = 0$ and (2.45b) yields

$$\sum_k \gamma_{ik} w_{lk} - \gamma_{il} w_i = 0 \qquad (i, l = 1, 2, \ldots \alpha) \qquad (2.46)$$

where the w_{lk} are known coefficients (cf. (2.42) and later (2.48)). This system of homogeneous linear equations determines both the perturbations w_i of the level and the coefficients γ_{ik}. The γ_{ik} are determined except for a multiplicative constant. In this kind of problem, one has to reduce the matrix (w_{lk}) to its diagonal form using "a change of axis" (2.44). This problem is similar to the problem arising from equation (2.35), but it is simpler since it does not involve the whole function space but only the subspace spanned by the α first axes ψ_i. In order that the system (2.46) has non-trivial solutions, its determinant must be zero. We obtain *the secular equation,*

$$\begin{vmatrix} w_{11}-w_i & w_{12} & \cdots & w_{1\alpha} \\ w_{21} & w_{22}-w_i & \cdots & w_{2\alpha} \\ \cdots & \cdots & \cdots & \cdots \\ w_{\alpha 1} & \cdots & \cdots & w_{\alpha\alpha}-w_i \end{vmatrix} = 0 \qquad (2.47)$$

whose α roots: $w_1, w_2, \ldots w_a$ are real, since the matrix (w_{lk}) is Hermitian (as will be proved below). To each root w_j there corresponds a system of coefficients $\gamma_{j1}, \ldots \gamma_{j\alpha}$ which are determined by the homogeneous equations (2.46) except for a multiplicative constant. One takes this opportunity to normalize the eigenfunctions ψ_j. If the secular equation has multiple roots the degeneracy is not completely removed and some γ_{ik} remain undetermined.

The practical difficulties are purely analytical. One must calculate first the elements w_{lk} of the perturbation matrix: according to (2.39a). These elements are given by the integrals

$$w_{lk} = \int \psi_l^* \mathrm{W} \psi_k \, d\tau, \qquad l \leqq \alpha \qquad (2.48)$$

the evaluation of which is generally difficult. One then has to solve the algebraic equation (2.47).

(ii) Now let us suppose $l > \alpha$, $w_{lk} = \gamma_{il} = 0$. According to (2.45b) we have

$$c_{il} = \sum_{k=1}^{\alpha} \frac{\gamma_{ik} v_{lk}}{E-E_l} = \frac{\omega_{il}}{E-E_l}, \tag{2.49}$$

with (cf. (2.39a))

$$\begin{cases} v_{lk} = \int \psi_l^* \mathrm{W} \psi_k \mathrm{d}\tau, & l > \alpha \\ \omega_{il} = \sum_k \gamma_{ik} v_{lk} = \int \psi_l^* \mathrm{W} \psi_i' \mathrm{d}\tau, \end{cases} \tag{2.48a}$$

according (2.44a) and neglecting the terms in λ. Since the perturbing function W is real, (2.48) and (2.48a) show that the perturbation matrix W is Hermitian. The preceding calculation does not involve any restrictive hypothesis about the nature of the perturbing function. The perturbation can even depend explicitly on time: this happens in dispersion theory whose starting equations are (2.49) and (2.48a).

7.4. QUASI DEGENERACY

Let us suppose that all the levels are distinct and that two of them E_1 and E_2 are close†. Let us write

$$\frac{E_2 - E_1}{\lambda} = \eta \tag{2.49}$$

$E_2 - E_1$ being fixed by the nature of the unperturbed system, the auxiliary parameter η is very large when the perturbation λW is negligible, and is of the order of unity when this perturbation becomes of the same order as the difference $E_2 - E_1$. In this last case the perturbing action will almost couple those two levels as if they were degenerate. As in (2.44) one is led to put

$$\psi_1' = \gamma_{11}\psi_1 + \gamma_{12}\psi_2 + \lambda \sum_{l=3}^{\infty} c_{1l}\psi_l; \qquad \psi_2' = \gamma_{21}\psi_1 + \gamma_{22}\psi_2 + \lambda \sum c_{2l}\psi_l;$$
$$\psi_3' = \psi_3 + \lambda \sum_{j=1}^{\infty} c_{3j}\psi_j \ldots . \tag{2.44a}$$

The calculation is similar to the preceding one: we choose as the initial

† There can be more than two.

level $E = E_1$ and in order that the eigenvalues w_i are 0 and η in the absence of the perturbation, we replace w_i by $w_i - \eta$ in the lower right-hand corner of the secular matrix. The equations (2.46) which determine the coefficients γ_{ik} can be written

$$\left.\begin{array}{ll}\gamma_{i1}(w_{11}-w_i)+ & \gamma_{i2}w_{12} = 0\\ \gamma_{i1}w_{21} & +\gamma_{i2}(w_{22}+\eta-w_i) = 0\end{array}\right\} \quad i = 1, 2 \tag{2.50}$$

from which follows the secular equation

$$\begin{vmatrix} w_{11}-w & w_{12} \\ w_{21} & w_{22}+\eta-w \end{vmatrix} = 0 \tag{2.51}$$

the two roots w_1 and w_2 are given by

$$w = \tfrac{1}{2}[w_{11}+w_{22}+\eta\pm\sqrt{(w_{22}+\eta-w_{11})^2+4w_{12}w_{21}}]. \tag{2.51a}$$

We have finally

$$E_1' = E_1+\lambda w_1, \qquad E_2' = E_1+\lambda w_2.$$

We find for $l > 2$ the equations (2.49) again, i.e.

$$c_{1l} = \frac{\gamma_{11}v_{l1}+\gamma_{12}v_{l2}}{E_1-E_l}; \qquad c_{2l} = \frac{\gamma_{21}v_{l1}+\gamma_{22}v_{l2}}{E_1-E_l}.$$

If the perturbation is very weak, η is large and by expanding (2.51a) we obtain again in first approximation the formulae of § 7.2,

$$\begin{array}{ll} E_1' = E_1+\lambda w_{11} & \gamma_{11} \approx 1 \\ E_2' = E_1+\lambda(\eta+w_{22}) = E_2+\lambda w_{22} & \gamma_{22} \approx 1 \\ & \gamma_{12} \approx \gamma_{21} \approx 0. \end{array} \tag{2.52}$$

We will first discuss the special case in which both w_{12} and w_{21} happen to be zero. The equations (2.52) remain valid no matter what the order of magnitude is for η and equations (2.51) reduce to a system of two equations of the first degree. *No coupling between the two close levels is established,* and when the perturbation becomes stronger the two curves $E_1' = E_1'(\lambda)$, $E_2' = E_2'(\lambda)$ cross without mutual modification. Their intersection is at the point where

$$w_{11} = \eta+w_{22} \quad \text{i.e.} \quad \lambda = \frac{E_2-E_1}{w_{11}-w_{22}}.$$

(It is necessary that $w_{11}-w_{22}$ and E_2-E_1 are both positive.)

On the contrary if $w_{12} \neq 0$ this intersection cannot occur. The two roots are always distinct since the radical involves the sum of two squares. The two energy curves come closer, the energy difference becomes a minimum which is equal to $2\sqrt{w_{12}w_{21}}$ when $\lambda = (E_2-E_1)/(w_{11}-w_{22})$, and then the two curves diverge.

It is also interesting to examine the behavior of the wave functions during this process. It is now a simple matter to calculate the coefficients γ. We find:

$$\gamma_{11}^2 = \gamma_{22}^2 = \tfrac{1}{2}[1+(w_{11}-w_{22}-\eta)/D]$$
$$\gamma_{12}^2 = \gamma_{21}^2 = \tfrac{1}{2}[1-(w_{11}-w_{22}-\eta)/D]$$

where D is the (positive) square root in equation (2.51a). For large positive $\eta : \gamma_{11} = \gamma_{22} = 1$ and $\gamma_{21} = \gamma_{12} = 0$. If we let E_1 and E_2 "cross over", η changes sign and the result is that for large negative η we have $\gamma_{11} = \gamma_{22} = 0$ and $\gamma_{12} = \gamma_{21} = 1$. Hence if the system was originally in the state E_1, the lower state, it will now be in the lower state (E_2) again. This is only true if $w_{12} \neq 0$ and as we saw above the energy levels do not actually cross over but do only approach each other closely. During the closing in, the wave functions get mixed and after separation turn out to be interchanged.

The preceding considerations can be applied to the theory of complex atoms where the perturbing function represents the mutual interaction between the electrons.

They can be applied to the theory of molecules which involves also the nuclei. This picture is also the basis for the existence of a gap in the energy surfaces at the Brillouin zones in a periodic structure. An important example is the following:

Let us consider two different atoms which are coupled to form a molecule. When the distance R between the nuclei is large, it is possible that two states of the system have nearly the same energy. This happens when the ionization potential of the electro-positive atom is small and partially compensated by the affinity of the electro-negative atom for the electron, since a slight increase of the energy $\lambda\eta$ is then sufficient to transform the homopolar state where the two atoms are neutral into the heteropolar state where they are ionized[1].

A similar case occurs when one of the atoms has an excited level near the

[1] Nevertheless the bond remains homopolar (Fig. 2.1) but the number $|\gamma_{21}|^2$ gives, as a function of R, the percentage of ionic state contained in the actual state ψ_1' and consequently its contribution to the electric moment of the molecule in the ground state.

ground level. If we are describing, for instance, the interaction energy we can choose for the parameter λ a suitable negative power of the distance R.

The precise theory of molecular systems is more complex. Added to the quasi degeneracy there is an essential degeneracy (Chapter 4, § 2.2) which comes from the indistinguishability of the electrons; this is called exchange degeneracy. It is the following: to any stationary state in which the electrons 1 and 2 play a definite role, 1 being bound to the atom a and 2 to the atom b, there corresponds another state with the same energy obtained by permuting the two electrons, 1 being bound to b, and 2 to a.[1] This degeneracy plays a fundamental role in the calculation of the levels by the perturbation method. In the case of two electrons one gets a secular equation of the second degree which is raised to the fourth degree if a quasi degeneracy occurs. Then the difference between the two interacting levels (cf. (2.49)) depends on the distance R and one is obliged to use some rough approximations. Nevertheless one can estimate the general features of the four curves which represent the energy versus distance for the four perturbed states.

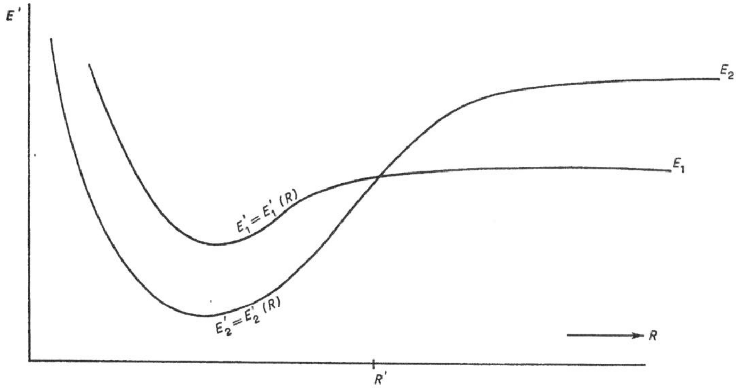

Fig. 2.1. Two energy curves as a function of the parameter R. Off-diagonal elements are zero at the point of intersection.

To illustrate the theory of the quasi degeneracy as clearly as possible, we have drawn in each of the two figures only two of those curves, those which correspond to the two lowest, i.e. more stable states. Figure 2.1 corresponds to the case where $w_{12} = 0$, Fig. 2.2 to $w_{12} \neq 0$.

[1] When n electrons can be permuted the degeneracy is of order $n!$ One knows that the electrons which are permuted can be connected to the same nucleus but in different states.

According to (2.42) we get $w_{12} = 0$ whenever ψ_1 is symmetrical, ψ_2 anti-symmetrical (Chapter 4, § 3.1).

7.5. APPLICATION: DIATOMIC MOLECULE

If one tries to apply the general perturbation theory to cases (analogous to the preceding one) where the electrons which are permuted are bound to different nuclei, one finds a difficulty which is worth pointing out. Let us consider for example, as Heitler and London did, two hydrogen atoms which are brought together to form a molecule. The permutation of the two electrons changes neither the expression of the energy nor its value, *but only the partition of the Hamiltonian into a principal term and a perturbing function.* As an example the interaction potential between the proton *a* and the electron 1 belongs to the principal term if one considers that these two charges form a neutral atom and the same interaction potential belongs to the perturbing function after the permutation of the electrons 1 and 2 has been performed.

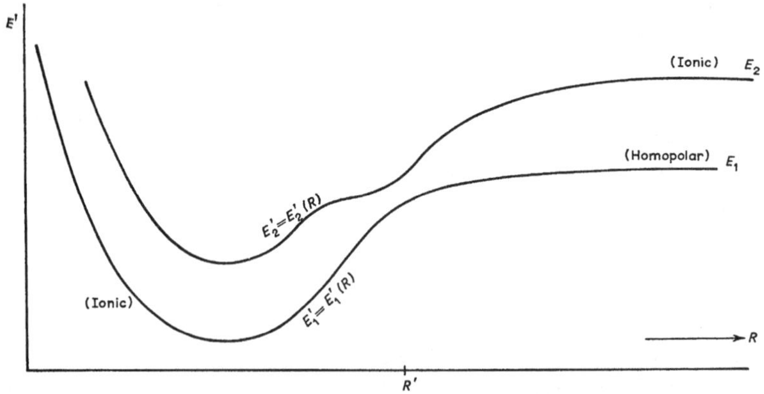

Fig. 2.2. Illustration of the non-crossing rule: Off-diagonal elements are non-zero. Notice the labelling of the curves.

It follows that the zeroth approximation wave functions are orthogonal only if the nuclei are infinitely distant from each other. For a finite but large distance the orthogonality is only approximative, which is not surprising because these wave functions are not eigenfunctions of the same Hamiltonian. At the same time the operator W which represents the perturbing function changes when the electrons are permuted and thus it depends on the wave function ψ_k on which it acts. Taking into account these two facts in the definition (2.39a) of the matrix w_{ik}, one can easily perform a first order calculation, following the calculation done in § 7.3 of this section. The further approximations are more difficult.

Let us make a last remark: we have the study of non-degenerate, degenerate and quasi-degenerate problems divided in three parts, §§ 7.2, 7.3 and 7.4, but the method which is used to solve these problems is always the same, the last case is the bridge between the two others.

CHAPTER 3

GROUP THEORY

1. The Role of Group Theory in Quantum Mechanics

In general, group theory is instrumental in solving those problems in quantum mechanics where we have degeneracy because usually the degeneracy is due only to the fact that the Schrödinger equation allows certain groups, i.e. the Schrödinger equation remains invariant if the system under consideration undergoes a set of transformations on the variables that enter into the wave function and this set of transformations forms a group.

Take for instance an atom with the nucleus at its center: because the field is spherically symmetric, the Hamiltonian function H is invariant under an arbitrary rotation around the center. For this reason the energy levels depend only on the quantum numbers n and l, whereas the wave functions (leaving out the spin) depend also, on the magnetic quantum number m. This means a degeneracy of orientation.

We consider as an example the $n = 2$ level of a non-relativistic hydrogen atom. It has m, l values: $l = 0$ $(m = 0)$ and $l = 1$ $(m = 0, \pm 1)$ and is therefore four-fold degenerate since the energy of such a system depends only on n: a part of this degeneracy is accidental, which means that it has nothing to do with symmetry, since any radial potential different from the Coulomb law will give rise to a splitting of the $l = 0$ and $l = 1$ level. The other part is due to spherical symmetry, since the three levels $m = 0, \pm 1$ of $l = 1$ stay, of course, degenerate as long as the potential has spherical symmetry.

Application of a linear electric field to this system will split these four levels into a pair with $m = \pm 1$ and two levels which have both $m = 0$, but different l values. This is explained in the following way. Take the axis of quantization (that is the direction in which the angular momentum operator is diagonal) along the field. Under the combination of both the radial and linear fields there remains still some symmetry, since reflection with respect to a plane perpendicular to the direction of the linear field leaves the system unchanged.

A magnetic field, however, has the symmetry properties of a helix and hence will remove also the $m = \pm 1$ degeneracy.

If we deal with diatomic molecules, the symmetry of the field (cylindrical or conical) permits rotations around the molecular axis and reflections about the symmetry planes. This leads to another type of degeneracy which will be discussed in Chapter 4, § 4. In general, in the case of more complicated molecules, in order to have degeneracies it is sufficient that the molecule presents certain symmetries. These symmetries are comparable with the groups in crystallography.

Finally and particularly the indistinguishability of particles of the same kind, as for example, electrons around the same nucleus, means that the Hamiltonian H must be invariant under a permutation between particles of the same kind. This is the origin of a new degeneracy unforeseeable from classical theory: the exchange degeneracy as was pointed out in the papers of Heisenberg on the helium molecule.

To each of these types of degeneracy corresponds a group of transformations. The study of each of these groups containing rotations, reflexions, and permutations allows one to determine separately all the properties of the wave functions which are of kinematic origin and permits one to solve the dynamical problem completely. In general this is very difficult.

Finally quantum theory defines physical quantities as operators and these belong in general to certain groups (compare Chapter 2, § 3.2). These definitions, which are simple in one-electron systems, become more sophisticated in the case of complex systems particularly when the spin has to be taken into account. It is obvious that group theory is the most general and essentially the simplest method to define all the physical quantities in complicated cases.

2. Examples

2.1. GENERAL CONSIDERATIONS

(i) Let a collection of n objects be placed in a certain order defined by the indices 1, 2, . . . n. If this order is upset, resulting in, for instance, 3, 7, 6, . . . k, we have performed a *permutation*. We indicate this operation by the symbol

$$\mathrm{P}_1 = \begin{pmatrix} 1 & 2 & 3 \ldots n \\ 3 & 7 & 6 \ldots k \end{pmatrix}.$$

In order to specify the sense of the preceding symbol, we assume that the indices are attached to the objects as labels: P_1 is an exchange of the objects 1 and 3, 2 and 7, etc.[1]

To let the collection of objects undergo two successive permutations, $P_2 = \binom{3\ 7\ 6 \ldots}{5\ 4\ 9 \ldots}$, amounts obviously to the effect of a single permutation

$$S = P_2 P_1 = \begin{pmatrix} 1 & 2 & 3 \ldots \\ 5 & 4 & 9 \ldots \end{pmatrix}.$$

We say that S is the *product* of the two permutations P_1 and P_2, using *the convention that the first operation will be written at the right-hand side.* This will permit us (if necessary) to place a symbol representing the collection of objects on which they operate at the right[2]. *This rule is general,* that is we will use it for arbitrary operators acting on different mathematical objects: vectors, functions, etc.

Among the permutations defined in this way, there is one that does not change the place of the objects. This is the identity, E, which obviously satisfies the relations:

$$EP = PE = P; \qquad E = \begin{pmatrix} 1 & 2 \ldots n \\ 1 & 2 \ldots n \end{pmatrix}.$$

Permutations do not commute: $P_1 P_2 \neq P_2 P_1$ this can easily be verified. However, to every permutation $P = \binom{1\ 2 \ldots n}{5\ 8 \ldots k}$ corresponds an inverse $P^{-1} = \binom{5\ 8 \ldots k}{1\ 2 \ldots n}$, which puts the objects back into their place and which consequently satisfies $P^{-1}P = PP^{-1} = E$.

The set of $n!$ permutations of n objects form a *group, the symmetric group* of order $n!$, which we indicate with the symbol $\mathscr{S}_n$.

(ii) If we let n go to infinity and if we distribute our objects on a straight line, on a plane, or in a space of 3 or more dimensions and if we go to the limit so that now the redistributions are continuous, then there will correspond a point to each object and a continuously variable index m to each point. In three dimensions all these points form the set specified by the three coordinates x, y and z.

If we now permute our objects, it will give rise to a well-determined correspondence between an arbitrary point P of the space under consideration

[1] It is also possible to number the boxes in which the objects are placed. This makes a second type of permutation possible, i.e. the exchange of numbers between the boxes. The same formulas apply to this case. Quantum theory considers both types of permutations simultaneously (comp. DIRAC, section 65, first edition only).

[2] One can make the opposite convention. The one we adopt is the most convenient with regard to applications in physics.

and its image point P′, thus we have effected a *transformation* A of the space onto itself. We will express this by the notation

$$P \rightarrow P' = AP.$$

The identity is defined by $EP = P$.

From two successive transformations $P \rightarrow P' = AP$, $P' \rightarrow P'' = BP'$ results a unique transformation,

$$P \rightarrow P'' = CP = BAP,$$

their *product*, which can be abbreviated to $C = BA$.

We will occupy ourselves only with reversible transformations, that is one-to-one correspondences. Hence we assume that after a transformation A it is always possible to bring the space back into its original order by a well-defined transformation A^{-1}. The transformation A^{-1} is the inverse of A and we have,

$$P' \rightarrow P = A^{-1}P', \quad \text{where} \quad A^{-1}A = E.$$

This is obviously only possible if the correspondence $P \leftrightarrows P'$ established by the transformation A is one-to-one.

(iii) Special cases of spatial transformations are: translations and/or rotations around an axis or a point (these operations are established in the concept of a crystalline solid), the motions of an arbitrary fluid (see BIRKHOFF [1950]), images formed in optics, etc.

Linear and reversible transformations of affine or unitary spaces are the most important groups.

The permutations themselves are transformations of a discontinuous space consisting of n points.

2.2. GROUP POSTULATES

From these examples we can deduce by abstraction the general definition of a group:

Let there be a collection $\mathscr{G}$ of operators or elements: A, B, . . . finite or infinite in number; they will form a group if the following four conditions are fulfilled:

I. The *product* BA of two arbitrary elements is an element of the collection $\mathscr{G}$,

$$BA = C, \quad C \text{ belongs to } \mathscr{G}. \tag{3.1}$$

II. The identity E is a part of the collection $\mathscr{G}$ and satisfies $EA = AE = A$.

III. To each element A corresponds another operation A^{-1}, called the inverse of A, belonging to $\mathscr{G}$ and defined by:

$$A^{-1}A = AA^{-1} = E. \tag{3.2}$$

IV. The product of these operations satisfies the associative law,

$$(AB)C = A(BC). \tag{3.3}$$

But in general we have $AB \neq BA$. A group in which the operations *are* commutative is called *Abelian*. If the number g of operations belonging to $\mathscr{G}$ is finite, the group is finite and g is its *order*.

2.3. FURTHER EXAMPLES OF GROUPS

1. The collection or set of rational numbers form a group under the operation of multiplication:

I. The product of two rational numbers is a rational number,
II. the identity is represented by the number 1,
III. the inverse of (A/B) is $(B/A) = (A/B)^{-1}$.

This group is Abelian and of infinite order.
From *the point of view of multiplication* the collection of integers (positive and negative) do not form a group.

2. However, they form a group if we consider the *operation of addition.* The operation resulting from the combination of two elements A and B is then written $A+B$ instead of AB.

I. The sum of two integers is an integer: $A+B = C$,
II. the identity is represented by the symbol *zero*,

$$A+0 = 0+A = A,$$

III. the inverse of A is $-A$.

This group is Abelian. It is an *additive group* of infinite order.

3. Let us go back to the space of three or n dimensions. A vector $\boldsymbol{a}$ represents an operation, a displacement for example.

I. The sum of two vectors is a vector $\boldsymbol{a}+\boldsymbol{b} = \boldsymbol{c}$,
II. $\boldsymbol{a}+\boldsymbol{o} = \boldsymbol{a}$, the operation E is the null vector,
III. the inverse of $\boldsymbol{a}$ is $-\boldsymbol{a}$.

Hence the set of displacements or vectors form an additive Abelian group since $\boldsymbol{a}+\boldsymbol{b} = \boldsymbol{b}+\boldsymbol{a}$.

The displacements are a special case of the projections of the space on itself, one of the few in which the operations do commute. Affine geometry, i.e., the metric used in vector spaces, is the study of the group just mentioned with one additional property: each operator or vector $\boldsymbol{a}$ may be multiplied with a real number if we deal with a real space, or with a complex number if we deal with a complex or unitary space. This is called *an additive group with multiplicators*. The mapping of a vector space upon itself results from the combination of two processes, the process of addition and the process of multiplication by numbers. These numbers are operators acting on the vectors. These vectors are themselves elements of an additive group. The operators could be of a more general form than simple multiplicators; for instance matrices. Hence a vector-space, on which a system of matrices acts can be considered as an *additive group submitted to a system of operators.* The extension of the theorems for the theory of groups to this particular case is of tremendous value.

2.4. GROUP TABLE

In summary, *a group consists of a set of symbols satisfying the postulates* I *to* IV *for which the "multiplication rules" are given* a priori. These symbols represent operations which are not specified by the abstract group theory. An abstract group outlines the properties of a certain number of concrete groups, which are *realizations* of the abstract group, in a kind of Pythagorean table (CAYLEY [1854]), usually called group table. For example: Each of the two Tables 3.1 and 3.2 represents an abstract group whose properties are those of the group $\mathscr{S}_3$ of permutations of three objects and those of the group of operations which bring the equilateral triangle into coincidence with itself including the possibility of turning it over on its face. We can give the following significance to the symbols in the table: A is a cyclic permutation $\binom{1\ 2\ 3}{2\ 3\ 1}$ or a rotation of $\frac{2}{3}\pi$ around an axis normal to the center of the triangle; $\mathrm{B} = \mathrm{A}^2$ is the inverse of the previous permutation or rotation of $\frac{4}{3}\pi$ around the same axes $\binom{1\ 2\ 3}{3\ 1\ 2}$; C, D, F the transposition of two objects or rotations around the medians of the triangle [Compare Fig. 3.3].

Table 3.1 looks like an ordinary multiplication table. In Table 3.2 the sequence of operations in the first column is not in the order E, A, B, C, D, F, but in the order of their inverses E, $\mathrm{A}^{-1} = \mathrm{B}$, $\mathrm{B}^{-1} = \mathrm{A}$, $\mathrm{C}^{-1} = \mathrm{C}$, $\mathrm{D}^{-1} = \mathrm{D}$, $\mathrm{F}^{-1} = \mathrm{F}$.[1]

[1] At the intersection of the row C and the column A we find the operation $\mathrm{F} = \mathrm{AC}$ obtained by effecting first C and then A.

TABLE 3.1

	E	A	B	C	D	F
E	E	A	B	C	D	F
A	A	B	E	D	F	C
B	B	E	A	F	C	D
C	C	F	D	E	B	A
D	D	C	F	A	E	B
F	F	D	C	B	A	E

TABLE 3.2

	E	A	B	C	D	F
E	E	A	B	C	D	F
A^{-1} = B	B	E	A	F	C	D
B^{-1} = A	A	B	E	D	F	C
C^{-1} = C	C	F	D	E	B	A
D^{-1} = D	D	C	F	A	E	B
F^{-1} = F	F	D	C	B	A	E

3. Subgroups

3.1. DEFINITION

In the groups of space rotations around a center O those operations which are rotations around an axis Oz form a *subgroup*, because they satisfy the postulates I to IV. The identity E is a null-rotation around Oz as well as around any other axis. One can also form a subgroup in the group $\mathscr{S}_3$ which we just examined. The operations E, A and B form a subgroup $\mathscr{H}$, the subgroup of the rotations around an axis normal to the triangle. This subgroup forms a *cyclic group* because it can be generated by powers of a single operation A ($B = A^2$, $E = A^3$).

From these considerations we form the following general definition: *If we can find in a group $\mathscr{G}$ a set of operations $\mathscr{H}$, finite or infinite, such that:*

1. *The product of two of these operations belongs to $\mathscr{H}$,*

2. *if $\mathscr{H}$ contains the elements* A, B, ... *it also contains their inverses* $A^{-1}, B^{-1}, \ldots$ *and as a result of this the element* E; *then $\mathscr{H}$ is a subgroup of $\mathscr{G}$.*

We have defined at the same time the *cyclic group generated by iterating a unique operation* A, i.e., formed by the "powers" of A. In order that a cyclic group be finite, it is necessary that a power p of A reproduces the identity: $E = A^p$ where p is the order of the group.

In a vector space of n dimensions let us put aside all the vectors of a sub-space with $m < n$ dimensions. These constitute a subgroup of the main additive group. In this way a plane $\mathfrak{R}_2$ going through the origin constitutes a subgroup of the ordinary space $\mathfrak{R}_3$ and the vectors lying on a straight line $\mathfrak{R}_1$ in this plane form a subgroup of this subgroup. The line must go through the origin since a subgroup must contain the unit element, i.e. the null-vector.

We want to study the influence of operators on the subgroup. If we subject the space to a system of projection matrices, these operators will, in general, bring the vectors belonging to $\mathfrak{R}_2$ out of this plane, i.e., the subgroup $\mathfrak{R}_2$ does not allow the same operators as the main group $\mathfrak{R}_3$. If the operators are multiplicators it is always true that the projected elements of the subgroup belong to the subgroup. An interesting case is the case of the additive subgroups, which allow the same operators as the main group. For example the operation could consist of adding constant vector $\boldsymbol{a}$ lying in the plane $\mathfrak{R}_2$. Thus these additive subgroups are subspaces which are *invariant with respect to the operators considered.* Hence it is natural to adapt the axes to this division of $\mathfrak{R}_3$; if $\mathfrak{R}_2$ happens to be an invariant plane, we designate it as the xOy plane and take Oz normal to it. We will come back to this point in § 7.

3.2. COSETS OR COMPLEXES ASSOCIATED WITH A SUBGROUP

Let $\mathscr{G}$ be a group of order g, $\mathscr{H}$ one of the subgroups of order h. Both g and h may be infinite, but we will first consider the case of finite groups.

Let E, A, B, C, ... F be the elements of $\mathscr{H}$.

1°. Multiply all these elements at the left side by the same element D of $\mathscr{H}$. As a result of the definition of the subgroups all products thus obtained: DE, DA, DB, ... are again belonging to $\mathscr{H}$. On the other hand they are all different (DA will not be equal to DB if $\text{A} \neq \text{B}$). Hence this multiplication amounts to writing all elements of $\mathscr{H}$ in a different order. If a subgroup $\mathscr{H}$, or an arbitrary group, undergoes a *transformation from the left* by multiplication by one of its elements this means a permutation of its elements.

2°. We take now an element S_2 of $\mathscr{G}$ that does not belong to $\mathscr{H}$ and form the products S_2A, S_2B ... S_2F. This collection constitutes a *complex of elements or coset*, all different from each other. We will call this *a left coset or a complex associated at the left* to the subgroup $\mathscr{H}$ and indicate it by $\text{S}_2\mathscr{H}$.

No one of these elements belongs to $\mathscr{H}$ because if S_2A would belong to $\mathscr{H}$, then $\text{S}_2\text{A}\text{A}^{-1} = \text{S}_2$ would also belong to $\mathscr{H}$ according to the definition of a subgroup.

These elements do not form a subgroup: the product $\text{S}_2\text{A}\text{S}_2\text{B}$ is certainly not a part of the subgroup since we know that $\text{A}\text{S}_2\text{B}$ does not belong to $\mathscr{H}$. If S_3 is an element of $\mathscr{G}$ and is neither a part of $\mathscr{H}$ nor of $\text{S}_2\mathscr{H}$, then we can form with S_3 a second coset $\text{S}_3\mathscr{H}$ and we can show in the same way that it does not contain an element of $\mathscr{H}$ nor of $\text{S}_2\mathscr{H}$.

If $\mathscr{G}$ is a finite group we can exhaust this procedure. As a result we have the equation:

$$\mathscr{G} = \mathscr{H} + \text{S}_2\mathscr{H} + \text{S}_3\mathscr{H} \dots \text{S}_l\mathscr{H}. \tag{3.4}$$

Each of these cosets contains the same number h of different elements as $\mathscr{H}$, hence we have:

$$g = lh. \tag{3.5}$$

The order of a subgroup of a finite group is an integral divisor of the order of the group. The integer l is the index of the subgroup.
One defines in the same way right cosets to a subgroup $\mathscr{H}$ and these will, in general, not coincide with the left cosets. We have

$$\mathscr{G} = \mathscr{H} + \mathscr{H}s_2 + \mathscr{H}s_3 + \ldots \mathscr{H}s_l. \tag{3.4a}$$

The nature of the cosets does not depend on the choice of the generating elements $s_2, s_3 \ldots$. They are uniquely determined by the structure of $\mathscr{G}$ and $\mathscr{H}$. To show this proposition let us consider an arbitrary element s'_2 of $s_2\mathscr{H}$. We claim that $s'_2\mathscr{H} = s_2\mathscr{H}$. As a matter of fact s'_2 can be written as $s_2\text{D}$ as it belongs to $s_2\mathscr{H}$, where D is a certain element of $\mathscr{H}$. Hence $s'_2\mathscr{H}$ consists of the collection of elements $s_2\text{DA}$, $s_2\text{DB}$, ... $s_2\text{DF}$, i.e. those of $s_2\mathscr{H}$ written in a different order (compare 1°).
In the example of § 2.4 the subgroup $\mathscr{A}_3$ (A, B, E) possesses a single coset $\text{C}\mathscr{A}_3 = \text{D}\mathscr{A}_3 \equiv \text{F}\mathscr{A}_3$.

The same considerations apply for infinite groups. The index l of $\mathscr{H}$ could then be finite or infinite.

Let $\mathscr{R}$, for example, be the set of vectors in a three-dimensional space, considered as an additive group, $\mathscr{H}$ the set of vectors $\boldsymbol{a}$ of the xOy plane and $\boldsymbol{s}$ a vector not located in $\mathscr{H}$. Then $\boldsymbol{s}\mathscr{H}$ is the set of all the vectors $\boldsymbol{s}+\boldsymbol{a}$ from which all the points S′ are in a plane Z parallel to xOy. Each of these planes correspond to a coset associated with $\mathscr{H}$. There are a continuous infinite number of these planes (compare Fig. 3.1). (See p. 69.)

4. Conjugated Elements. Classes

4.1. THE CASE OF LINEAR SUBSTITUTIONS

Let us consider in an n-dimensional vector space, the group (of infinite order) of all the linear vector transformations with complex coefficients. One of them will be

$$y_i = \sum_k a_{ik} x_k, \tag{1.4}$$

or in matrix language

$$\mathsf{Y} = \mathsf{AX}. \tag{1.4b}$$

We can consider this from two points of view: either the axes $\boldsymbol{e}_i$ stay fixed and we let a transformed vector $\boldsymbol{y} = \text{A}\boldsymbol{x}$ correspond to the vector $\boldsymbol{x}$ (A represents a mapping of the space upon itself), or the vectors $\boldsymbol{x}$ and $\boldsymbol{y}$ are identical in space and the axes are rotated. If (1.4) is considered as a change in basis vectors we find according to (1.3a) that the basis vectors underwent the transformation A^{-1}.

Let us take now two transformations A and S and

$$\mathsf{B} = \mathsf{SAS}^{-1}, \tag{3.6}$$

then by using the first interpretation, i.e. the axes $\boldsymbol{e}_i$ are kept fixed, we say that the matrix B produces a mapping of the space upon itself, which in general differs from A: *we say that the two transformations* A *and* B *are conjugated.*

With the similarity transformation (1.7), or the second interpretation used for the matrix S, we know that they go over into each other by a change of coordinates S. Hence two conjugated linear transformations are in general different but *equivalent* in the sense that either one can be obtained from the other by a transformation of axes.

4.2. GENERALIZATIONS. INVARIANT SUBGROUPS

Two elements A and B of a group $\mathscr{G}$ are conjugated if one can find a third element S *of the same group*, such that

$$\mathrm{B} = \mathrm{SAS}^{-1}. \tag{3.6a}$$

A *class* is a collection of all operations conjugate to a given operation A: we obtain the class from this element by considering S in (3.6a) as representing successively all the elements of the group $\mathscr{G}$.

In an Abelian group $\mathrm{SA} = \mathrm{AS}$ or $\mathrm{SAS}^{-1} = \mathrm{A}$ holds for every S. Hence each element forms a class in itself.

In each group the element E forms a class.

Let $\mathscr{H}$ be a subgroup of $\mathscr{G}$. Then $\mathrm{S}\mathscr{H}\mathrm{S}^{-1}$ is again a subgroup (S is a fixed element), the conjugated subgroup of $\mathscr{H}$. Indeed, if $\mathrm{AB} = \mathrm{C}$ and all three belong to $\mathscr{H}$, then $\mathrm{SAS}^{-1}\mathrm{SBS}^{-1} = \mathrm{SABS}^{-1} = \mathrm{SCS}^{-1}$.
Hence a subgroup $\mathscr{H}$ has as many conjugates as there are operations S in the group; but not all of these are distinct from each other.

It often happens that $\mathscr{H}$ is identical with all its conjugates, that is to say that SFS^{-1} belongs to $\mathscr{H}$, whatever the element F of $\mathscr{H}$ is, for every S belonging to $\mathscr{G}$. In this case $\mathscr{H}$ is an *invariant subgroup* or normal divisor. For example the subgroup $\mathscr{A}_3$ of the group $\mathscr{S}_3$ (see § 3.2) is invariant.

4.3. FACTOR GROUP

If $\mathscr{H}$ is an *invariant* subgroup in $\mathscr{G}$, let us decompose $\mathscr{G}$ into $\mathscr{H}$ and its cosets,[1]

$$\mathscr{G} = \mathscr{H} + \mathrm{s}_2\mathscr{H} + \mathrm{s}_3\mathscr{H} + \ldots \mathrm{s}_l\mathscr{H}. \tag{3.4}$$

We claim that each term of this sum may be considered in its turn as an element of a new subgroup, called the *factor group* of $\mathscr{G}$ and designated by the symbol $\mathscr{G}/\mathscr{H}$.

Indeed if we take the product $\mathrm{s}_i\mathscr{H}\mathrm{s}_j\mathscr{H}$ we obtain a coset containing all of the elements of $\mathscr{G}$ which are of the form $\mathrm{s}_i\mathrm{A}\mathrm{s}_j\mathrm{B}$, A and B belonging to $\mathscr{H}$. We claim that this product is one of the cosets associated with $\mathscr{H}$ in (3.4), that it is the one which contains the element $\mathrm{s}_k = \mathrm{s}_i\mathrm{s}_j$ and which we designate by $\mathrm{s}_k\mathscr{H}$.

Indeed

$$\mathrm{s}_i\mathrm{A}\mathrm{s}_j\mathrm{B} = \mathrm{s}_i\mathrm{s}_j\mathrm{s}_j^{-1}\mathrm{A}\mathrm{s}_j\mathrm{B},$$

and as $\mathscr{H}$ is invariant (a necessary and sufficient condition) $\mathrm{s}_j^{-1}\mathrm{A}\mathrm{s}_j = \mathrm{D}$ which is an element of $\mathscr{H}$ as well as $\mathrm{DB} = \mathrm{C}$. Hence we have

$$\mathrm{s}_i\mathrm{A}\mathrm{s}_j\mathrm{B} = \mathrm{s}_i\mathrm{s}_j\mathrm{C} \equiv \mathrm{s}_k\mathrm{C}.$$

If we take for A and B all other elements in $\mathscr{H}$, keeping s_k fixed, C will vary but remain in $\mathscr{H}$ and hence we will have

$$\mathrm{s}_i\mathscr{H}\mathrm{s}_j\mathscr{H} = \mathrm{s}_k\mathscr{H}. \tag{3.7}$$

If we put $\mathscr{F}_1 = \mathscr{H}$, $\mathscr{F}_2 = \mathrm{s}_2\mathscr{H}, \ldots \mathscr{F}_l = \mathrm{s}_l\mathscr{H}$, the symbols $\mathscr{F}_1$, $\mathscr{F}_2, \ldots \mathscr{F}_l$ may be considered as the elements of a new group, *the factor group* $\mathscr{G}/\mathscr{H}$, using this notation (3.7) takes the form of the usual definition of a group product $\mathscr{F}_i\mathscr{F}_j = \mathscr{F}_k$.

Evidently we have $\mathscr{H}\mathscr{H} = \mathscr{H}$, which simply results from the fact that the product of two elements of $\mathscr{H}$ is also an element of this subgroup. Hence $\mathscr{H}$ is the unit element of $\mathscr{G}/\mathscr{H}$.

The definition of a factor group amounts to blurring out the individuality of the elements of $\mathscr{H}$. They blend into a single element as is also the case for all the elements of each coset associated with $\mathscr{H}$.

All preceding propositions can be extended to infinite groups. It could happen that the factor groups are infinite.

[1] The left and right cosets are the same in the case of an invariant subgroup $\mathscr{H}$, as follows immediately from the definitions.

4.4. ABELIAN GROUPS

If AS = SA then $SAS^{-1} = A$ for every S. Hence each subgroup is invariant and can be used to define a factor group.

Let us take for example a three-dimensional vector space $\mathfrak{R}_3$. The vectors in the plane $\mathfrak{R}_2 = xOy$ form an invariant subgroup $\mathscr{H}$ ($SAS^{-1} = S + A - S = A$). As we have seen in § 4.2, a coset associated with $\mathscr{H}$ consists of the set of vectors which have their points in a plane Z perpendicular to Oz the z-axis (see Fig. 3.1); to each of these planes, i.e. to each value of z, corresponds a coset. They form a continuous series.

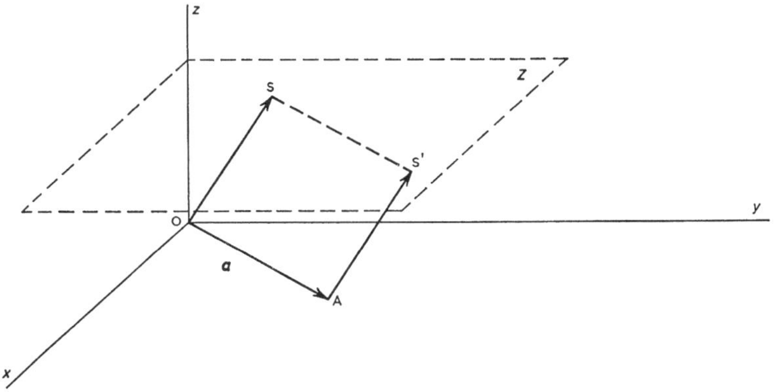

Fig. 3.1. Coset associated with $\mathscr{H} = xOy$.

The element F_z of the factor group can be obtained by disregarding the differences between the different vectors $\boldsymbol{v}$, $\boldsymbol{v}'$, $\boldsymbol{v}''$ . . . which have their points in the plane Z, that is by projecting them all on the axis Oz. This is expressed in mathematics by the symbol of *congruency*. (For this notation see also problem 3.11):

$$\boldsymbol{v} \equiv \boldsymbol{v}' \equiv \boldsymbol{v}'' \pmod{\mathfrak{R}_2}$$

The notation means that $\boldsymbol{v}$ and $\boldsymbol{v}'$ are identical if one disregards the (vector) difference lying in the plane $\mathfrak{R}_2$. In the same way we treat angles in trigonometry where 45° and 405° are considered identical disregarding the difference of 360°. The latter is irrelevant in certain cases as for instance the calculation of the trigonometric functions.

The figure accompanying the previous formula is identical with (3.1). The plane Z is the space $\mathfrak{R}_2$; all vectors $\boldsymbol{v}$, $\boldsymbol{v}'$, $\boldsymbol{v}''$ etc. are vectors between the origin and a certain point in Z.

To each value of z corresponds an element F_z of the factor group $\mathscr{G}/\mathscr{H} = \mathfrak{R}_3/\mathfrak{R}_2$. This group is continuous and represented by a one-dimensional vector space, the axis Oz. More generally (the proof is the same) if $\mathscr{G}$ is an n-dimensional vector space, $\mathscr{H}$ has m dimensions ($m < n$); each element of $\mathscr{F}_i$ of $\mathscr{G}/\mathscr{H}$ consists of a set of vectors $\boldsymbol{v}_i, \boldsymbol{v}_i', \boldsymbol{v}_i'' \ldots$ that satisfy,

$$\boldsymbol{v}_i \equiv \boldsymbol{v}_i' \equiv \boldsymbol{v}_i'' \pmod{\mathscr{H}}$$

any $\mathscr{G}/\mathscr{H}$ is a vector space with $n-m$ dimensions obtained by projection on an axis, a plane or a hyperplane.

5. Some Properties of the Group of Permutations of n Objects $\mathscr{S}_n$ (Symmetric group)

5.1. NOTATION WITH CYCLES

If $P = \begin{pmatrix} 1 & 2 & 3 & 4 & 5 \\ 2 & 3 & 4 & 5 & 1 \end{pmatrix}$ is a cyclical permutation, we could write behind each term the one we have to substitute for it. We obtain the following notation:

$$P = (1\ 2\ 3\ 4\ 5).$$

In a similar way we write

$$\begin{pmatrix} 1\,2\,3\,4\,5 \\ 5\,4\,1\,2\,3 \end{pmatrix} = (1\ 5\ 3)(2\ 4);$$

the permutation can be separated into two *cycles*, each set of elements forms an independent cycle that terminates in itself. An object that does not move from its place will form a cycle on itself. In this notation one encloses it between brackets or sometimes one omits it altogether. The *order* of a cycle is the number of objects contained in it.

Only three things are important in this notation: the number of cycles, the objects contained in each of them, and the *succession* or order in which they appear in the cycle. One can arbitrarily choose the order of cycles: $(1\ 5\ 3)(2\ 4) = (2\ 4)(1\ 5\ 3)$ or the object that appears first in a cycle $(1\ 5\ 3) = (5\ 3\ 1) = (3\ 1\ 5)$.

A *transposition* is a permutation of two objects. The following permutation is equivalent to a single transposition:

$$\begin{pmatrix} 1\,2\,3\,4\,5 \\ 1\,4\,3\,2\,5 \end{pmatrix} = (2\ 4)(1)(3)(5) = (2\ 4).$$

Each permutation is obviously a product of transpositions:

$$(1\ 5\ 3)(2\ 4) = (1\ 5)(5\ 3)(2\ 4) \quad \text{or} \quad (2\ 4)(5\ 3)(1\ 3)$$

but in the cycle notation we usually arrange the objects such that the different cycles do not have an object in common. A permutation is *even* or *odd* if it is the result of an even or odd number of transpositions. This is called the parity of a given permutation.

One can show — and it is almost obvious — that the parity of a permutation is unambiguous, i.e. independent of the manner in which one decomposes it into transpositions.

The inverse permutations of a given permutation can be obtained, in the cycle notation, by reversing the order of the terms in each bracket. The order of the brackets as a whole is irrelevant since they are independent.

$$(1\,5\,3)(2\,4)^{-1} = (3\,5\,1)(4\,2) = (1\,3\,5)(2\,4).$$

5.2. CONJUGATED PERMUTATIONS

If we have

$$\mathrm{S} = \begin{pmatrix} 1 & 2 & \dots & n \\ i_1 & i_2 & \dots & i_n \end{pmatrix}; \qquad \mathrm{T} = \begin{pmatrix} 1 & 2 & \dots & n \\ k_1 & k_2 & \dots & k_n \end{pmatrix} = \begin{pmatrix} i_1 & i_2 & \dots & i_n \\ l_1 & l_2 & \dots & l_n \end{pmatrix};$$

$$\mathrm{T}^{-1} = \begin{pmatrix} k_1 & k_2 & \dots & k_n \\ 1 & 2 & \dots & n \end{pmatrix}$$

then the conjugate of S with respect to T is $\mathrm{S}' = \mathrm{TST}^{-1}$, *where we have to read the operations from the right to the left*, so we obtain

$$\mathrm{S}' = \mathrm{TST}^{-1} = \begin{pmatrix} k_1 & k_2 & \dots & k_n \\ l_1 & l_2 & \dots & l_n \end{pmatrix}.$$

Hence to obtain the conjugate S′ of S with respect to T, we have to carry out the permutation T on both lines of S. This rule can be transferred easily to the cycle notation: suppose the S requires several cycles. It is then obviously sufficient to carry out the permutation T on the terms of each cycle in order to obtain TST^{-1}, i.e. we neither move the parenthesis nor change the number of cycles, nor the order in which they appear. Take for example, $\mathrm{S} = (1\,5\,3)(2\,4)$, $\mathrm{T} = (1\,2\,3\,4\,5)$, then, forming the conjugate, we have $\mathrm{TST}^{-1} = (2\,1\,4)(3\,5)$. In brief, to go from a permutation to its conjugate means to change the labels of the objects which one permutes without modifying the manner in which they are permuted.

From this we see that a *class of permutations* (compare § 4.2) is completely determined by the number of cycles and the order of each of them.

5.3. ALTERNATING GROUP $\mathscr{A}_n$ OF n VARIABLES

This is the group formed by all the even permutations of n objects, a subgroup of index 2 of $\mathscr{S}_n$. Its coset is a set of odd permutations; these permutations do not form a subgroup since the product of two odd permutations is even. $\mathscr{A}_n$ is an invariant group of $\mathscr{S}_n$, according to the rule of § 5.2.

Most of the preceding results were obtained by Cauchy.

6. Isomorphism and Homomorphism

6.1. DEFINITION

If we have two groups $\mathscr{G}$ and $\mathscr{G}'$ such that:

1. To each element A of $\mathscr{G}$ corresponds one *and only one* element A′ of $\mathscr{G}'$ and vice versa.

2. If $AB = C$ then $A'B' = C'$ for every element in the group.

The group tables of these two groups are the same and actually they differ only in the designation of the elements. In this case the two groups are called isomorphic (or have a holohedral isomorphism). From the point of view of abstract group theory the groups are identical. However, the elements refer to different mathematical objects. We have seen an example in 2.4; the group $\mathscr{S}_3$ and the group of operations which describe the coincidence of an equilateral triangle with itself are identical.

6.2. GENERAL THEOREMS

Let us suppose that the correspondence between $\mathscr{G}$ and $\mathscr{G}'$ is not one-to-one, that means that to an element A′ of $\mathscr{G}'$ there will correspond several different elements $A_1, A_2, \ldots A_p$ of $\mathscr{G}$. Any element A of $\mathscr{G}$, for example A_i, multiplied by one of the elements of $\mathscr{G}$ corresponding to B′ in $\mathscr{G}'$ should lead to an element C_k which corresponds to $A'B' = C'$. (see Fig. 3.2). In this case, the groups are called homomorphic, that is the isomerism is merohedral

These concepts are used in crystallography and one can find a number of examples there. A simple arithmetic example is the following. Let $\mathscr{S}_n$ be the group of permutations of n objects. With the even permutations we associate the number $+1$ and to the odd permutation the number -1. The multiplicative group $\mathscr{G}'$ consists of two elements $+1$ and -1 and is homomorphic to $\mathscr{S}_n$. The element $+1$ corresponds to the alternating subgroup and -1 to its coset.

In general if a group $\mathscr{G}$ possesses an invariant subgroup $\mathscr{H}$, then the group $\mathscr{G}$ is homomorphic to the factor group $\mathscr{G}/\mathscr{H}$. All the elements of $\mathscr{H}$ correspond to the unit element F_1 of $\mathscr{G}/\mathscr{H}$ and those of the complex $S_i\mathscr{H}$ to F_i. The converse of this obvious statement is also true and forms a fundamental theorem.

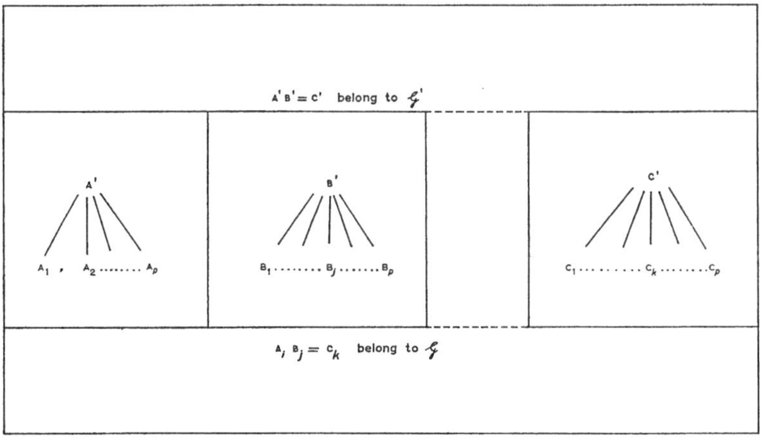

Fig. 3.2. Diagrammatical representation of the homomorphism between two groups.

Theorem: If $\mathscr{G}$ is homomorphic to $\mathscr{G}'$ and letting E′ *be the unit element of $\mathscr{G}'$, then:*

1. *The set of elements of $\mathscr{G}$ that correspond to* E′ *forms an invariant subgroup of $\mathscr{G}$;*
2. *$\mathscr{G}'$ is isomorphic to the factor group $\mathscr{G}/\mathscr{H}$.*

The proof is as follows. 1. If the elements J_1 and J_2 of $\mathscr{G}$ correspond to E' then $J_3 = J_1 J_2$ corresponds to $E'E' = E'$. We conclude that the set of elements $J_1 J_2 \ldots J_p$ of $\mathscr{G}$ which correspond to E' forms a subgroup $\mathscr{H}$ of $\mathscr{G}$. This subgroup is invariant because using an arbitrary element X of $\mathscr{G}$, then XJX^{-1} corresponds in $\mathscr{G}'$ to

$$X'E'X'^{-1} = X'X'^{-1} = E'$$

and we see that XJX^{-1} belongs to $\mathscr{H}$ for an arbitrary X.

2. If S_1 and S_2 are two elements of $\mathscr{G}$, not belonging to $\mathscr{H}$ and corresponding to the same element S' in $\mathscr{G}'$, then $S_1^{-1}S_2$ corresponds to $S'^{-1}S' = E'$ and

belongs to $\mathscr{H}$: $s_1^{-1}s_2 = J_m$. Hence $s_2 = s_1 J_m$ lies in the coset $s_1\mathscr{H}$ associated with $\mathscr{H}$.

Inversely two elements $s_1 J_m$ and $s_1 J_n$ of this coset correspond to the same element s′ in $\mathscr{G}'$. Hence the element s′ corresponds to the complex $s_1\mathscr{H}$, just as E′ corresponds to $\mathscr{H}$. We find that $\mathscr{G}'$ is isomorphic to $\mathscr{G}/\mathscr{H}$.

6.3. REPRESENTATIONS OF A GROUP

A special example of isomorphisms are the so-called representations of an abstract group. Let us suppose that $\mathscr{G}'$ is a group of *linear transformations.* To each operation A of $\mathscr{G}$ corresponds a matrix **A** created by the linear substitutions of a set of variables called the basis. We say that $\mathscr{G}'$ is a *representation* of $\mathscr{G}$. If the connection is isomorphic the representation is faithful.[1] For example, if every group element is represented by the number 1, the multiplication rules are preserved but the relation is not isomorphic, hence not faithful. One is unable to recognize the structure of the group table from this representation. We will designate representations of groups by bold face capitals $\boldsymbol{\mathscr{G}}$ or also by $\mathscr{R}$(G) or $\mathscr{D}$(G). In particular $\mathscr{D}$ stands for "Darstellung".[2]

The order of the representation matrices is equal to the dimensionality of the *representation space* $\mathfrak{R}$, i.e. number of variables employed in the substitution.

For instance each abstract group has an obvious representation, the *identity representation* in which we let correspond to each element of the group the substitution Y = X i.e. a one-dimensional matrix. To the product AB = C corresponds in this representation the product of the matrices $1 \cdot 1 = 1$.

6.4. EQUIVALENT REPRESENTATIONS

If $\mathscr{G}$ is a group with elements E, A, B, C . . ., and if $\mathscr{G}_1$ and $\mathscr{G}_2$ are two of its representations consisting of matrices $\boldsymbol{E}_1, \boldsymbol{A}_1, \boldsymbol{B}_1, \boldsymbol{C}_1 \ldots$, and $\boldsymbol{E}_2, \boldsymbol{A}_2, \boldsymbol{B}_2, \boldsymbol{C}_2 \ldots$ then we will call $\mathscr{G}_1$ and $\mathscr{G}_2$ *equivalent* if we can convert one into the other by a change of coordinates, i.e. if we can find a matrix S such that $\boldsymbol{A}_2 = S^{-1}\boldsymbol{A}_1 S$; $\boldsymbol{B}_2 = S^{-1}\boldsymbol{B}_1 S$; . . . shortly $\mathscr{G}_2 = S^{-1}\mathscr{G}_1 S$.
Obviously the two representation spaces $\mathfrak{R}_1$ and $\mathfrak{R}_2$ are identical.

[1] Also used: "true", which is an incorrect translation from the German "treu" meaning faithful.

[2] The reader will have noticed that we used mainly lower case bold face letters for vectors in order to avoid confusion.

7. Reducibility of Representations

7.1. INVARIANT SUBSPACE

If we have a representation $\mathcal{G}$ in an n-dimensional vector space, it often happens that there exists a vectorial subspace $\mathfrak{R}_1$ which has only $m < n$ dimensions and which is *invariant with respect to the transformations of* $\mathcal{G}$, i.e. such that all mappings A of $\mathcal{G}$ transform the vectors of $\mathfrak{R}_1$ into vectors of $\mathfrak{R}_1$.

It is natural to adapt the coordinate system to this invariance. We span the space $\mathfrak{R}_1$ by the first set of m unit vectors and we leave the remaining $n-m$ outside this space. A vector $\boldsymbol{x}$ of $\mathfrak{R}_1$ will have in this case the components $x_{m+1} = x_{m+2} = \ldots = x_n = 0$. Since after the transformation $\boldsymbol{x}' = \boldsymbol{A}\,\boldsymbol{x}$ remains in $\mathfrak{R}_1$, the preceding equations lead to: $x'_{m+1} = x'_{m+2} = \ldots = x'_n = 0$. The mapping $\boldsymbol{x}' = \boldsymbol{A}\boldsymbol{x}$ can be written

$$\begin{array}{ll} x'_1 & = a_{11}x_1 + \ldots a_{1\,m+1}x_{m+1} + \ldots \quad a_{1n}x_n, \\ \cdot & \cdot\ \cdot\ \cdot\ \cdot\ \cdot\ \cdot\ \cdot\ \cdot\ \cdot\ \cdot\ \cdot\ \cdot\ \cdot\ \cdot \\ x'_m & = a_{m1}x_1 + \ldots a_{m\,m+1}x_{m+1} + \ldots \quad a_{mn}x_n, \\ x'_{m+1} & = \qquad\qquad a_{m+1\,m+1}x_{m+1} + \ldots a_{m+1\,n}x_n, \\ \cdot & \cdot\ \cdot\ \cdot\ \cdot\ \cdot\ \cdot\ \cdot\ \cdot\ \cdot\ \cdot\ \cdot\ \cdot\ \cdot\ \cdot \\ x'_n & = \qquad\qquad a_{n\,m+1}x_{m+1} + \ldots \quad a_{nn}x_n. \end{array}$$

Hence the matrices A of $\mathcal{G}$ have the form

$$A = \begin{pmatrix} \boldsymbol{P}_a & \boldsymbol{Q}_a \\ 0 & \boldsymbol{S}_a \end{pmatrix} \tag{3.8}$$

where $\boldsymbol{P}_a$ and $\boldsymbol{S}_a$ are square matrices[1] of the order m and $(n-m)$, $\boldsymbol{Q}_a$ is a rectangular matrix, and 0 a null matrix. Hence, the matrices $\boldsymbol{P}_a, \boldsymbol{P}_b \ldots$ generate a representation of $\mathcal{G}$ in the space $\mathfrak{R}_1$ with m dimensions.

In case we can find such a space $\mathfrak{R}_1$ and its corresponding axes, the representation $\mathcal{G}$ is reducible. *If we cannot find such a space the representation is irreducible.* These definitions are independent of the concept of groups and of representations because they apply to an arbitrary set of matrices.

The matrices $\boldsymbol{S}_a, \boldsymbol{S}_b \ldots$ also form a representation of the group $\mathcal{G}$ in a space which one obtains by ignoring the components $x_1, x_2 \ldots x_m$, $x'_1, x'_2 \ldots x'_m$ of the vectors in $\mathfrak{R}$. This space is geometrically obtained by

[1] The word matrix is used here in the loose sense of "array of numbers" not in the usual strict definition of "array of numbers that transform like a direct product of vector components" since the latter obviously does not hold for subsections of a matrix.

projecting the vectors of $\mathfrak{R}$ *parallel to* $\mathfrak{R}_1$, i.e. by projecting all those that differ only by their m first components into one single vector. We will designate these by the symbol $\mathfrak{R}/\mathfrak{R}_1$.

Indeed, if we consider the vector space $\mathfrak{R}$ as an additive group with operators defined by the matrices of the system $\mathscr{G}$, $\mathfrak{R}_1$ is an invariant subgroup with the same operators (invariant in the double sense of the word with respect to the transformations of $\mathscr{G}$ and also because the group is Abelian). $\mathfrak{R}/\mathfrak{R}_1$ is the factor space as defined in § 4.4 of this Chapter. Hence, *the matrices* S_a *form a representation of* $\mathscr{G}$ *in the factor* subspace $\mathfrak{R}/\mathfrak{R}_1$.

7.2. COMPLETE REDUCTION OR DECOMPOSITION

The most interesting cases are those in which one succeeds, by a convenient choice of axes, to decompose the space $\mathfrak{R}$ into two independent invariant subspaces $\mathfrak{R}_1$ and $\mathfrak{R}_2$. All the rectangular matrices Q_a become zero and the matrices of the system $\mathscr{G}$ take on the form of a *step-wise matrix.*

$$\begin{pmatrix} P_a & 0 \\ 0 & S_a \end{pmatrix}. \tag{3.8a}$$

Thus the representation, or more generally the system of matrices $\mathscr{G}$ decomposes into two representations or into two separate matrix systems, one with m, the other with $n-m$ dimensions. This is called a *complete reduction* or *decomposition.* This is indicated by a *symbolic* plus sign:

$$\mathscr{G} = \mathscr{G}_1 + \mathscr{G}_2, \tag{3.9}$$

$$\mathfrak{R} = \mathfrak{R}_1 + \mathfrak{R}_2 \tag{3.9a}$$

hence $\mathfrak{R}_1$ is *isomorphic with* $\mathfrak{R}/\mathfrak{R}_2$.

Theorem: *Every system of reducible unitary matrices is decomposable*, i.e. for a unitary system we can always go from (3.8) to (3.8a) by a suitable change of coordinates.

It is sufficient to take in the unitary space $\mathfrak{R}$ — in which the transformations of the system $\mathscr{G}$ are performed — a set of axes such that the first m span the space $\mathfrak{R}_1$ and the remaining $n-m$ span a space $\mathfrak{R}_2$ that is orthogonal to $\mathfrak{R}_1$, i.e. a space that contains all the vectors "perpendicular" to $\mathfrak{R}_1$. Since the transformations $\mathscr{G}$ are unitary they conserve the orthogonality relations among vectors. They transform $\mathfrak{R}_1$ into itself and will do the same to $\mathfrak{R}_2$ as this space is orthogonal to $\mathfrak{R}_1$. The result is that the second one is invariant too.

7.3. REDUCTION OF THE UNITARY MATRICES OF A GROUP INTO THEIR IRREDUCIBLE PARTS

An isolated unitary matrix can always be brought into the diagonal form. For a *system* of unitary matrices this is only possible if these matrices commute with each other (compare § 3.3). (We assume of course that all the matrices of the system are transformed with the help of the *same* transformation matrix.) If the matrices are not commuting, we can only hope for the simultaneous reduction of all the matrices of the system to "boxes" of the size $m_1, m_2 \ldots m_p (m_1+m_2+ \ldots m_p = n)$ along the diagonal. If this reduction is carried out as far as possible we say that the system of matrices is reduced and the parts, each corresponding to one set of boxes (one for each matrix element of $\mathbf{\mathscr{G}}$), are called irreducible.

When we deal with the representation $\mathbf{\mathscr{G}}$ of an abstract group $\mathscr{G}$, it is important to decompose this representation into *irreducible representations* since, as we will see below, only these have a fundamental meaning in physics and mathematics. Once we have found these p irreducible constituents $\mathbf{\mathscr{G}}_1$, $\mathbf{\mathscr{G}}_2 \ldots \mathbf{\mathscr{G}}_p$ we can again use the notation introduced in equation (3.9).

$$\mathbf{\mathscr{G}} = \mathbf{\mathscr{G}}_1 + \mathbf{\mathscr{G}}_2 + \ldots \mathbf{\mathscr{G}}_p .$$

None of these preceding definitions implies that the group $\mathscr{G}$ is finite.

7.4. EXAMPLE

We will again use the group $\mathscr{S}_3$ since this is the simplest example of a non-Abelian group. The elements are the six permutations (1), (1 2), (2 3), (3 1), (1 2 3) and (1 3 2). In order to make use of the group table in § 2.4, Chapter 3 we indicate the isomorphism: (1 2 3) ≡ A; (3 2 1) ≡ B; (1 2) ≡ D; (2 3) ≡ C and (1 3) ≡ F.

It is easy to find a representation of this group. Let us take for the objects which permuted the three variables x_1, x_2 and x_3, i.e. the three projections of a vector $\boldsymbol{x}$ on three rectangular axes $\boldsymbol{e}_1$, $\boldsymbol{e}_2$ and $\boldsymbol{e}_3$. The different elements of this group are represented by the system of mapping equations:

$$(1\ 2) \rightarrow \begin{aligned} y_1 &= x_2 \\ y_2 &= x_1 \\ y_3 &= x_3 \end{aligned} \qquad (1\ 2\ 3) \rightarrow \begin{aligned} y_1 &= x_2 \\ y_2 &= x_3 \\ y_3 &= x_1 \end{aligned} \qquad \text{etc.}$$

We obtain in this way *one* 3-dimensional representation of $\mathscr{S}_3$ from which the matrices can be written as:

$$(1\,2) \to \begin{pmatrix} 0&1&0\\ 1&0&0\\ 0&0&1 \end{pmatrix} \qquad (1\,2\,3) \to \begin{pmatrix} 0&1&0\\ 0&0&1\\ 1&0&0 \end{pmatrix} \quad \text{etc.}$$

The following simple geometrical remarks permit us to reduce these matrices. The permutations of $\mathscr{S}_3$ leave the sum $x_1+x_2+x_3$ invariant. Hence the plane $x_1+x_2+x_3 = 0$ (plane $\mathfrak{R}_2$) and the normal to this plane $x_1 = x_2 = x_3$ are two invariant orthogonal subspaces of the representation space $\mathfrak{R}_3$. Let us take an axis e'_1 along the normal and two axes e'_2 and e'_3 in the plane $\mathfrak{R}_2$.

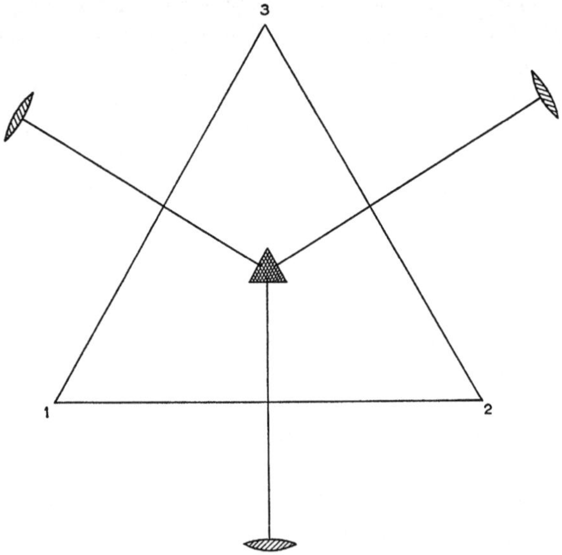

Fig. 3.3. An object that allows the six operations of the group $\mathscr{S}_3$. Indicated are the three-fold axis (perpendicular to the plane) and the three two-fold axes (in the plane).

It is convenient, in the sense of obtaining more symmetrical formulas, to take these last two axes neither perpendicular to each other nor of unit length. We will take e'_2 in the plane of e_1 and e_2 and e'_3 in the plane of e_2 and e_3. We take:

$$e'_1 = \tfrac{1}{3}(e_1+e_2+e_3); \qquad e'_2 = \tfrac{1}{3}(e_1-e_2); \qquad e'_3 = \tfrac{1}{3}(e_2-e_3).$$

(e'_2 is directed along the bisectrix of e_1 and $-e_2$, etc.)

The corresponding change in variables is found by taking the transposed inverse of the transformation of the basis vectors (compare equations (1.2) and (1.3)):

$$\tilde{\mathsf{S}} = \tfrac{1}{3}\begin{pmatrix} 1 & 1 & 1 \\ 1 & -1 & 0 \\ 0 & 1 & -1 \end{pmatrix} \to \mathsf{S}^{-1} = \begin{pmatrix} 1 & 1 & 1 \\ 2 & -1 & -1 \\ 1 & 1 & -2 \end{pmatrix}.$$

Hence the new variables are

$$x_1' = x_1+x_2+x_3, \qquad x_2' = 2x_1-x_2-x_3, \qquad x_3' = -2x_3+x_2+x_1$$

and the transformation formulas for the mapping become

$$(1\,2) \to \begin{cases} y_1' = x_1' \\ y_2' = 2y_1-y_2-y_3 = 2x_2-x_1-x_3 = x_3'-x_2'; \quad \text{etc.} \\ y_3' = y_1+y_2-2y_3 = x_1+x_2-2x_3 = x_3' \end{cases}$$

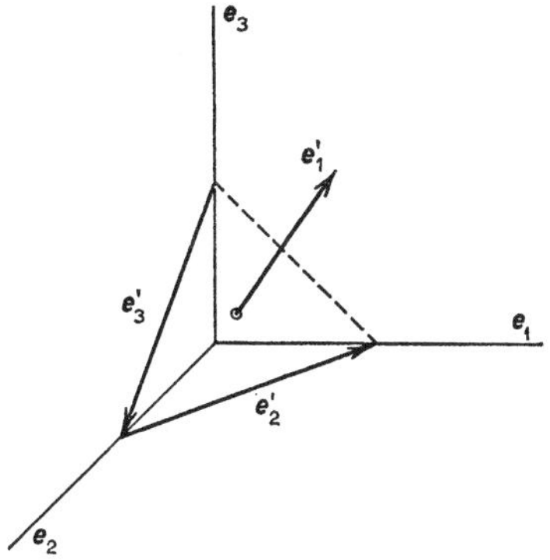

Fig. 3.4. The choice of unit vectors that establishes a decomposition of the representation into two irreducible representations.

From which we find the matrices

$$(1\,2) \to \begin{pmatrix} 1 & 0 & 0 \\ 0 & -1 & 1 \\ 0 & 0 & 1 \end{pmatrix}; \qquad (1\,3) \to \begin{pmatrix} 1 & 0 & 0 \\ 0 & 0 & -1 \\ 0 & -1 & 0 \end{pmatrix}; \qquad (2\,3) \to \begin{pmatrix} 1 & 0 & 0 \\ 0 & 1 & 0 \\ 0 & 1 & -1 \end{pmatrix}$$

$$(1\,2\,3) \to \begin{pmatrix} 1 & 0 & 0 \\ 0 & -1 & 1 \\ 0 & -1 & 0 \end{pmatrix}; \qquad (1\,3\,2) \to \begin{pmatrix} 1 & 0 & 0 \\ 0 & 0 & -1 \\ 0 & 1 & -1 \end{pmatrix}; \qquad (1) \to \begin{pmatrix} 1 & 0 & 0 \\ 0 & 1 & 0 \\ 0 & 0 & 1 \end{pmatrix}.$$

The reduction cannot be performed any further. There is no invariant axis in the plane $\mathfrak{R}_2$.

Hence we have found a three-dimensional representation of the group $\mathscr{S}_3$ and we reduced it to two irreducible representations. One is the identity representation, consisting of the matrices 1, and the other a two-dimensional representation in the plane $\mathfrak{R}_2$.

The group $\mathscr{S}_3$ has a third irreducible representation. It is one-dimensional and can be found very easily. The alternating group $\mathscr{A}_3$ is an invariant subgroup of $\mathscr{S}_3$. It consists of the even permutations: (1), (1 2 3) and (1 3 2) and we may write (compare § 3.2)

$$\mathscr{S}_3 = \mathscr{A}_3 + (1\ 2)\mathscr{A}_3.$$

$(1\ 2)\mathscr{A}_3$ is the coset associated with $\mathscr{A}_3$ and consists of the odd permutations (1 2), (2 3) and (3 1). If we let the number -1 correspond to the elements of the coset and if we let the number $+1$ correspond to the elements of the subgroup, we obtain the antisymmetric representation of $\mathscr{S}_3$. The identical representation could be called the symmetric representation. As a matter of fact any symmetry group $\mathscr{S}_n$ must have these two representations as is obvious from the explanation in § 6.2. It is possible to prove that for $\mathscr{S}_3$ only these three irreducible representations exist. As a last remark only the two-dimensional representation is faithful (compare problem 3.1).

7.5. FINITE GROUPS

Theorem: The representations of a finite group can all be considered to be unitary and therefore completely reducible. Let $\mathscr{G}$ be an arbitrary representation of order n of a finite group $\mathscr{G}$. Let us take an arbitrary Hermitian form, for instance, the unit form $F = x_1^* x_1 + x_2^* x_2 + \ldots x_n^* x_n$ constructed with the variables of the representation $\mathscr{G}$. We submit this form to all the substitutions of $\mathscr{G}$ and add all the results. This way we obtain an Hermitian form that stays invariant under all these substitutions, since they only permute the terms in the sum. By a convenient choice of coordinates we can bring this form on its main axis and by a "choice of units", into the unit form itself (compare problem 3.12).

The transformation just described is then used to transform all the matrices of the representation $\mathscr{G}$ into $\mathscr{G}' = \mathrm{S}\mathscr{G}\mathrm{S}^{-1}$. This new representation, equivalent with the first one, is unitary since it has the property that it leaves "the length of a vector" $(x_1')^* x_1' + (x_2')^* x_2' + \ldots (x_n')^* x_n'$ invariant.

It is often convenient not to limit oneself to unitary representations, as in the example of § 7.4.

8. Uniqueness Theorem: *The decomposition of a given representation $\mathcal{G}$ from a group $\mathcal{G}$ into irreducible constituents is only possible in one way*

More precisely, if one finds two decompositions

$$\mathcal{G} = \mathcal{G}_1 + \mathcal{G}_2 + \dots \mathcal{G}_p \quad \text{and} \quad \mathcal{G} = \mathcal{G}'_1 + \mathcal{G}'_2 + \dots \mathcal{G}'_{p'},$$

we must have $p = p'$ and the two sequences are formed by irreducible representations which are one by one equivalent after changing the order in a proper way.

In modern algebra which deals only indirectly with the notion of representations this proposition is connected to a more abstract theorem due to Jordan, Hoelder, and Noether.[1]

In order to give an idea of the proof of the theorem without going into details we will show it for only three dimensions. This gives the possibility of specifying the precise meaning of the process of decomposition with the help of a simple geometric example. In the case of a three-dimensional space two hypotheses are possible, either the irreducible invariant subspaces are a plane $\mathfrak{R}_2 = xOy$ and a line $\mathfrak{R}_1 = Oz$. (If the representation is unitary the line is perpendicular to the plane) or they are the three axes $\mathfrak{R}_1 = Oz$, $\mathfrak{R}'_1 = Ox$ and $\mathfrak{R}''_1 = Oy$.

1. Suppose the first assumption is true:

$$\mathfrak{R}_3 = \mathfrak{R}_1 + \mathfrak{R}_2$$

i.e. each vector of $\mathfrak{R}_3$ that goes through the origin can be decomposed unambiguously into a component lying in $\mathfrak{R}_1$ and a component in $\mathfrak{R}_2$ and each of these will stay in its subspace under the transformations of the representations $\mathcal{G}$ of the group $\mathcal{G}$. It is impossible to find an invariant plane $\mathfrak{R}'_2$ that does not coincide with $\mathfrak{R}_2$. Indeed if it existed, it would cut $\mathfrak{R}_2$ along a line L. This line would be an invariant subspace of $\mathfrak{R}_2$ since it is the *intersection* between two invariant subspaces. But $\mathfrak{R}_2$ is irreducible, the line L cannot exist and each invariant plane $\mathfrak{R}'_2$ has to coincide with $\mathfrak{R}_2$. In the same way it is impossible to find an invariant line $\mathfrak{R}'_1$ outside the axes $Oz = \mathfrak{R}_1$ because if it existed it would determine with this axis an invariant plane different from $\mathfrak{R}_2$.

2. In the second case we have

$$\mathfrak{R}_3 = \mathfrak{R}_1 + \mathfrak{R}'_1 + \mathfrak{R}''_1;$$

[1] Compare for instance Speiser [1937] section 11.

the irreducible representations are one-dimensional, i.e. the matrices of $\mathscr{G}$ are diagonal. In this case there is no *irreducible* invariant plane since the intersection of such a plane with the plane $\mathfrak{R}_1 \mathfrak{R}'_1$ (which is itself an invariant plane) is invariant. Finally there could be a line L different from the three coordinate axes that forms an invariant subspace on itself. In order that the vectors $\boldsymbol{v}$ on this line stay on it after the transformations of the representation $\mathscr{G}$, it would be necessary that all components of the vector v_x, v_y and v_z be multiplied with the same number. In this case the representation would have one dimension instead of three. Hence the only invariant subspaces are the three coordinate planes which are each reducible to their axes.

We see from this example the uniqueness of the decomposition of a three-dimensional representation and a similar proof can be given in the general case of n dimensions.

9. Schur's Lemma and Related Theorems

The theorems discussed below are crucial in the theory of representations.

Consider two vector spaces $\mathfrak{R}$ and $\mathfrak{S}$, one with m, the other with n dimensions. Let:

1. $\mathscr{G}_R$ be a system of mappings or bilinear transformations of $\mathfrak{R}$ onto itself, consisting of the matrices $\boldsymbol{A}_R$, $\boldsymbol{B}_R$ ....

2. $\mathscr{G}_S$ be a system of mappings of $\mathfrak{S}$ onto itself, consisting of the matrices $\boldsymbol{A}_S$, $\boldsymbol{B}_S$ ....

The matrices of the two systems correspond one by one to each other. $\mathscr{G}_R$ and $\mathscr{G}_S$ could be two representations of the same group, but it is not necessary to introduce the concept of a group at all.

3. Finally let there be a mapping T of $\mathfrak{R}$ into $\mathfrak{S}$. T is a rectangular matrix such that to every vector $\boldsymbol{x}$ of $\mathfrak{R}$ there corresponds a vector $\boldsymbol{y}$ of $\mathfrak{S}$;

$$\boldsymbol{y} = \mathsf{T}\boldsymbol{x}. \tag{3.10}$$

The converse in general is not true. To a null-vector of $\mathfrak{S}$ corresponds a subspace $\mathfrak{R}'$ in $\mathfrak{R}$ (this is an invariant subspace of the additive group $\mathscr{R}$, comp. § 4.4) and to a vector $\boldsymbol{y}$ which is different from zero corresponds a coset associated with $\mathfrak{R}'$ (compare § 6.2).

Hence T established a homomorphism of $\mathfrak{R}$ into $\mathfrak{S}$, or at least between a part of the space $\mathfrak{S}$ and the space $\mathfrak{R}$, since it may be possible that there exits vectors in $\mathfrak{S}$ which are not used at all in our homomorphism, that is they do not correspond to any vector of R.[1]

Having supposed all this we assume:

1. that the system $\mathscr{G}_R$ is irreducible;

2. that the matrix T establishes between the vectors $\boldsymbol{A}_R\boldsymbol{x}$ and $\boldsymbol{A}_S\boldsymbol{y}$, $\boldsymbol{B}_R\boldsymbol{x}$ and $\boldsymbol{B}_S\boldsymbol{y}$, etc. . . . *the same correspondence* as between $\boldsymbol{x}$ and $\boldsymbol{y}$, i.e.

$$A_S y = \mathsf{T} A_R x; \qquad B_S y = \mathsf{T} B_R x; \qquad \ldots \tag{3.10a}$$

or, by taking into account (3.10) and leaving out the vector symbol $\boldsymbol{x}$

$$A_S \mathsf{T} = \mathsf{T} A_R; \qquad B_S \mathsf{T} = \mathsf{T} B_R; \qquad \ldots. \tag{3.11}$$

From these two hypotheses we will show the following theorems (compare Fig. 3.5).

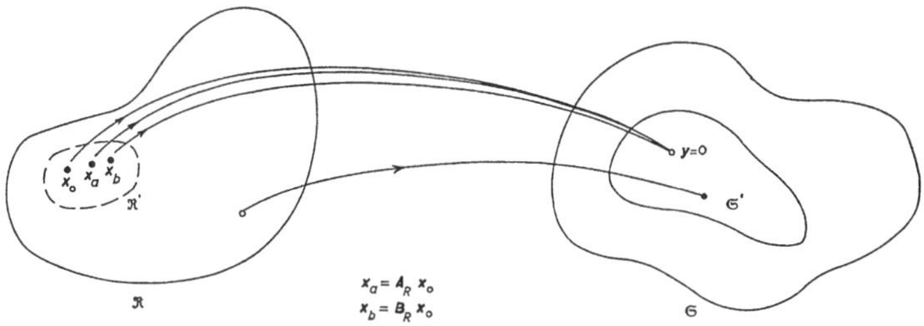

Fig. 3.5. Symbolic representation of the assumptions on Schur's Lemma. Spaces $\mathfrak{R}$ and $\mathfrak{S}$ are represented by point sets. The mapping T by connecting lines. $\mathfrak{S}'$ is that part of $\mathfrak{S}$ actually used in the mapping T. Since $\mathsf{T}x_0$ is equal to zero, then $\mathsf{T}x_a$, $\mathsf{T}x_b$, etc. are also equal to zero. We prove that either $\mathfrak{R}' = 0$ or $\mathfrak{R}' = \mathfrak{R}$.

Theorem I. The relation T *existing between* $\mathfrak{R}$ *and* $\mathfrak{S}$ *is either an isomorphism and* Det $\mathsf{T} \neq 0$, *or* T *is identically zero.* Indeed if we consider the subspace $\mathfrak{R}'$ of $\mathfrak{R}$ that corresponds to the null-vector of $\mathfrak{S}$ and we let $\boldsymbol{x}_0$ be an arbitrary vector of this space then by hypothesis we have $\boldsymbol{y} = \mathsf{T}\boldsymbol{x}_0 = 0$.

[1] For instance the equations

$$y_1 = t_{11}x_1 + t_{12}x_2 + t_{13}x_3;$$
$$y_2 = t_{21}x_1 + t_{22}x_2 + t_{23}x_3,$$

establish a relation such that to each vector x of a three-dimensional space $\mathfrak{R}_3$ there corresponds a vector y of the plane y_1Oy_2 in the space $\mathfrak{S}$ and if this space has more than two dimensions, its vectors lying outside this plane do not correspond to any vector in $\mathfrak{R}$.

The vectors in $\mathfrak{R}$ that correspond to the null-vector in $\mathfrak{S}$ are those which lie on a line whose equations are determined by putting the left-hand sides of the preceding equations equal to zero. To a given vector $y \neq 0$ there corresponds a set of vectors in $\mathfrak{R}$ having their origins at O and their end points on a line parallel to the line mentioned above.

The matrix T makes the vectors $A_R x_0$, $B_R x_0, \ldots$ correspond to $A_S y_0$, $B_S y_0, \ldots$, according to (3.10a). Because $y_0 = 0$ the latter are all zero. Hence $A_R x_0$; $B_R x_0$; ... all belong to the subspace $\mathfrak{R}'$ which appears to be invariant with respect to the transformation of the system $\mathscr{G}_R$. But we have supposed that this system is irreducible and we are left with the following alternative: either $\mathfrak{R}' = \mathfrak{R}$, and $\mathsf{T}x_0 = 0$ for every x_0, i.e. $\mathsf{T} \equiv 0$; or $\mathfrak{R}' = 0$ and the null-vector of $\mathfrak{S}$ uniquely corresponds to the null-vector of $\mathfrak{R}$. In the latter case we know as a result of the fundamental theorem of § 6.2 that the relation between $\mathfrak{R}$ and $\mathfrak{S}$ established by T is a one-to-one correspondence or *isomorphism*. The collection of vectors y of $\mathfrak{S}$ which correspond to vectors x in $\mathfrak{R}$ do not have to fill up the whole space $\mathfrak{S}$, but only a subspace $\mathfrak{S}'$ which is isomorphic to $\mathfrak{R}$ and invariant under the transformations of $\mathscr{G}_S$. This isomorphism has as a consequence the reversibility of T, i.e. the existence of T^{-1} or the mapping of $\mathfrak{S}'$ upon $\mathfrak{R}$.

Theorem II. If $\mathscr{G}_S$ is also irreducible, $\mathfrak{S}'$ is identical to $\mathfrak{S}$ and $\mathfrak{R}$ is isomorphic to $\mathfrak{S}$. They have the same number of dimensions and (3.11) can be written as

$$A_S = \mathsf{T}A_R\mathsf{T}^{-1}; \quad B_S = \mathsf{T}B_R\mathsf{T}^{-1} \quad \ldots \quad \mathscr{G}_S = \mathsf{T}\mathscr{G}_R\mathsf{T}^{-1}.$$

The two systems $\mathscr{G}_S$ and $\mathscr{G}_R$ are equivalent. They transform into each other through a change of coordinates.

Theorem III. Suppose this identification is made $\mathscr{G}_R = \mathscr{G}_S = \mathscr{G}$, then (3.11) can be written

$$A\mathsf{T} = \mathsf{T}A; \quad B\mathsf{T} = \mathsf{T}B; \ldots.$$

The matrix T *commutes with all the matrices of the irreducible system* $\mathscr{G}$ *and we will show below that this matrix is necessarily a multiple of the unit matrix in n dimensions,* or $\mathsf{T} = \lambda \mathrm{I}$, λ being a number.

Let us consider the equation $\mathrm{Det}(\mathsf{T} - \lambda \mathrm{I}) = 0$. This equation will have at least one root which is unequal to zero. Let us use this value for λ. The matrix $\mathsf{T} - \lambda \mathrm{I}$ will commute with all the matrices of the system $\mathscr{G}$ for every value of λ since T commutes with all of them. The preceding theorems confront us with the following alternative, either $\mathsf{T} - \lambda \mathrm{I}$ establishes also a one-to-one projection of $\mathfrak{R}$ on $\mathfrak{S}$ and the $\mathrm{Det}\,(\mathsf{T} - \lambda \mathrm{I}) \neq 0$, which is impossible, or $\mathsf{T} - \lambda \mathrm{I} = 0$. This establishes the theorem.

Summarizing we have: *A matrix* T, *which commutes with all the matrices of the irreducible representation of a group* $\mathscr{G}$, *is necessarily the multiple of a unit matrix. If the matrix relates two non-equivalent irreducible representations* like the relations (3.11), *it is identically zero*, a statement of major importance in the theory of groups and quantum mechanics.

If the representation is reducible we can easily construct a matrix which will commute with all matrices of the representation and which is not a constant times the unit matrix.

Suppose the reducible representation is transformed by a similarity transformation S to a set of step-wise matrices. We construct a diagonal matrix T that has elements λ_1, at the places which correspond to the diagonal positions of the first box of the step-wise matrices, elements λ_2 on places corresponding to the diagonal elements of the second box, etc. This matrix will, according to the previous theorem, commute with all matrices in the representation presumed above. If we bring the reducible representation back to its original form by the transformation S^{-1} and if we transform T simultaneously, the commutation relation will be maintained and the matrix T will not be a constant times the unit matrix, provided of course we take $\lambda_1 \neq \lambda_2 \neq \ldots$.

Hence we find that if matrices exist which commute with all the matrices of a certain representation and if they are not proportional to the unit matrix then the representation is reducible. If they are proportional to the unit matrix, the representation is irreducible. This forms a simple criterion about irreducibility and it will finally lead to a prescription for finding the irreducible parts of a reducible matrix system (§ 12.3).

10. Characters of a Representation

10.1. DEFINITION

Let $A, B, C \ldots$ be the matrices of a representation $\mathscr{G}$ of a group $\mathscr{G}$. The characters of the representations are the traces of the matrices.

Equivalent representations have the same system of characters, i.e. (compare § 3.5) if $A' = SAS^{-1}$; $B' = SBS^{-1}$; ... then

$$\mathrm{Tr}\, A' = \mathrm{Tr}\, A; \quad \mathrm{Tr}\, B' = \mathrm{Tr}\, B; \ldots.$$

The trace of a product of two matrices is independent of the order in which the matrices appear. A similar statement for an arbitrary number of matrices holds only if the order is changed cyclically, which gives

$$\mathrm{Tr}\, SAS^{-1} = \mathrm{Tr}\, S^{-1}SA = \mathrm{Tr}\, A.$$

This proof is only valid for finite matrices; in the case of infinite matrices we have to consider the convergence of the sums.

We will designate the character of the matrix A by $X(\text{A})$. Its value depends on the particular matrix we have selected, just as a function depends on the chosen value of the variable x.
The characters of the matrices representing operations belonging to the same class (in the sense of § 4.2) are identical as they can all be represented in the form SAS^{-1} (S runs through all the representation matrices of the group) and $\text{Tr}\, SAS^{-1} = \text{Tr}\, A$. For this reason the *character* is said to be *a function of a class* instead of a function of an element as was suggested above.
If the representation is irreducible the character is called *primitive*.[1]

Let $\mathscr{G}$ be a representation which is decomposed into its irreducible elements

$$\mathscr{G} = m_0 \mathscr{G}_0 + m_1 \mathscr{G}_1 + \ldots, \tag{3.12}$$

where the integers m_1 indicate how many times a particular representation is contained in $\mathscr{G}$. The representations $\mathscr{G}_0, \mathscr{G}_1, \ldots$ are not equivalent. Obviously the characters have a similar relation

$$X = m_0 X_0 + m_1 X_1 + \ldots. \tag{3.12a}$$

The symbols $X_0, X_1, \ldots$ designate the character systems of the irreducible representations, i.e. the set of characters $X_0(\text{A})$, $X_0(\text{B}), \ldots$ $X_1(\text{A})$, $X_1(\text{B}), \ldots$ etc. They are different only if A and B are elements belonging to different classes. Therefore the number of relations (3.12a), which are distinct, is the same as the number of classes. We will see below (§ 11) that these character sets satisfy a number of relations between themselves.

10.2. THE NUMBER OF IRREDUCIBLE REPRESENTATIONS OF A FINITE GROUP

We have seen the importance of the notion of irreducibility in the preceding paragraphs. (A crucial theorem which we will prove in § 11 brings this out very clearly.)
We know that an arbitrary representation of a given group $\mathscr{G}$ can be decomposed according to (3.12) into its irreducible parts. These irreducible parts may differ from one arbitrary representation to another and there seems to be no limit to the number of possible irreducible representations resulting from these decompositions. Actually this is not the case.

[1] Some authors, comp. e.g. VAN DER WAERDEN [1949] reserve the word character for the irreducible representation only and use trace or spur otherwise.

The number of non-equivalent irreducible representations of a finite group $\mathcal{G}$ is equal to the number of classes into which we can divide its elements. We establish this theorem by studying a special representation of the group, *the regular representation*, which is one of the most natural ways of representing a given abstract group.

10.3. REGULAR REPRESENTATION OF A GROUP

Before we introduce the regular representation in a more formal way, we want to point out that multiplication of all group-elements with a certain fixed group element induces a permutation among the group elements. As we can see from the group table, all these permutations are distinct from each other.

An arbitrary permutation of elements can be represented by a matrix in the following way:

$$P = \begin{pmatrix} 1\,2\,3\,4 \\ 2\,3\,4\,1 \end{pmatrix} \rightarrow \begin{pmatrix} 0 & 1 & 0 & 0 \\ 0 & 0 & 1 & 0 \\ 0 & 0 & 0 & 1 \\ 1 & 0 & 0 & 0 \end{pmatrix} \begin{pmatrix} a_1 \\ a_2 \\ a_3 \\ a_4 \end{pmatrix} = \begin{pmatrix} a_2 \\ a_3 \\ a_4 \\ a_1 \end{pmatrix}.$$

This matrix contains only one unit element per row and per column. The permutation induced by the multiplication by a given element and represented by this type of matrix is called the regular representation.

In order to establish this idea in a more formal way we let a variable x_s and a basis vector $\boldsymbol{s}$ correspond to each operation S of a group. The collection of vectors $\boldsymbol{s}$ spans a new space ρ, the so-called *group space*, which has g dimensions in the case of a finite group of order g. A vector ξ of this space can be written as follows:

$$\xi = \sum_s x_s \boldsymbol{s} \tag{3.13}$$

the sum being extended over the g symbols $\boldsymbol{s}$.

We *define* the *product* of two vectors ξ and $\eta = \sum y_t \boldsymbol{t}$ by the rule:

$$\xi\eta = \sum_{st} x_s y_t \boldsymbol{st} \tag{3.14}$$

$\boldsymbol{st}$ represents the basis vector that corresponds with the operation ST of the group. The expression (3.14) may be read either as a double sum or after replacing the products by the proper group elements as a single sum in which each element is repeated g times.

It is often practical to refrain from the vector description altogether. The ξ and η are then considered as *hypercomplex numbers*[1] and defined as linear combinations of group elements with (real or complex) coefficients x_s and y_t. Hence there is no longer any need to use bold face letters and we will refrain from doing so from now on. The symbols S, T, ... are the *basis* of the hypercomplex number system and all the quantities obtained this way form the *group algebra*.[2] The structure of the algebra is determined by the rules that define the products U = ST, i.e. by the multiplication table of the group (§ 2.4).

The two expressions "group space" and "group algebra" as well as the notions of "vector in the group space" and "hypercomplex number of the algebra" are equivalent. We will use the latter expressions if we want to stress the multiplication rules (3.14).

From the last equation we find that an arbitrary operation A of the group generates a projection from the space ρ upon itself,

$$\xi \to \xi' = \mathsf{A}\xi = \sum_s x_s as = \sum_t x_{a^{-1}t} t.$$

($t = as$ corresponds with the product T of the operations A and S and we have $\text{S} = \text{A}^{-1}\text{T}$).

The collection of projections or mappings A *forms the regular representation of the group.*

Let us designate the components of ξ' by x'_s then the preceding equation is equivalent with the set of substitutions

$$x'_s = x_{a^{-1}t} \qquad t = a, b, c, \ldots .$$

Hence the matrix A can be written

$$\mathsf{A} = (a_{ts}) = (\delta_{s,\, a^{-1}t}) = (\delta_{a,\, ts^{-1}}) \tag{3.15}$$

in which the rows and columns are labeled with the help of the elements of the group themselves: E, A, B, ... S ... T ... and $\delta_{a,ts^{-1}}$ is equal to 1 if $\text{A} = \text{TS}^{-1}$ and zero otherwise. The matrix contains mainly zeros and only a single 1 per row and per column.

If we return to the example of the group $\mathscr{S}_3$ of permutation of three objects, referring to its group table displayed in § 2.4 (second form), we see that the matrix A that represents the abstract element A is the following

[1] Compare problem no. 3.

[2] The word algebra is suggested because both sum and product are defined.

arrangement of 0 and 1

$$\mathsf{A} = \begin{pmatrix} 0&0&1&0&0&0 \\ 1&0&0&0&0&0 \\ 0&1&0&0&0&0 \\ 0&0&0&0&1&0 \\ 0&0&0&0&0&1 \\ 0&0&0&1&0&0 \end{pmatrix}.$$

The importance of the regular representation is due to the following theorem. *All irreducible representations of a group $\mathscr{G}$ can be obtained by reducing its regular representation $\mathscr{G}_r$. The number of times an irreducible representation appears in the regular representation is equal to the dimensionality of the matrices of that particular representation.*

The proof of this theorem can be established directly from the result in § 11.2. [Compare problem 3.8 where it is explicitly indicated that one can obtain this proof from (3.23).]

Taking up again our example of $\mathscr{S}_3$; if one would reduce the regular representation of this group, one would find the unit representation once, the alternating one-dimensional representation once, and the irreducible two-dimensional representation mentioned in § 7.4 twice.

11. Orthogonality Relations (Finite Groups)

11.1. GENERAL FORMULAS

Let $\mathscr{G}$ and $\mathscr{G}'$ be two irreducible representations of the finite group $\mathscr{G}$ and (a_{ik}), (a'_{ik}) their representation matrices; n and n' the order of these matrices. An important case is the one in which $\mathscr{G}'$ is unitary, but first we will deal with the general case. Consider a rectangular matrix $\mathsf{S} = (s_{l\mu})$ with n rows and n' columns and let us form the sum

$$\mathsf{T} = \sum_{\mathsf{A}} ASA'^{-1}, \tag{i}$$

extended over all operations A of $\mathscr{G}$. (The number is of course equal to the order of the group g.) T will be, like S, a rectangular matrix with n rows and n' columns. Let us write this expression explicitly:

$$t_{ik} = \sum_{l,\,\mu} (a_{il} s_{l\mu} a'^{-1}_{\mu k} + b_{il} s_{l\mu} b'^{-1}_{\mu k} \ldots).$$

Let C and C' be two matrices of $\mathscr{G}$ and $\mathscr{G}'$ corresponding to an arbitrary element C of $\mathscr{G}$. We claim that for an arbitrary C the following expression holds

$$\mathsf{C}\mathsf{T}C'^{-1} = \mathsf{T}. \tag{ii}$$

Indeed

$$C\mathsf{T}C'^{-1} = \sum_{\mathrm{A}} CASA'^{-1}C'^{-1},$$

if we call $CA = D$, then $A'^{-1}C'^{-1} = (C'A')^{-1} = D'^{-1}$ and D and D' correspond in $\mathscr{G}$ and $\mathscr{G}'$ to the same element D = CA of $\mathscr{G}$.
If A runs through the group $\mathscr{G}$, i.e. represents successively all the operations of $\mathscr{G}$, D = CA goes through the same group, but in a different order (the same idea was used in § 3.2).
Hence

$$C\mathsf{T}C'^{-1} = \sum_{\mathrm{D}} DSD'^{-1} = \sum_{\mathrm{A}} ASA'^{-1} = \mathsf{T},$$

which proves (ii).
For arbitrary C the result is

$$C\mathsf{T} = \mathsf{T}C'. \tag{iii}$$

If we apply now Schurs' lemma we find that either C and C', i.e. $\mathscr{G}$ and $\mathscr{G}'$ are non-equivalent and hence $\mathsf{T} \equiv 0$, or $\mathscr{G}$ and $\mathscr{G}'$ are equivalent and T is a constant times the unit matrix.

First Case: $\mathscr{G}$ and $\mathscr{G}'$ are non-equivalent; (i) gives

$$\sum_{\mathrm{A}} ASA'^{-1} \equiv 0.$$

The matrix S is arbitrary. If we take a matrix that is everywhere zero except for the component $s_{k\lambda}$ we obtain

$$\sum_{\mathrm{A}} a_{ik} a'^{-1}_{\lambda\nu} = 0 \tag{3.16}$$

and in the unitary case this means

$$\sum_{\mathrm{A}} a_{ik} a'^{*}_{\nu\lambda} = 0. \tag{3.16a}$$

In the relations (3.16) and (3.16a) the indices i, k, ν and λ are arbitrary, i.e. we take an arbitrarily located element in the collection of matrices, and keep the position fixed in going from one matrix to the other. These relations are important since they characterize the inequivalent representations.

Second Case: $\mathscr{G}$ and $\mathscr{G}'$ are equivalent. We can choose a new coordinate system for the representation $\mathscr{G}'$ such that $\mathscr{G}' = \mathscr{G}$. In that case our theorem can be written

$$\sum_{A} ASA^{-1} = \alpha \mathrm{I}$$

where α is a constant depending only on S. Writing this out, we have

$$\sum_{jk} \sum_{A} a_{ij} s_{jk} a_{kl}^{-1} = \alpha \delta_{il}.$$

Since the matrix S is arbitrary, let us suppose that all the elements are zero except one of them $s_{jk} = 1$; the preceding equation then becomes

$$\sum_{A} a_{ij} a_{kl}^{-1} = \alpha \delta_{il} \tag{iv}$$

where the constant α, determined by the choice of S, i.e. by the indices j and k, is independent of i and l. In order to determine α we have by definition,

$$\sum_{i=1}^{n} a_{ki}^{-1} a_{ij} = \delta_{kj}. \tag{v}$$

(The reader should not confuse this summation, which is extended over the n dimensions of the representation space, with the summation $\sum_{A}$ extended over the g operations of the group $\mathscr{G}$.)

In equation (iv) let $i = l$, and if we perform successively the two summations with respect to the indices A and i we find, taking into account (v),

$$\sum_{i=1}^{n} \sum_{A} a_{ij} a_{ki}^{-1} = n\alpha = g\delta_{kj}$$

or finally,

$$\sum_{A} a_{ij} a_{kl}^{-1} = 0. \tag{3.17}$$

in case $i \neq l$ or $j \neq k$. Otherwise we have:

$$\sum_{A} a_{ik} a_{ki}^{-1} = g/n. \tag{3.18}$$

In case the representation matrices are unitary the equations take the form

$$\sum_{A} a_{ik} a_{ji}^{*} = 0 \quad \text{and} \quad \sum_{A} a_{ik} a_{ik}^{*} = g/n. \tag{3.19}$$

The equations (3.16), (3.17), (3.18), and (3.19) are *the basic orthogonality relations between irreducible representations*. The use of the term orthogonality will be explained in the remaining part of this section.

11.2. APPLICATION TO THE CHARACTERS OF IRREDUCIBLE REPRESENTATIONS

Restricting ourselves to unitary matrices, we have for (3.19) putting $i = k$,

$$\sum_{\mathrm{A}} a_{ii} a_{ii}^* = g/n.$$

Hence using the right side of (3.19), we have the following relation between the characters of a unitary irreducible representation

$$\sum_{i=1} \sum_{\mathrm{A}} a_{ii} a_{ii}^* = \sum_{\mathrm{A}} X(\mathrm{A}) X^*(\mathrm{A}) = g. \tag{3.20}$$

Similarly if we deal with two non-equivalent irreducible representations we find from (3.16):

$$\sum_{\mathrm{A}} X'(\mathrm{A}) X^*(\mathrm{A}) = 0. \tag{3.21}$$

$X(\mathrm{A})$, $X'(\mathrm{A}) \ldots$ are functions of the variable A representing an element of the group (compare § 10.1). These functions are defined only at g discrete points. (Moreover the "shape" of this function is further simplified by the fact that its functional value is the same for all elements belonging to the same class. Hence the characters are actually functions of a class instead of functions of the elements.) The function space of this function is identical with the group space mentioned in § 10.3. Or to formulate it differently we may construct a hypercomplex number by using the character of the element as the coefficients of every element. Using the geometric interpretation we use the vector symbol

$$\boldsymbol{X} = \sum_{\mathrm{A}} X(\mathrm{A}) \; .$$

In order to write the preceding equations in the form of an orthogonality relation it is sufficient to normalize them with the help of a factor $1/\sqrt{g}$:

$$\sum_{\mathrm{A}} \frac{1}{\sqrt{g}} X^{(p)}(\mathrm{A}) \frac{1}{\sqrt{g}} X^{(g)*}(\mathrm{A}) = \delta_{pg} \tag{3.22}$$

where the superscript labels the irreducible representation.

The normalized primitive characters of the inequivalent irreducible representations of a group $\mathcal{G}$ form in the group space a system of orthogonal functions or a system of orthogonal vectors.

One may express the equations (3.19) in a similar way. The result is that according to well-known theorems $X(\mathrm{A})$, $X'(\mathrm{A}), \ldots$ are linearly independent of each other. The same holds for the matrix elements a_{ik}, a_{jl} . . . or, according to (3.16) for a_{ik}, $a'_{\nu\lambda}$ . . ..[1]

Similar statements can be made for continuous i.e. (infinite) groups. For example if we have the group of rotations around an axis, equation (3.22) becomes the well-known relation

$$\frac{1}{2\pi}\int_0^{2\pi} \exp(im'\varphi)\exp(-im\varphi)\mathrm{d}\varphi = \delta_{mm'}.$$

Finally let us consider a reducible representation. Equation (3.21) applied to (3.12a) gives

$$\frac{1}{g}\sum_{\mathrm{A}} X^{(\mu)}(\mathrm{A})X^*(\mathrm{A}) = m_\mu \tag{3.23}$$

and

$$\frac{1}{g}\sum_{\mathrm{A}} X(\mathrm{A})X^*(\mathrm{A}) = m_0^2+m_1^2+\ldots, \tag{3.23a}$$

where m_μ is the number of times the representation μ is contained in the reducible representation. From this we conclude that:

1. (3.20) is the necessary condition for the irreducibility of a representation $\mathcal{G}$ and (3.23) shows that this condition is sufficient.

2. *The necessary and sufficient condition for the equivalence of two irreducible representations $\mathcal{G}$ and $\mathcal{G}'$ of the group $\mathcal{G}$ is the identity of their character system.* Indeed this condition is necessary because a change of axes will not change the traces of the matrices. It is sufficient because, if it were fulfilled we would have

$$\frac{1}{g}\sum_{\mathrm{A}} X'(\mathrm{A})X^*(\mathrm{A}) = 1,$$

contradicting (3.21) the necessary condition for non-equivalence. Hence the characters completely determine the irreducible representations and the irreducible representations completely determine the characters.

11.3. CLASS-SPACE

We have mentioned in §§ 10.1 and 10.2 that the characters are functions of a class. We will indicate the p classes into which we can subdivide the group $\mathcal{G}$ by $C_1, C_2, C_3, \ldots C_i \ldots C_p$. Each class contains $h_1, h_2, \ldots h_i, \ldots, h_p$ elements of the group.

[1] Compare problem 5.

If we introduce $\tilde{X}_i = \sqrt{h_i/g}X(C_i)$ the equations (3.20) and (3.21) can be written

$$\sum_{i=1}^{p} \tilde{X}_i \tilde{X}_i^* = 1 \quad \text{and} \quad \sum_{i=1}^{p} \tilde{X}_i' \tilde{X}_i^* = 0. \tag{3.24}$$

We are led in this way to consider a *class space* of p dimensions which is derived from the group space by ignoring the differences between the elements of the same class. The successive irreducible representations $\mathscr{G}$, $\mathscr{G}'$ ... correspond to the vectors $\tilde{X}$, $\tilde{X}'$ ... of this space with components $\tilde{X}_i$ ..., $\tilde{X}_i'$ .... According to (3.24) these components are orthogonal. Because there can only be p orthogonal vectors in a p-dimensional space, *the number of irreducible representations of the group $\mathscr{G}$ must be less than or equal to the number of classes.* The theorem in § 10.2 states that they would be equal, i.e. the vectors $\tilde{X}$ form in the class space a *complete* system of orthogonal axes. This theorem is also true of the vectors X with components $X(\mathrm{A})$ in the group space.

12. Sum of a Class; Projection Operators

12.1. DEFINITION OF THE SUM OF A CLASS; STRUCTURE COEFFICIENTS

Suppose we take the sum of all elements belonging to a given class i:

$$\mathrm{C}_i = \mathrm{A}_1^{(i)} + \mathrm{A}_2^{(i)} + \dots \mathrm{A}_{h_i}^{(i)} \tag{3.25}$$

where h_i is the number of elements in the class. If these symbols represent abstract group elements, C_i is a hypercomplex number (see § 10.3). Most of the time we will apply this definition to a representation of the group. In this case we write:

$$\boldsymbol{C}_i = \boldsymbol{A}_1^{(i)} + \boldsymbol{A}_2^{(i)} + \dots \boldsymbol{A}_{h_i}^{(i)} \tag{3.25'}$$

and the matrices $\boldsymbol{C}_i$ are obtained by adding the corresponding elements of the matrices $\boldsymbol{A}_1$, $\boldsymbol{A}_2$, etc.

If we take a product of two of these quantities, the group elements at the right-hand side can be grouped into classes, that is, the result can be written as a linear combination of sums of classes, the coefficients cited being either zero or integers:

$$\mathrm{C}_i \mathrm{C}_k = \sum_{l}^{h_i} \sum_{m}^{h_k} \mathrm{A}_l^{(i)} \mathrm{A}_m^{(k)} = \sum_{l} c_{ikl} \mathrm{C}_l. \tag{3.26}$$

To show this we take the conjugate of the left-hand side with an arbitrary element x which will only permute the elements in each sum of a class. As a result the right-hand side will never contain a set of elements which do not fill a whole class, because all possible conjugations would generate the missing members of that class, and because the left-hand side must be invariant under any conjugation.

The coefficients c_{ikl}, which can be determined from the group table, reflect the structure of the group. With the help of these *structure coefficients* we can determine the character of all possible irreducible representations of a given group.

The hypercomplex numbers c_i formed by the sum of a class commute with all elements of the group, since

$$\mathrm{c}_i\mathrm{x} = \mathrm{x}\mathrm{x}^{-1}\mathrm{c}_i\mathrm{x} = \mathrm{x}\mathrm{c}_i \tag{3.27}$$

where x is an arbitrary element of the group; $\mathrm{x}^{-1}\mathrm{c}_i\mathrm{x} = \mathrm{c}_i$ because conjugation only permutes the terms in the sum according to the definition of a class.

If we now take the representation by *matrices* of (3.26) and apply one of the theorems of § 9, we conclude that these matrices are constants times the unit matrix if the representation is irreducible. (It is not difficult to obtain the explicit value of this constant but its value is not needed for the following argument.) If two matrices are multiples of the unit matrix the product of their traces is equal to the trace of their product times the dimensionality of the matrices.

Applying this to the matrix representation of (3.26) we find that

$$h_i X_i h_k X_k = n \sum_{l=1}^{r} c_{ikl} h_l X_l \tag{3.28}$$

because the character of the sum-of-a-class matrix is equal to the character of that class X_i times the number of elements in the class h_i. n is the dimensionality of the matrices of the particular irreducible representation we are referring to. (This number is of course equal to the character of the unit element X_1.) Note that the relation (3.28) only holds for primitive characters.

12.2. CHARACTER TABLES

In this subsection we will give the complete proof of a statement made in § 10.2, namely: The number of irreducible representations is equal to the number of classes. This was partially verified in § 11.3. However, the following proof is independent of the arguments used in that section.

In equation (3.28) the representation was not specified. If we sum over all irreducible representations we have:

$$\sum_{\mu=1}^{r'} h_i X_i^{(\mu)} h_k X_k^{(\mu)} = \sum_{\mu=1}^{r'} n^{(\mu)} \sum_{l=1}^{r} c_{ikl} h_l X_l^{(\mu)}, \tag{3.29}$$

where r' is the number of irreducible representations. The right-hand side contains a factor, which we will evaluate separately:

$$\sum_{\mu=1}^{r'} n^{(\mu)} X_l^{(\mu)} = \sum_{\mu=1}^{r'} X^{(\mu)}(\mathrm{E}) X^{(\mu)}(\mathrm{C}_l).$$

This is simply a change in notation since n is the dimension of the representation and $X(\mathrm{E})$ the character of the unit representation. However, according to problem 3.8 the characters thus obtained are the characters of the regular representation, and these are all zero, except of course if $\mathrm{c}_l = \mathrm{E}$. We find

$$\sum_{\mu}^{r'} X^{(\mu)}(\mathrm{E}) X^{(\mu)}(\mathrm{C}_l) = X^{\mathrm{reg}}(\mathrm{C}_l) = g\delta_{l1},$$

since the dimension of the regular representation is the order of the group. Substituting this into (3.29) we obtain:

$$\sum_{\mu=1}^{r'} X^{(\mu)}(\mathrm{C}_i) X^{(\mu)}(\mathrm{C}_i^{-1}) = g/h_i. \tag{3.30}$$

The symbol C_i^{-1} refers to the elements of the class that contains the inverse elements of the elements of class C_i. This may be either the same class or a different one. In both cases it will contain the same number of elements.

If the representation is unitary, we have

$$\sum_{\mu=1}^{r'} X^{*(\mu)}(\mathrm{C}_i) X^{(\mu)}(\mathrm{C}_i) = g/h_i. \tag{3.30$'$}$$

This orthogonality relation looks similar to (3.20), which can be written as follows

$$\sum_{i=1}^{r} h_i X^{*(\mu)}(\mathrm{C}_i) X^{(\mu)}(\mathrm{C}_i) = g, \tag{3.20a}$$

but is entirely different in nature. Equation (3.20a) contains a sum over all classes $i = 1, \ldots r$ while (3.30′) contains a sum over all irreducible representations $\mu = 1 \ldots r'$.

The theorem is now easily proved if we consider the double sum

$$\sum_{i=1}^{r} \sum_{\mu=1}^{r'} h_i X^{*(\mu)}(\mathrm{C}_i) X^{(\mu)}(\mathrm{C}_i) = gr = gr'$$

depending on which sum was taken first.

As an example we give the character table of the group mentioned in § 2.4. It has three classes C_i: E the unit element, C_2 the two rotation elements around the axis normal to the center, and C_3 the three turnover elements. Using the results of § 7.4 the following character table is obtained:

	E	C_2	C_3
$\mu = 1$	1	1	1
$\mu = 2$	1	1	−1
$\mu = 3$	2	−1	0

For other character tables see for instance KOSTER [1957].

12.3. PROJECTION OPERATORS (IDEMPOTENT ELEMENTS)

A matrix that generates a given irreducible representation is called a projection matrix. Consider a set of matrices already in box form and multiply by a diagonal matrix as indicated below.

$$\left(\begin{array}{cc|cc|ccc} 0 & & & & & & \\ & 0 & & & & & \\ \hline & & 1 & & & & \\ & & & 1 & & & \\ \hline & & & & 0 & & \\ & & & & & 0 & \\ & & & & & & 0 \end{array}\right)\left(\begin{array}{cc|cc|ccc} x & x & & & & & \\ x & x & & & & & \\ \hline & & x & x & & & \\ & & x & x & & & \\ \hline & & & & x & x & x \\ & & & & x & x & x \\ & & & & x & x & x \end{array}\right)=\left(\begin{array}{cc|cc|ccc} 0 & 0 & & & & & \\ 0 & 0 & & & & & \\ \hline & & x & x & & & \\ & & x & x & & & \\ \hline & & & & 0 & 0 & 0 \\ & & & & 0 & 0 & 0 \\ & & & & 0 & 0 & 0 \end{array}\right).$$

This matrix has zero elements except at those places on the diagonal which correspond with the box we want to "project". The irreducible representations can be considered as "components" of the reducible representation. The projection matrices ε are also referred to as idempotents because they have the property $\varepsilon^n = \varepsilon$ for any power n.

The result at the right-hand side is a set of matrices that are again isomorphic (or homomorphic) with the group. They form an irreducible representation if we omit all superfluous rows and columns (those that contain nothing but zeros).

If the representation was not in box form we could bring it into this form by a transformation $A \to \mathrm{S}^{-1}A\mathrm{S} = A^{\mathrm{b}}$. The representation in box form is designated by a superscript b. Conversely if we consider the inverse transformation applied to the equation

$$\varepsilon^{\mathrm{b}(\mu)} A^{\mathrm{b}} = A^{\mathrm{b}(\mu)} \tag{3.31'}$$

representing the matrices depicted above, we have

$$\mathrm{S}\varepsilon^{\mathrm{b}(\mu)}\mathrm{S}^{-1}\mathrm{S}A^{\mathrm{b}}\mathrm{S}^{-1} = \mathrm{S}A^{\mathrm{b}(\mu)}\mathrm{S}^{-1} \quad \text{or} \quad \varepsilon^{(\mu)}A = A^{(\mu)}. \tag{3.31}$$

It is possible to construct a priori the matrices $\varepsilon^{(\mu)}$. If we form all products $\varepsilon^{(\mu)}A$ the result will be a set of matrices $A^{(\mu)}$ which can be reduced in size. This is done in this case not only by omitting rows and columns that are zero (as mentioned above) but also by noticing that certain rows or columns are linear combinations of other rows or columns and keeping only those that are independent.

The procedure can be summed up as follows:

Projecting a certain irreducible representation out of a reducible one consists of three steps. (i) Convert the matrices to box form; (ii) Multiply by the diagonal matrix indicated above; (iii) Omit the unnecessary rows and columns in the result. It turns out, however, that steps two and three can be taken first, making the first step, which is in practical cases of course the most elaborate one, unnecessary.

After this introduction we display the projection operators or idempotent elements. They are

$$\varepsilon^{(\mu)} = (n_\mu/g)\sum_{\mathrm{A}} X^{(\mu)}(\mathrm{A}^{-1})A = (n_\mu/g)\sum_i X_\mu^*(i)\mathrm{C}_i \tag{3.32}$$

where g is the number of elements of the (finite) group, n_μ the dimensionality of the irreducible representation μ, X_μ the characters of this representation. The sum is taken either over all the matrices that form the reducible representation of the group, or as the right-hand side of (3.32) indicates the sum over all C_i. Since the characters of all elements are equal as long as they belong to the same class it is obvious that the second sum is identical with the first, provided the representation is unitary.

The matrices $\varepsilon^{(\mu)}$ can be constructed in a straightforward way from a given representation if the characters of the irreducible representations are known. These can be calculated from the structure coefficients as mentioned in § 12.1 and these in turn follow directly from the abstract group table. Usually, however, the necessary character tables can be found in the literature. If a calculation is performed, it is of course advisable to investigate first whether representation μ is really contained in the irreducible representation. This can be done with equation (3.23).

Finally we have to demonstrate that the products $\varepsilon^{(\mu)}A$ really form for all A an irreducible representation μ. The necessary and sufficient condition is that the characters of $\varepsilon^{(\mu)}A$ are equal to the characters of the irreducible representation μ

$$X(\varepsilon^{(\mu)}A) = X^{(\mu)}(\mathrm{A}). \tag{3.33}$$

In order to show this we first point out that $\varepsilon^{(\mu)}$ commutes with all A:

$$A\varepsilon^{(\mu)}A^{-1} = \varepsilon^{(\mu)}, \qquad \text{(for every } A \text{ belonging to } \mathcal{G}\text{)}$$

since $\varepsilon^{(\mu)}$ is made up of class-sums which are invariant under conjugation as mentioned in § 12.1.

Let A be, for the moment, an irreducible representation, for example ν. Then from theorem III, § 9 we conclude that

$$\varepsilon^{(\mu)} = c\,\mathbf{I}.$$

We calculate the constant c by taking the trace of this equation

$$X(\varepsilon^{(\mu)}) = \frac{n_\mu}{g}\sum X^{(\mu)}(\mathrm{A}^{-1})X(A^\nu) = n_\mu \delta_{\mu\nu}. \tag{3.34}$$

(We must assume that A contains the representation ν only once.) From this we conclude that

$$c = \delta_{\mu\nu}. \tag{3.35}$$

If A is reducible and in box form, the matrix $\varepsilon^{(\mu)}$ will be a diagonal matrix with the diagonal elements zero except at the places corresponding to the box μ, according to (3.35). If the representation is in box form equation (3.33) holds. But then it should hold for any representation, because the characters are invariants.

The prescription for performing a reduction is the following. We have a set of basis functions and operate with the prescribed operations of the group in order to create matrices. Now, depending on how complicated the problem is, it might be worthwhile to see which representations are contained by taking the trace of each matrix (one of each class is enough, if the classes were not known this is the moment to find out) and determining the coefficients in eq. (3.12a)

If a certain irreducible representation is contained one constructs the projection operator (3.32) and inspection of these matrices usually tells us which rows or columns are superfluous.

If not obvious we multiply the projection matrix with the basis. The resulting linear combinations are either zero (in case a certain row contains nothing but zeros) or partially dependent on each other. We take from each set of dependent linear combinations one and drop the others.

As an epilogue we should like to point out that the idea of projection operators is solely based on the theory of characters. First, the only reason that the projections $\varepsilon^{(\mu)}A$ are the matrices we want them to be is that they have the proper characters, and that fact is sufficient for irreducibility. Second, the way the operators $\varepsilon^{(\mu)}$ are constructed is to make use of the orthonormality of character systems, and this was used (3.34).

If a certain irreducible representation is contained several times in a reducible representation, one has to proceed with more caution. In this case the projection will result in a linear combination of the multiple occurring representations. It is necessary to perform an additional orthogonalization in order to obtain the box form. There is an arbitrariness in this procedure similar to the arbitrariness in the choice of orthogonal wave functions in a degenerate eigenvalue problem.

In the important case that the reducible representation is the product of two irreducible representations of the rotation group, the method does not apply, because these are continuous groups. This case can be treated in an entirely different way (Chapter 5 : § 5.3).

13. Representations of the Permutation Group

13.1. YOUNG-TABLEAUX

In the special case in which we are dealing with the permutation group it is possible to obtain irreducible representations in a clear and concise way. The ideas explained below are originally due to A. Young, who published a number of papers on group theory around 1900.[1]

If we consider an element of the permutation group and if we employ the notation with cycles for instance,

$$(1\ 5\ 3)(2\ 4)(7\ 8)(9) \tag{3.36}$$

then we can write these numbers in a so-called Young tableau as follows:

1	5	3
2	4	
7	8	
9		

(3.37)

[1] A. Young was a country clergyman.

Every row corresponds to a cycle and hence the numbers in a row can always be cyclically permuted. It is the custom to draw the different cycles in declining order, that is to say the first row is the longest cycle, the second is either the same length or shorter, etc. If there are two rows of the same length the order in which they are written is irrelevant. The shape or contour of this scheme determines a class of equivalent permutations. (Compare the end of § 5.2.) This is a one-to-one correspondence. If two schemes are of the same shape, they belong to the same class of permutations and if they are not the same shape they belong to a different class of permutations. By the shape we mean, of course, the empty box, i.e., the tableau without the numbers. Instead of describing the shape or the class of permutations by indicating the length of each row or the length of each cycle, one can do just as well by telling how many cycles there are of order one, how many cycles there are of order two, how many cycles there are of order three, etc. The number of cycles of a given order, that is, the number of rows of a given length, is sometimes called the rank number. If n is the total number of elements which we are permuting, we have the following equality:

$$n = 1 \cdot a_1 + 2a_2 + 3a_3 \dots ka_k \tag{3.38}$$

where the rank number a_k indicates the number of cycles of order k.

The number of permutations that belong to a class $C_{\{a_i\}}$ characterized by the set of rank numbers $a_1, a_2, a_3, \dots a_k \equiv \{a_i\}$ is equal to:

$$h_{\{a_i\}} = \frac{n!}{1^{a_1} a_1!\, 2^{a_2} a_2! \dots k^{a_k} a_k!}. \tag{3.39}$$

The derivation of this formula is easily demonstrated with the help of example (3.36). If we take the corresponding Young shape we have h, i.e. in this case 8 open places and hence can fill in the numbers in 8! (h!) different ways. The number of rows of length k is a_k and they can be permuted in any order. Hence we have to divide the total number of ways $h!$ by $a_k!$ in order to get the total number of classes. Now we can cyclically permute all the numbers in each one of these rows which gives a factor k^{a_k} since each row allows k cyclical permutations. That means $k^{a_k} a_k!$ in our example (3.37) 3 for the first row and $2^2 \cdot 2! = 8$ for the next two rows. The number expressed by (3.39) is found in different places in mathematics and physics. In the theory of numbers it is called the partitio numerorum, i.e., the number of ways one can assort a certain number n, of objects into piles with a_1 piles of one object, a_2 piles of two objects each, etc. To physicists this factor is well known from the work of Mayer in the development of the partition function in statistical mechanics. (MAYER and MAYER [1940] p. 437.)

13.2. THE P · Q-OPERATIONS; IDEMPOTENTS

We will introduce a special set of elements which will eventually lead to the construction of an idempotent.

Conjugation of an arbitrary element T with respect to P (i.e. $T' = P^{-1}TP$) will have the effect that the Young tableau of T will maintain its shape. The numbers in the shape will be permuted in the way described by P. We consider only the elements P which will create a horizontal permutation in T. That is to say, the numbers in one row will be permuted but will never leave that particular row. Such elements form a subgroup. We will call this subgroup $\mathscr{P}$ and its order is equal to $\Pi_k a_k!$ (Compare Fig. 3.6 for the general Young scheme.) Not all elements represent different permutations. In a similar way one can introduce those permutations, which, if used to conjugate given permutations, create permutations only within a given column. We call this subgroup $\mathscr{Q}$. Any rearrangement of numbers in a given tableau, i.e. keeping the shape the same, corresponds to a similarity transformation

$$P_{op}T \equiv P^{-1}TP. \tag{3.40}$$

This holds for both P and Q.

We introduce now P · Q, which indicates that we are dealing with an operation which consists of a vertical permutation followed by a horizontal permutation. Such elements, which do not form a subgroup, will be used to characterize a hypercomplex number ζ by the following convention: Suppose the conjugation of the Young scheme cannot be obtained by the operation P · Q, then the coefficient ζ will be equal to zero. If, however, it can be obtained by the operation P · Q then ζ will be either plus 1 if Q is an even permutation, or minus 1 if Q is an odd permutation. In formula we have

$$\zeta = \sum_{R} \zeta(R)R \tag{3.41}$$

R is an element of $\mathscr{S}_n$ either a reducible or irreducible representation of this element, or, in the true sense of group algebra, an abstract element. The numbers $\zeta(R)$ are:

$$\zeta(R) = \begin{cases} 0 \text{ if } R \neq PQ \\ 1 \text{ if } R = PQ \text{ and } Q \text{ is an even permutation} \\ -1 \text{ if } R = PQ \text{ and } Q \text{ is an odd permutation.} \end{cases} \tag{3.42}$$

We want to study the set of elements P · Q more carefully. First a word of warning. It is not true that P · Q corresponds to a similarity transformation with Q followed by one with P. The second similarity transformation is $Q^{-1}PQ$ instead. This is easy to see. From (3.40) we have: $(QP)_{op}T \equiv (QP)^{-1}$ $TQP = (P^{-1}QP)_{op}P_{op}T$. Second, it is interesting and useful to have a criterion that tells which elements of R belong to the set P · Q and which do not. The elements R belong to the set P · Q if and only if, two numbers of a given column never arrive in the same row. The condition is necessary, as one can easily see from an example.

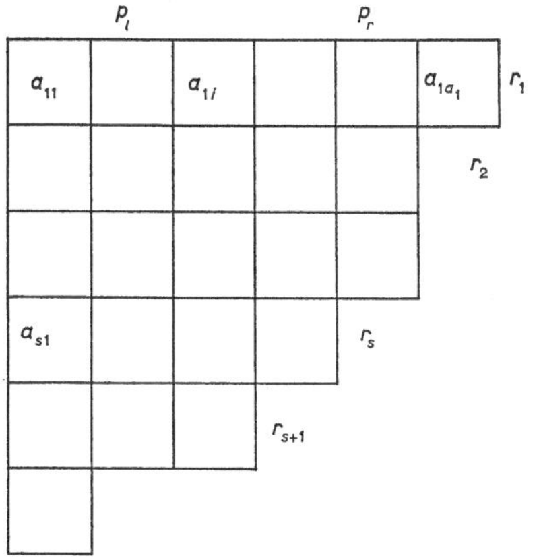

Fig. 3.6. Tableau.

To show that the condition is sufficient we separate the Young tableau into two parts. The left part p_l contains the first i columns, the right part p_r the columns labelled by $i+1, \ldots, a_1$. The operation R never brings two numbers of the same column into the row r_1, hence the a_1 numbers of r_1 come from different columns. The numbers in the positions $i+1, \ldots, a_1$ of the row r_1 originated from the columns in p_r, hence these numbers can never come from the rows r_{s+1}, $r_{s+2}, \ldots$ etc.

The second criterion is; if R does not belong to P · Q, the elements $R^{-1}PR$ and Q will have at least one transposition in common.

The hypercomplex number ζ defined in this way has the property that its square is proportional to the number itself, as will be shown below. Such a number is called an essential idempotent because it is possible to make it idempotent by multiplication with an ordinary number. Hence if we take the square of (3.41) we find by using the multiplication rules for hypercomplex numbers of § 10.3

$$\zeta^2 = \sum \zeta(\text{R})\zeta(\text{R}^{-1}\text{T})\text{T}. \tag{3.43}$$

With the help of the result (3A.6) of the Appendix to this Chapter we find $\zeta^2 = \rho\zeta$ and with

$$\rho^{-1}\zeta = \varepsilon \tag{3.44}$$

we have $\varepsilon^2 = \varepsilon$. The number ρ is a real integer as can be seen from the derivation in the Appendix.

The fact that the ζ is an essential idempotent allows us to use this operator as a projection operator. Any idempotent ε fulfills the equation $\varepsilon^2 - \varepsilon = 0$, hence the idempotent must have eigenvalues 1 and 0. If we multiply the matrix ε with a set of basis functions, all components that correspond to a zero in the diagonal of the ε-matrix will be removed. Hence the number of independent basis vectors is reduced. If the ε were not in diagonal form the last arguments would still hold as we have shown in § 12.3. An obvious aspect of projection, viz. projecting twice gives the same result as projecting once, is automatically fulfilled since $\varepsilon^2 = \varepsilon$.

13.3. IRREDUCIBLE REPRESENTATIONS

We return to the regular representation described in § 10. The basis of the regular representation can be considered in two ways. Either we consider the basis E, A, B, . . . S, . . . as abstract elements, or we consider the basis represented by matrices ***E***, ***A***, ***B***, . . . ***S*** . . .. These matrices are operating in turn on a "subbasis" which is usually not written in the formulas. (This is reminiscent of second quantization or field theory where the operator wave function operates on nothing "visible" either.) The distinction between these two cases is irrelevant but helpful in order to make the following point.

We want to use the projection operator or idempotent ε to induce a new representation in the regular representation. We do this by operating with ε on the "subbasis" and hence the elements E, A, B, . . . are replaced by Eε, Aε, Bε, . . ., Sε . . .. Again one can either consider this as a set of abstract elements as above, or as a set of matrices ***E***ε, ***A***ε, ***B***ε, . . ., ***S***ε, . . . where ε is a matrix which has the property that the eigenvalues are either zero or one. Again, this distinction is irrelevant but we wanted to make it clear that the projection operators are written *on the right.*[1]

[1] In the language of group algebra all hypercomplex numbers obtained by multiplying ε at the left by the hypercomplex numbers of the algebra form a subset of the algebra called the left ideal. In this language the minimal (simple) left ideal corresponds to our irreducible representation.

We want to calculate the characters of the representation induced in this way, since we finally want to show that the special choice of idempotents resulting from the Young shapes leads to irreducible representations and irreducibility is easily demonstrated if the characters are known.
The representation matrices induced by the idempotent are

$$\mathrm{S} \to \mathrm{AS}\varepsilon = \mathrm{A}\frac{\mathrm{S}}{\rho}\sum_{\mathrm{R}} \zeta(\mathrm{R})\mathrm{R}$$

$$= \frac{1}{\rho}\sum_{\mathrm{T}} \zeta(\mathrm{S}^{-1}\mathrm{A}^{-1}\mathrm{T})\mathrm{T}. \tag{3.45}$$

That is, according to § 10.3 the element A is represented by a matrix:

$$\mathrm{A} = (a_{st}) \quad \text{where} \quad a_{st} = \rho^{-1}\zeta(\mathrm{S}^{-1}\mathrm{A}^{-1}\mathrm{T}).$$

Hence the diagonal elements of this matrix are obtained by taking $\mathrm{S} = \mathrm{T}$ and the trace is

$$X(\mathrm{A}) = \frac{1}{\rho}\sum_{\mathrm{S}} \zeta(\mathrm{S}^{-1}\mathrm{A}^{-1}\mathrm{S}). \tag{3.46}$$

We indicate with $\{a_i\}$ a certain partitio, hence a certain class of elements and hence a certain shape. The hypercomplex numbers $\zeta^{\{a\}}$ or $\varepsilon^{\{a\}}$ derived from this shape will induce a representation of which the element A has a character $X^{\{a\}}(\mathrm{A})$. In order to see whether this representation is irreducible we calculate the sum (3.23a) and show that the right-hand side is equal to one. Using (3A.7) we find indeed

$$g^{-1}\sum_{\mathrm{A}} X^{\{a\}}(\mathrm{A}^{-1})X^{\{a\}}(\mathrm{A}) = (g\rho^2)^{-1}\sum_{\mathrm{A}}\sum_{\mathrm{S}}\sum_{\mathrm{T}} \zeta^{\{a\}}(\mathrm{S}^{-1}\mathrm{A}^{-1}\mathrm{S})\zeta^{\{a\}}(\mathrm{T}^{-1}\mathrm{AT})$$

$$= \rho^{-2}\sum_{\mathrm{A}}\sum_{\mathrm{S}} \zeta^{\{a\}}(\mathrm{S}^{-1}\mathrm{A}^{-1}\mathrm{S})\zeta^{\{a\}}(\mathrm{A}) = 1. \tag{3.47}$$

The result obtained implies that all irreducible representations of thc symmetric group can be obtained by a straight forward method. (Compare problem 3.13.) The tableaux are not only useful for this purpose but are also of great help in the decomposition of representations that were irreducible with respect to the symmetry group of n elements but are reducible with respect to a symmetry group of $n-1$ elements. If, for instance, we have the permutation group of three elements, the Young tableaux are:

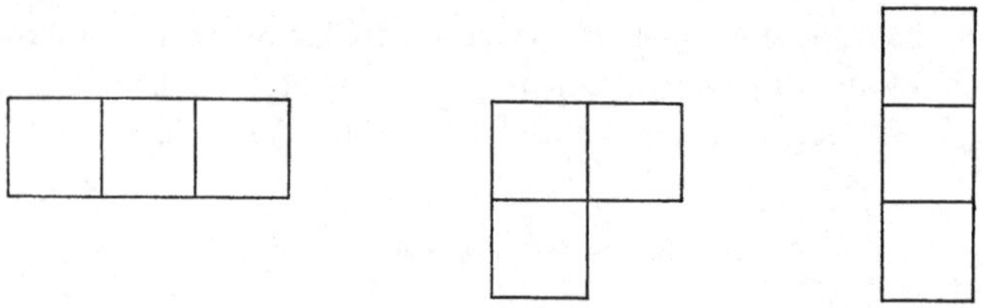

We can decompose very simply each one of these irreducible representations into those which are the irreducible representations of the permutation group of two elements. According to the following scheme:

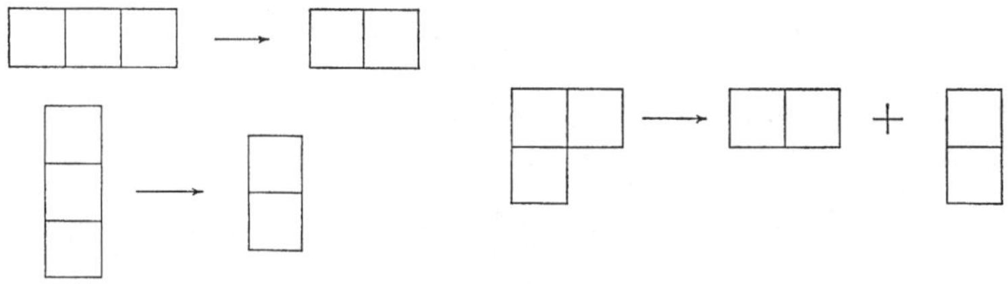

The plus sign indidates a direct sum of two permutations. This idea is extensively used in electronic and nuclear spectroscopy. If one wants to find the energy levels of a n-electron system from the knowledge of the wave function and energy levels of a $n-1$ electron system. In such a procedure it is necessary to form products of linear combinations of wave functions but not all of these products obey the Pauli principle. Hence one is interested in only certain representations of the permutation group. In this case the permutation is with respect to the electrons.

Appendix 3.I

In order to prove the equation mentioned in § 13.3, we study the properties of the hypercomplex numbers ζ (compare Molenaar [1930]). First of all we notice that

$$\zeta(\mathrm{E}) = 1; \quad \zeta(\mathrm{P}) = 1; \quad \zeta(\mathrm{Q}) = \pm 1$$

$$\zeta(\mathrm{PQ}) = \zeta(\mathrm{Q}) \quad (3.\mathrm{A}.1); \quad \zeta(\mathrm{Q}_1\mathrm{Q}_2) = \zeta(\mathrm{Q}_1)\zeta(\mathrm{Q}_2) \qquad (3\mathrm{A}.2)$$

which follows simply from the definition of ζ. It is also easy to see that

$$\zeta(\text{PR}) = \zeta(\text{R}) \tag{3A.3}$$

since either $\text{R} = \text{P}_1\text{Q}_1$ and according to (3A.1) we have:

$$\zeta(\text{PR}) = \zeta(\text{PP}_1\text{Q}_1) = \zeta(\text{Q}_1) \text{ and } \zeta(\text{R}) = \zeta(\text{P}_1\text{Q}_1) = \zeta(\text{Q}_1),$$

or $\text{R} \neq \text{P}_1\text{Q}_1$. In this case the right-hand side is zero. The left-hand side is also zero because $\text{PR} \neq \text{PP}_1\text{Q}_1$ is also not a product of a P and a Q. Hence the second possibility satisfies the equation as well.

Following the same line of reasoning one can show the relation

$$\zeta(\text{RQ}) = \zeta(\text{R})\zeta(\text{Q}) \tag{3A.4}$$

since either $\text{R} = \text{PQ}$, then on the one hand we have

$$\zeta(\text{RQ}) = \zeta(\text{P}_1\text{Q}_1\text{Q}) = \zeta(\text{Q}_1)\zeta(\text{Q})$$

using (3A.3) and (3A.2). On the other hand

$$\zeta(\text{R})\zeta(\text{Q}) = \zeta(\text{P}_1\text{Q}_1)\zeta(\text{Q}) = \zeta(\text{Q}_1)\zeta(\text{Q})$$

according to (3A.1). Or if $\text{R} \neq \text{PQ}$ then both sides are zero since also $\text{RQ} \neq \text{P}_1\text{Q}_1\text{Q} = \text{P}_1\text{Q}_2$.

We have now enough material to show the first summation formula:

$$\sum_{\text{P}} \zeta(\text{APB}) = \sum_{\text{P}} \zeta(\text{AP})\zeta(\text{B}). \tag{3A.5}$$

Either $\text{B} = \text{P}_1\text{Q}_1$ and we have, according to (3A.4)

$$\sum_{\text{P}}\zeta(\text{APB}) = \sum_{\text{P}}\zeta(\text{APP}_1)\zeta(\text{Q}_1) = \sum_{\text{P}}\zeta(\text{AP})\zeta(\text{Q}_1),$$

since P and PP_1 are both summed over all elements. The right-hand side, using (3A.1) can be written

$$\sum_{\text{P}} \zeta(\text{AP})\zeta(\text{P}_1\text{Q}_1) = \sum_{\text{P}} \zeta(\text{AP})\zeta(\text{Q}_1).$$

Or if $\text{B} \neq \text{P}_1\text{Q}_1$ the right-hand side is equal to zero and since according to the second criterion B^{-1}PB and Q have a transposition in common: $\text{Q}_2 = \text{B}^{-1}\text{P}_2\text{B}$. Hence replacing P by PP_2 in the sum we have

$$\sum_{\text{P}} \zeta(\text{APB}) = \sum_{\text{P}} \zeta(\text{APP}_2\text{B}) = \sum_{\text{P}} \zeta(\text{APBQ}_2) = \sum_{\text{P}} \zeta(\text{APB})\zeta(\text{Q}_2)$$

according to (3A.4). Since Q_2 is a transposition, we have $\zeta(\text{Q}_2) = -1$ and the sum is zero because

$$\sum_{\text{P}} \zeta(\text{APB}) = -\sum_{\text{P}} (\text{APB}).$$

The next formula we want to prove is,

$$\sum_{\mathrm{R}} \zeta(\mathrm{R})\zeta(\mathrm{R}^{-1}\mathrm{T}) = \rho\zeta(\mathrm{T}), \tag{3A.6}$$

where ρ is an integer. Call the left-hand side $\sum$ and replace R by $\mathrm{P}^{-1}\mathrm{R}$. If we now sum over all elements P then we have:

$$p\sum = \sum_{\mathrm{P}}\sum_{\mathrm{R}} \zeta(\mathrm{P}^{-1}\mathrm{R})\zeta(\mathrm{R}^{-1}\mathrm{PT}) = \sum_{\mathrm{P}}\sum_{\mathrm{R}} \zeta(\mathrm{R})\zeta(\mathrm{R}^{-1}\mathrm{P})\zeta(\mathrm{T})$$

using (3A.3) and (3A.5).

The number p represents the total number of elements that have the property that they only create horizontal permutations in the tableau. If we replace R by PR then we have

$$p\sum = \sum_{\mathrm{P}}\sum_{\mathrm{R}} \zeta(\mathrm{PR})\zeta(\mathrm{R}^{-1}\mathrm{P}^{-1}\mathrm{P})\zeta(\mathrm{T}) = p\sum_{\mathrm{R}} \zeta(\mathrm{R})\zeta(\mathrm{R}^{-1})\zeta(\mathrm{T})$$

according to (3A.3). Hence we find (3A.6) where the integer ρ is defined by $\rho = \sum_{\mathrm{R}} \zeta(\mathrm{R})\zeta(\mathrm{R}^{-1})$.

The last formula we want to prove is the following double sum:

$$\sum_{\mathrm{R}}\sum_{\mathrm{S}} \zeta(\mathrm{S}^{-1}\mathrm{R}^{-1}\mathrm{S})\zeta(\mathrm{R}) = \rho^2. \tag{3A.7}$$

If we again call the left-hand side $\sum$ and replace R by PR and take the sum over P, we obtain:

$$\begin{aligned} p\sum &= \sum_{\mathrm{P}}\sum_{\mathrm{R}}\sum_{\mathrm{S}} \zeta(\mathrm{S}^{-1}\mathrm{R}^{-1}\mathrm{P}^{-1}\mathrm{S})\zeta(\mathrm{PR}) \\ &= \sum_{\mathrm{P}}\sum_{\mathrm{R}}\sum_{\mathrm{S}} \zeta(\mathrm{S}^{-1}\mathrm{R}^{-1}\mathrm{P}^{-1})\zeta(\mathrm{S})\zeta(\mathrm{R}), \end{aligned}$$

according to (3A.5) and (3A.3).

If now we replace R by $\mathrm{P}^{-1}\mathrm{R}$ then we have:

$$\begin{aligned} p\sum &= \sum_{\mathrm{P}}\sum_{\mathrm{R}}\sum_{\mathrm{S}} \zeta(\mathrm{S}^{-1}\mathrm{R}^{-1}\mathrm{PP}^{-1})\zeta(\mathrm{S})\zeta(\mathrm{P}^{-1}\mathrm{R}) \\ &= p\sum_{\mathrm{R}}\sum_{\mathrm{S}} \zeta(\mathrm{S}^{-1}\mathrm{R}^{-1})\zeta(\mathrm{S})\zeta(\mathrm{R}) \end{aligned}$$

again using (3A.3). The result is, that by replacing S by PS and summing over P, we will have:

$$\begin{aligned} p\sum &= \sum_{\mathrm{P}}\sum_{\mathrm{R}}\sum_{\mathrm{S}} \zeta(\mathrm{S}^{-1}\mathrm{P}^{-1}\mathrm{R}^{-1})\zeta(\mathrm{PS})\zeta(\mathrm{R}) \\ &= \sum_{\mathrm{P}}\sum_{\mathrm{R}}\sum_{\mathrm{S}} \zeta(\mathrm{S}^{-1}\mathrm{P}^{-1})\zeta(\mathrm{R}^{-1})\zeta(\mathrm{S})\zeta(\mathrm{R}) \end{aligned}$$

using (3A.5) and (3A.3). The sum over P can be performed by replacing S by PS and we obtain, using again (3A.3):

$$p\sum = \sum_{P}\sum_{R}\sum_{S} \zeta(S^{-1}PP^{-1})\zeta(R^{-1})\zeta(P^{-1}S)\zeta(R)$$

$$= p\sum_{R}\sum_{S} \zeta(S^{-1})\zeta(R^{-1})\zeta(S)\zeta(R) = p\rho^2.$$

This completes the proof.

CHAPTER 4

GENERAL APPLICATIONS TO QUANTUM MECHANICS; WIGNER'S THEOREM

1. Invariant Properties of the Schrödinger Equation

1.1. THE TWO GROUPS OF THE SCHRÖDINGER EQUATION

There are two important groups that leave the Schrödinger equation, i.e., the Hamilton operator H, invariant.

1. The permutation group — the exchange of positions in space — between identical particles, either electrons or nuclei. This group will always leave H invariant. The quantum theory makes the fullest use of the indistinguishability of identical objects.
2. The group of rotations and reflections. This group plays a role only if the potential energy has certain symmetries.

These groups are related to the space coordinates of the constituting particles only, i.e., their operators are *orthogonal linear substitutions in the configuration space* (the axes in the *ordinary space stay rectangular*). They occur in ordinary as well as in relativistic mechanics.

The Lorentz group which acts in space-time, does not leave the Schrödinger equation invariant, but only the Dirac equation.

1.2. TRANSFORMATIONS INDUCED IN FUNCTION SPACE BY THE TRANSFORMATIONS IN CONFIGURATION SPACE

If $x_1, x_2, \dots x_n$ are the coordinates of the particles, i.e., the coordinates in the configuration space Γ and if S represents an operation of one of the groups we mentioned (for instance, a rotation of the entire helium atom around its nucleus, or a permutation among its electrons, both of which can be expressed in six-dimensional configuration space), then S can be expressed by a system of N equations

$$x_i' = \sum_{k=1}^{n} \sigma_{ik} x_k , \tag{4.1}$$

where the matrices σ are always orthogonal in the real sense of the word.

We will use the following shorthand

$$x \to x' = \mathrm{s}x; \quad x = \mathrm{s}^{-1}x' \tag{4.1a; 1b}$$

where x represents the set of coordinates $x_1 \dots x_n$, i.e., a point in the configuration space.

What are the repercussions of the operator s in the function space or state-vector space $\mathfrak{R}$? Or according to the language of H. Weyl: "Which transformation induces s into the space $\mathfrak{R}$?"

Generalizing the statements of Chapter II § 3.2: The operation s substitutes in Γ-space the point $x' = \mathrm{s}x$ for the point x and *at the same time carries the value the function ψ had at the point x to the point x'*. We may use the following picture. The operation creates, so to speak, a redistribution of the wave function distribution in Γ-space as one redistributes the masses in ordinary space if one describes the motion of a fluid. The coordinate system being fixed, we obtain in this way a new wave function $\psi'(x)$ and we have by definition

$$\psi(x) \to \psi'(x) = \mathrm{s}\psi(x), \tag{4.2}$$

but, according to the statement in italics, we have for every value of x,

$$\psi'(x') = \mathrm{s}\psi(\mathrm{s}x) = \psi(x) \tag{4.2a}$$

or

$$\mathrm{s}\psi(x) = \psi(\mathrm{s}^{-1}x). \tag{4.2b}$$

In the equations (4.2) and (4.2b) the elements s of the group $\mathcal{G}$ can be considered as operators acting on the vectors or rays of the function space, as mappings of the space $\mathfrak{R}$ on itself, or as mappings induced in the space $\mathfrak{R}$ by the group $\mathcal{G}$.

We can easily see that these are linear. They are unitary since we have for two functions ψ_1 and ψ_2,

$$(\psi_1 . \psi_2) = (\mathrm{s}\psi_1 . \mathrm{s}\psi_2)$$

because the transformation (4.1) is equivalent with an orthogonal transformation of coordinates in Γ-space.

1.3. EXPRESSION OF THE INVARIANCE OF H

We suppose that the potential energy

$$V(x) = V(x_1, x_2, \dots x_n)$$

stays invariant under *all* operations s of $\mathcal{G}$. Hence we have

$$V(x_1' \dots x_n') = V(\mathrm{s}x) = V(x)$$

and since s is an arbitrary element from the group $\mathcal{G}$ we may write according to (4.2b)

$$\mathrm{s}V(x) = V(\mathrm{s}^{-1}x) = V(x) \tag{4.3}$$

i.e., *the operation* s *does not change the function* V. We shall say that the functions which satisfy the condition (4.3) *are symmetrical with regard to the group* $\mathcal{G}$ *or invariant under the operations of the group* $\mathcal{G}$.

If we consider the product of two functions, for example, V and ψ, we have according to (4.2b)

$$\begin{aligned} \mathrm{s}[V(x) \cdot \psi(x)] &= V(\mathrm{s}^{-1}x)\psi(\mathrm{s}^{-1}x) \\ &= \mathrm{s}V(x) \cdot \mathrm{s}\psi(x), \end{aligned}$$

and if V is invariant

$$\mathrm{s}[V(x) \cdot \psi(x)] = V(x)\mathrm{s}\psi(x).$$

More generally the invariance of an operator such as the Hamiltonian H, i.e., its invariance under the operations s of the group $\mathcal{G}$, is expressed by the equation

$$\mathrm{sH}\psi = \mathrm{Hs}\psi$$

or, as usually done in the theory of operators, we omit the object ψ on which they operate and write

$$\mathrm{sH} = \mathrm{Hs}. \tag{4.4}$$

Our hypothesis on the Schrödinger equation can be written

$$\mathrm{s}(\mathrm{H}-E)\psi = (\mathrm{H}-E)\mathrm{s}\psi. \tag{4.5}$$

1.4. CONSTANTS OF THE MOTION

The equation (4.4) shows that the operator s commutes with the Hamiltonian. If we remember the quantum mechanical meaning of H($\mathrm{H} = i\hbar\partial/\partial t$) we see that the operator s does not change in time. Considered as a physical quantity (for this it has to be Hermitian) this operator is a constant of the motion. The same conclusion follows immediately from the equation (2.27) of Born, Heisenberg and Jordan.

This remark is the basis for the quantum mechanical derivation of the classical theorems about momentum. (Chapter 5, § 3.)

2. Wigner's Theorem

2.1. THEOREM

Equation (4.5) shows that:

If ψ is an eigenfunction of the operator H *and corresponds to the eigenvalue E, then* sψ *is also an eigenfunction of* H *corresponding to the same eigenvalue E.* WIGNER [1927]-theorem.

From this fundamental remark follow some important consequences. First let us suppose that the Schrödinger problem is solved: the sequence of eigenvalues E and eigenfunctions ψ is known. They form a complete system of orthogonal functions. For the sake of simplicity we suppose that the energy level spectrum is discrete.

(i) If E is a *non-degenerate* eigenvalue: then sψ is equal to ψ except for a multiplicative constant μ of *modulus one*. In the special case that the group $\mathscr{G}$ has only two elements, the identity E and the elements s, the constant μ satisfies the equation $\mu^2 = 1$ and hence must have the two values ± 1. This occurs in the permutation group of the two electrons of the helium atom. If $\mu = +1$ the function ψ is symmetrical. If $\mu = -1$ the function ψ is antisymmetrical. For more precision we refer to section 3.1.

ii) If E is a *degenerate* eigenvalue of order α; $\psi_1, \psi_2 \dots \psi_\alpha$ are the orthogonal eigenfunctions describing the states of the energy level E. Let ψ_i be one of them, then sψ_i is an eigenfunction corresponding to the same value E for every element s of $\mathscr{G}$. Hence sψ_i is a linear combination of the functions,

$$\mathrm{s}\psi_i = \sum_{k=1}^{\alpha} \psi_k S_{ki}. \tag{4.6}$$

In the same way each element s of the group gives rise to a matrix $\boldsymbol{S} = (S_{ki})$ of order α the elements of which are generally complex numbers. The whole *set of the matrices or the whole set of the transformations* (4.6) *form a representation of the group $\mathscr{G}$.*

To prove this let T be another element of $\mathscr{G}$

$$\mathrm{ST}\psi_i = \mathrm{S}(\mathrm{T}\psi_i) = \mathrm{S}\sum_{l=1}^{\alpha} \psi_l T_{li} = \sum_{k,\,l} \psi_k S_{kl} T_{li} = \sum_{k} \psi_k (ST)_{ki}. \tag{4.7}$$

Hence to the element ST corresponds a matrix $\boldsymbol{ST}$.

If the functions ψ_i are orthogonal and the elements s are unitary operators, this representation is unitary because we have

$$(\mathrm{s}\psi_i \,.\, \mathrm{s}\psi_k) = (\psi_i \,.\, \psi_k) = \delta_{ik}$$

or according to (4.6),

$$\sum_{j,l} (\psi_j, \psi_l) S^*_{ji} S_{lk} = \sum_{jl} \delta_{jl} S^*_{ji} S_{lk} = \delta_{ik}.$$

Hence

$$\sum_l S^*_{li} S_{lk} = \delta_{ik} \qquad \tilde{S}^* S = \mathbb{I}.$$

It is known that the α fundamental functions which describe the state corresponding to the eigenvalue E are fixed except for an arbitrary unitary transformation A.
If we replace the functions ψ_i by linear combinations of these functions,

$$\psi'_i = \sum_{k=1}^{\alpha} \psi_k a_{ki}, \qquad (a_{ki}) = \mathsf{A}$$

the matrices S become

$$S' = \mathsf{A}^{-1} S \mathsf{A}$$

and the representation we obtained by a change of the axes in the "eigen space" of the level E is equivalent to the first one.

2.2. GENERAL SOLUTION BY SUCCESSIVE REDUCTIONS

Wigner's theorem is usually presented as formulated above. It seems suggestive to study the question in a more general way and in a certain way from the reverse point of view.

Consider an arbitrary complete set of orthogonal functions χ_i. They span the function space and we use them to represent any operator s or H as an infinite matrix, according to (1.25)

$$\mathrm{s}\chi_i = \sum_k \chi_k S_{ki} \qquad \mathrm{H}\chi_i = \sum_k \chi_k H_{ki}. \tag{4.8}$$

The elements $S_{ki} \ldots, H_{ki} \ldots$ are complex numbers and not dependent on time.

Let us suppose that $\mathrm{s}, \mathrm{T} \ldots$ are the elements of a group $\mathscr{G}$. The matrices $S' = (S_{ki})$, $T' = (T_{ki}), \ldots$ (primed in order to distinguish these matrices from the finite matrices S and T used in the first part of this section) form a representation $\mathscr{G}$ of $\mathscr{G}$ of infinite dimension. This representation is unitary if the group $\mathscr{G}$ is unitary. The proof is the same as above (compare (4.7)).

Let us reduce $\mathscr{G}$ in its irreducible elements

$$\mathscr{G} = \mathscr{G}_1 + \mathscr{G}_2 + \ldots. \tag{4.9}$$

In this series we will find several times the same irreducible representation, often even an infinite number of times. The reduction is obtained by a unitary change of coordinates, i.e., by a choice of new axes $\varphi_1, \varphi_2, \ldots$ which are suited to the problem. Thus each matrix takes the box form corresponding to (4.9)

$$S = \begin{pmatrix} S_1 & 0 & 0 \ldots \\ 0 & S_2 & 0 \ldots \\ 0 & 0 & S_3 \ldots \\ \ldots & \ldots & \ldots \end{pmatrix} \qquad T = \begin{pmatrix} T_1 & 0 & 0 \ldots \\ 0 & T_2 & 0 \ldots \\ 0 & 0 & T_3 \ldots \\ \ldots & \ldots & \ldots \end{pmatrix}, \tag{4.9a}$$

where $S_1, T_1 \ldots$ are matrices of the representation $\mathscr{G}_1$; $S_2, T_2 \ldots$ matrices of the representation $\mathscr{G}_2$, etc. ....
In other words the orthogonal functions φ_i are arranged in partial sets each of which span an invariant subspace of the function space $\mathfrak{R}$.

The Hamiltonian H is, however, not an operator of the group $\mathscr{G}$. The matrix which corresponds to this Hamiltonian in the system of axes can be written as follows: all the matrix elements connecting the functions of one subset to another are gathered in a unique symbol

$$H = \begin{pmatrix} H_{11} & H_{12} & H_{13} \ldots \\ H_{21} & H_{22} & H_{23} \ldots \\ H_{31} & H_{32} & H_{33} \ldots \\ \ldots & \ldots & \ldots \end{pmatrix}, \tag{4.9b}$$

where the H_{ik} are *submatrices*. Most of them are rectangular, for the different boxes $S, T \ldots$ do not necessarily have the same dimensions.

Let us form the products HS and SH (when multiplying two matrices the submatrices are dealt with as matrix elements).

$$HS = \begin{pmatrix} H_{11}S_1 & H_{12}S_2 \ldots \\ H_{21}S_1 & H_{22}S_2 \ldots \\ H_{31}S_1 & H_{32}S_2 \ldots \\ \ldots & \ldots \end{pmatrix}; \qquad SH = \begin{pmatrix} S_1H_{11} & S_1H_{12} & S_1H_{13} \ldots \\ S_2H_{21} & S_2H_{22} & S_2H_{23} \ldots \\ S_3H_{31} & S_3H_{32} & S_3H_{33} \ldots \\ \ldots & \ldots & \ldots \end{pmatrix}.$$

As a result of (4.4) all the matrices S of the group $\mathscr{G}$ satisfy:

$$HS - SH = 0$$

i.e.

$$H_{ik}S_k - S_iH_{ik} = 0. \tag{4.10}$$

Schurs' Lemma can now be applied and it leads to the following conclusion

1° When S_i and S_k are unequivalent $H_{ik} = 0$

2° When S_i is equivalent to S_k, H_{ik} is a multiple of the unit matrix and we have

$$H_{ik} = H'_{ik}\, \mathbb{I} \tag{4.11}$$

where H'_{ik} is a number.

It follows that $\boldsymbol{H}$ also has a box form but in a less restricted manner than the $\boldsymbol{S}$ matrices. Let us replace (4.9) by:

$$\mathcal{G} = n_0\mathcal{G}_0 + n_1\mathcal{G}_1 + \ldots n_l\mathcal{G}_l \tag{4.12}$$

where only the *non-equivalent* irreducible representations occur: $\mathcal{G}_0$, n_0 times, $\mathcal{G}_1$, n_1 times, etc. .... The matrices (4.9a) and (4.9b) take the form

$S =$

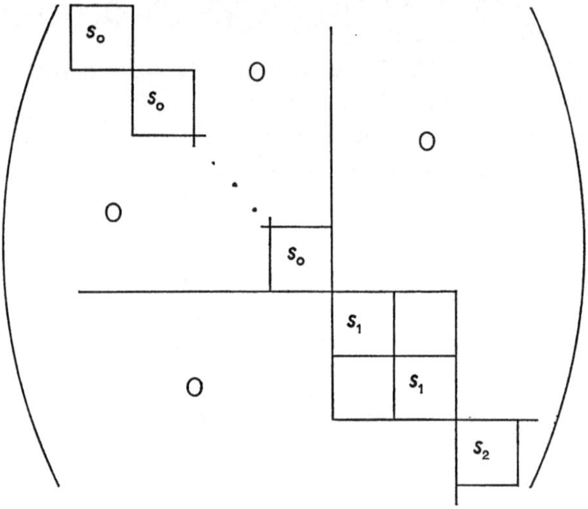

(4.13)

$$H = \left(\begin{array}{cccc|cccc} H_{11}(0) & H_{12}(0) & \cdots & & & & & \\ H_{21}(0) & H_{22}(0) & \cdots & & & \text{O} & & \\ \cdots & \cdots & \cdots & & & & & \\ \hline & & & & H_{11}(1) & H_{21}(1) & \cdots & \\ & \text{O} & & & H_{21}(1) & H_{22}(1) & \cdots & \\ & & & & \cdots & \cdots & \cdots & \end{array}\right) \qquad (4.1$$

Hence $\boldsymbol{H}$ is similarly decomposed as the matrix $\boldsymbol{S}$ of $\mathscr{G}$, i.e., in big square matrices situated along the principal diagonal and each corresponding to an irreducible representation $\mathscr{G}_i$ of $\mathscr{G}$. All the matrices $\boldsymbol{H}_{ik}(l)$ are multiples of the unit matrix and their dimension is the same as the dimension of the matrices $\mathscr{G}_i$

$$\boldsymbol{H}_{ik}(l) = H'_{ik}(l)\,\mathbb{I} \tag{4.11a}$$

$H'_{ik}(l)$ being a number.

Very often (we shall see this in examples below) the big square matrices in $\boldsymbol{S}$ and $\boldsymbol{H}$ are infinite. The difficulties which come from this are easily solved (cf. WEYL [1950]), since (4.13a) can be written by analogy with (4.12)

$$\boldsymbol{H} = \boldsymbol{H}_0 + \boldsymbol{H}_1 + \ldots \boldsymbol{H}(l) \ldots$$

$$\mathfrak{R} = \mathfrak{R}_0 + \mathfrak{R}_1 + \ldots \mathfrak{R}_l \ldots$$

The function space $\mathfrak{R}$ is decomposed in subspaces $\mathfrak{R}_l$. These subspaces are invariant with respect to the group $\mathscr{G}$ and the operator H *at the same time. Each of them corresponds to one of the non-equivalent irreducible representations of $\mathscr{G}$.*

This decomposition comes solely from the symmetry of H with respect to the group $\mathscr{G}$. It can be said that this decomposition is of kinematic nature. The axes or orthogonal functions which lead to this decomposition are represented by symbols with three indices φ_{lnm},

l labelling the big square matrices

n the small square matrices

m the rows and columns of the small square matrices.

These functions are not yet the Schrödinger wave functions for which H is diagonal. This is obvious considering the steps from (4.9b) to (4.13a). But these functions are not yet completely determined: we can still perform any unitary transformation in each of the subspaces $\mathfrak{R}_l$ and we shall use this transformation to make each of the big squares matrices of (4.13a) diagonal, i.e., the matrices obtained by replacing the submatrices $\boldsymbol{H}_{ik}(l)$ by the corresponding numbers $H'_{ik}(l)$. The indices i, k are the m-values, (l) is the angular momentum which characterizes a representation. Inside a box the elements are labelled by n, a label necessary because the representations l are multiple occurring irreducible representations.

The result so far obtained can be formulated without any reference to a particular quantum mechanical problem. We have a set of reducible matrices $\boldsymbol{S}$. The matrices $\boldsymbol{S}$ can be written as a sum of big boxes (labelled by l) each consisting of n_l small boxes of dimension m_l. If there is a matrix H that commutes with all the elements of the set or group, then H can be written as a sum of big boxes (again labelled by l) each consisting of n_l small boxes. These latter boxes each contain a different diagonal matrix of dimension m_l. By rearranging rows and columns we can also say that the big boxes (labelled by l) consist of m_l small boxes of dimension n_l:

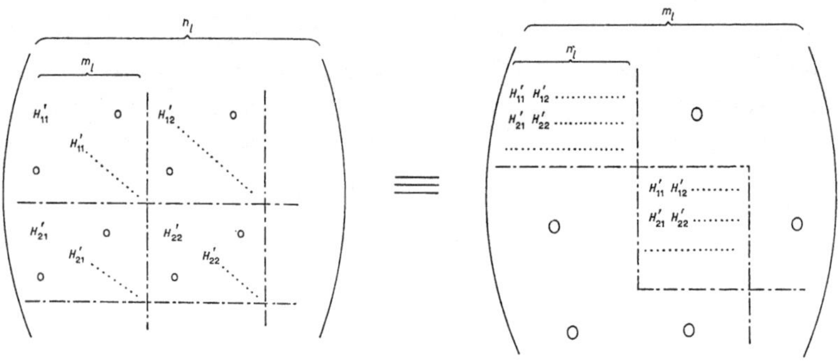

It is sufficient to choose in each subspace $\mathfrak{R}_l$, corresponding to the representation $\mathscr{G}_l$ a set of axes ψ_{nlm} which are eigenvectors of H. These are the eigenfunctions of the Schrödinger equation. The matrix $(H'_{ik}(l))$ of dimension n_l becomes diagonal. To each of its diagonal elements E_{nl} which are the energy levels of the system, there corresponds a submatrix in H, which is a multiple of the unit matrix $E_{nl}\mathbf{1}$, of the same dimension as $\mathscr{G}_l$. Thus we obtain after bringing rows and columns in their original arrangement,

$$H = \begin{pmatrix} E_{10}\mathbf{1} & O & & O & \\ O & E_{20}\mathbf{1} & & & \\ \hline & & E_{11}\mathbf{1} & O & \\ & O & & & \\ & & O & & E_{21}\mathbf{1} \end{pmatrix} \qquad (4.14)$$

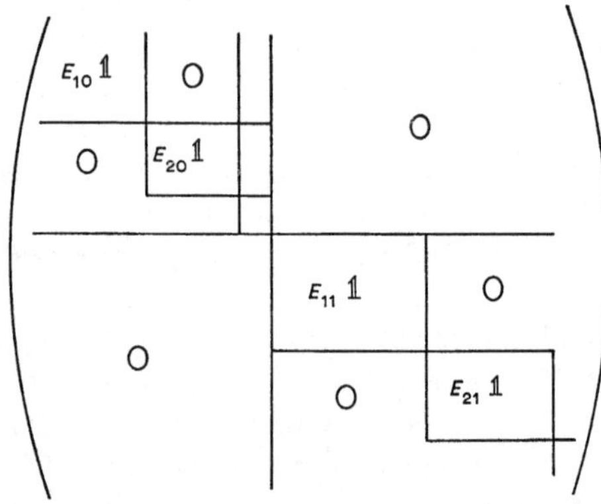

This matrix has clearly the same arrangement as the matrices S (4.13): each of the big square matrices corresponding to an irreducible representation $\mathscr{G}_l$ of $\mathscr{G}$ contains as many small square matrices as $\mathscr{G}$ contains $\mathscr{G}_l$ i.e. n_l. The set of the elements included in a big square matrix form a term system in the atomic case, i.e., a system with fixed l-value and different n-values. The combinations between the elements belonging to different term-systems are subject to selection rules which depend on the group $\mathscr{G}$ (cf. chapter 5, § 7). It can be proved that they are not allowed if $\mathscr{G}$ is a symmetrical group.

It is easily seen that one can reduce H to the diagonal form (4.14) without modifying the form (4.13) of the matrices S — (see Appendix 1).

The equality of the values E_{nl} of E in each small square matrix comes from the symmetry properties of H with respect to the group $\mathscr{G}$ and forms an *essential degeneracy*: it is impossible to separate the corresponding levels by a perturbation W unless this one alters the symmetry of H. If W is invariant only for *a subgroup* $\mathscr{H}$ of $\mathscr{G}$, the representations which were irreducible with respect to $\mathscr{G}$ cease to be so with respect to $\mathscr{H}$. Their reduction leads to a separation of the levels coalescing in E_n. The eigenvalue spectrum becomes finer.[1] The energy levels are labelled with 3 indices or more instead of only two. For example, this happens when a hydrogen atom is placed in an external magnetic or electric field. The spherical symmetry is replaced by a cylindrical or conical symmetry.

The preceding theory gives us valuable information on how to perform the perturbation calculation. Let α be the dimension of the representation $\mathscr{G}_l$ of $\mathscr{G}$, ψ_i $(i = 1, 2 \ldots \alpha)$ the eigenfunctions which form a basis for this representation, i.e., which describe all the states belonging to the α-fold degenerate level E_{nl} in the table (4.14). In order to calculate the first order perturbation W_i of this level it is generally sufficient (see Chapter 2, § 7.3) to solve a secular equation (2.47) of order α. In this equation only the matrix elements W_{ik} $(i, k = 1, 2 \ldots \alpha)$ resulting from the mutual interactions of the states ψ_i occur.

But if we know the subgroup $\mathscr{H}$ of $\mathscr{G}$ which leaves the perturbation function W invariant and the number β of the representations resulting from the reductions of $\mathscr{G}_l$ with respect to the subgroup $\mathscr{H}$, then we know the level E_{nl} is divided into only $\beta < \alpha$ distinct levels and the order of the secular equation is lowered from α to β. Besides we know *a priori* the basis of the new represen-

[1] The group $\mathscr{H}$ is formed with *certain* elements of $\mathscr{G}$. Then the invariant subspaces of the group $\mathscr{G}$ remain *a fortiori* invariant under the elements of $\mathscr{H}$ but these subspaces can be further subdivided.

tation i.e., the corresponding eigenfunctions. The theory of the Zeeman effect Chapter 5 (§§ 4 and 9) is a simple example of this method.

For a n-electron system, atom or molecule, one group of the Schrödinger equation is the group $\mathscr{S}_n$ of the permutations of n electrons. The function H is *always* invariant with respect to $\mathscr{S}_n$ because the electrons are physically identical. *It is impossible that the degeneracy corresponding to the group $\mathscr{S}_n$ can ever be lowered by any perturbation W.* This is called *exchange degeneracy.* That is one of the important results in the theory.

Sometimes it happens that the levels located in several different squares are the same. This is an *accidental degeneracy* which can be removed by any perturbation, even one having the same symmetry as the system.

NOTE - When H is invariant with respect to several groups (rotations, permutations . . .) one can consider each group isolated or combine them in a single group. The first method is more convenient if the elements of two different groups commute as this makes them absolutely independent.

In every important case and particularly in the case of the permutation groups the methods of the theory of groups give the possibility of building *a priori* the irreducible representations. Thus the structure of the matrix ***H*** is determined. Its decomposition into systems of elements, each system occupying a big square in the matrix, can be determined *before we calculate the elements and the wave functions.*

Historically the symmetry properties of H with respect to the group of rotations and reflections were used implicitly without reference to the theory of groups. But introduction of group theory allowed physicists to clear up the permutation problem and brought clarity and unity in the special symmetric problem.

2.3. EQUIVALENT DESCRIPTION

Let us make a last remark, one which is almost trivial but which may prevent misunderstandings. If we consider the equations (4.8) or (1.25) literally we can think about the matrices $\boldsymbol{S}_l$ of the irreducible representations of $\mathscr{G}$ as matrices of *rotation of the axes* in the subspaces $\mathfrak{R}_{nl}$ corresponding to the small square matrices of (4.13). We suppose that the basis functions ψ_{lnm} are known: i.e., we have:

$$\psi_{lnm} \rightarrow \psi'_{lnm} = \mathrm{S}\psi_{lnm} = \sum_{m'} \psi_{lnm'} S^{(l)}_{m'm} \tag{4.8a}$$

with

$$\boldsymbol{S}_l = (S^{(l)}_{mm'}).$$

We can also suppose the axes are fixed and consider these matrices *as mappings of the function space on itself.* Let ψ be a wave function expanded in a series of orthogonal basis functions,

$$\psi = \sum_{l,n,m} \beta_{lnm} \psi_{lnm} .$$

We have

$$\psi = \psi' = \mathrm{s}\psi = \sum_{lnm} \beta_{lnm} \sum_{m'} \psi_{lnm'} S^{(l)}_{m'm}$$
$$= \sum_{lnm'} \beta'_{lnm'} \psi_{lnm'} ,$$

with

$$\beta'_{lnm'} = \sum_{m} S^{(l)}_{m'm} \beta_{lnm} \tag{4.8b}$$

an equation that determines the linear transformation which the operation s exerts on the Fourier coefficients or components of ψ *along the fixed axes.* The functions ψ_{lnm} span the invariant subspaces $\mathfrak{R}_{nl}$. The two points of view are equivalent. We shall use either according to the case we are dealing with.

3. Abelian Groups

All the elements and all the matrices of the representations commute. So the matrices can be simultaneously diagonalized (Chapter 1, § 3.3). The irreducible representations are all one-dimensional.

Following below are two simple examples.

3.1. PERMUTATIONS OF TWO OBJECTS

This group $\mathscr{S}_2$ contains only two elements $\mathrm{E} = (1)$ and $\mathrm{s} = (1\ 2)$ with the only rule of multiplication $\mathrm{s}^2 = \mathrm{E}$. In order to represent this group let the element s be represented by a one-dimensional matrix such that the number $\boldsymbol{S}$ satisfies the equation $\boldsymbol{S}^2 = 1$. Thus one obtains two irreducible representations of $\mathscr{S}_2$: $\boldsymbol{E}, \boldsymbol{S} = +1$ and $\boldsymbol{E} = 1, \boldsymbol{S} = -1$.

It follows that in the space of the states for an atom with two electrons (helium) the matrix $\boldsymbol{S}$ of the equation (4.13) is decomposed in only *two* big square matrices: In the first one the number 1 occurs along the diagonal (obviously they are infinite in number). In the second square matrix the number -1 occurs in the same way. Once H has been put in the form (4.14) we obtain

$$S = \left(\begin{array}{c|c} \begin{matrix} 1 & 0 & 0 & \\ 0 & 1 & 0 & \\ 0 & 0 & 1 & \\ & & & \ddots \end{matrix} & \\ \hline & \begin{matrix} -1 & 0 & 0 & \\ 0 & -1 & 0 & \\ 0 & 0 & -1 & \\ & & & \ddots \end{matrix} \end{array}\right) \tag{4.15}$$

$$H = \left(\begin{array}{c|c} \begin{matrix} E_{10} & 0 & \\ 0 & E_{20} & \\ & & \ddots \end{matrix} & \\ \hline & \begin{matrix} E_{11} & 0 & \\ 0 & E_{21} & \\ & & \ddots \end{matrix} \end{array}\right). \tag{4.15a}$$

Hence, there exist two systems of eigenvalues corresponding to the two systems of eigenfunctions. One of them satisfies, after (4.15), the equation $\mathrm{s}\psi_k = \psi_k$, these are the *symmetric functions.* The second satisfies $\mathrm{s}\psi_k = -\psi_k$, these are the *antisymmetric functions.* (We delete: with respect to the group of permutations.)

3.2. PLANE ROTATIONS (AROUND A FIXED AXIS) (GROUP $\mathscr{D}_2$)

The operations commute, since any two arbitrary rotations of angle φ' and φ satisfy $\varphi'+\varphi = \varphi+\varphi'$.[1] Hence the irreducible representations are one-dimensional. To each element s_φ that is a rotation through an angle

[1] When φ is incommensurable with π we can consider that all the elements of the group are built by iteration of the rotation φ to any degree of approximation (two angles which differ by a multiple of 2π are considered identical). The group $\mathscr{D}_2$ can be considered from this point of view as a cyclic group of infinite order (closed group with one parameter).

φ, there corresponds a one-dimensional matrix, i.e., a number $X(\varphi)$ and because

$$\mathrm{s}_{\varphi'}\mathrm{s}_{\varphi} = \mathrm{s}_{\varphi+\varphi'},$$

we have

$$X(\varphi)X(\varphi') = X(\varphi+\varphi').$$

Since $X(\varphi)$ is a continuous function of φ and since s_0 is the identity $X(0) = 1$.

We shall suppose (this is not the only possible case) that the representation is faithful and unambiguous, i.e.,

$$X(2\pi) = X(0) = 1.$$

Let us put: $X(\varphi) = \mathrm{e}^{\mathrm{i}\lambda(\varphi)}$.
We have

$$\lambda(0) = 0, \quad \lambda(\varphi+\varphi') = \lambda(\varphi)+\lambda(\varphi').$$

A functional equation with the solution $\lambda = m\varphi$ with $\mathrm{e}^{2\pi m\mathrm{i}} = 1$, where m is a positive or negative integer. Finally:

$$X(\varphi) = \mathrm{e}^{\mathrm{i}m\varphi}, \qquad m = 0, \pm 1, \pm 2 \ldots.$$

An atom in an external magnetic field has the symmetry of the group $\mathscr{D}_2$. Every matrix $S(\varphi)$, representing in function space a rotation φ around the field, is decomposed according to the scheme (4.13) with

$$S_0(\varphi) = 1; \qquad S_1(\varphi) = \mathrm{e}^{\mathrm{i}\varphi}; \qquad S_2(\varphi) = \mathrm{e}^{2\mathrm{i}\varphi}\ldots;$$

$$S_1'(\varphi) = \mathrm{e}^{-\mathrm{i}\varphi}; \qquad S_2'(\varphi) = \mathrm{e}^{-2\mathrm{i}\varphi}; \ldots.$$

The elements can be ordered according to the value of m, *the magnetic quantum number* but this classification is generally insufficient. The study of the complete equation can only tell us how each value of m occurs in the final representation.

Let us, however, consider the case of a single electron with spherical coordinates r, φ, ϑ. The theory gives us immediately information about the wave function $\psi_m(r, \varphi, \vartheta)$ corresponding to the invariant subspace labelled by the number m. These functions satisfy the equation

$$\mathrm{s}_{\varphi}\psi_m(r, \varphi_0, \vartheta) = \psi_m(r, \varphi_0+\varphi, \vartheta) = \mathrm{e}^{\mathrm{i}m\varphi}\psi_m(r, \varphi_0, \vartheta).$$

In particular $\varphi_0 = 0$

$$\psi_m(r, \varphi, \vartheta) = \mathrm{e}^{\mathrm{i}m\varphi}\psi_m(r, 0, \vartheta) = \mathrm{e}^{\mathrm{i}m\varphi}\psi_m(r, \vartheta).$$

4. Non-Abelian Groups. Rotations and Reflections in a Plane

This group $\mathscr{D}'_2$ is obtained by adding to the rotations around an axis the reflections through planes containing this axis. This is the collection of diatomic molecules which have the symmetry of the cone or the cylinder.

Let us call the reflection through an arbitrary plane containing the axis T, and the rotations S_φ. The rotations commute with each other but not with T. We can see this immediately from Fig. 4.1.

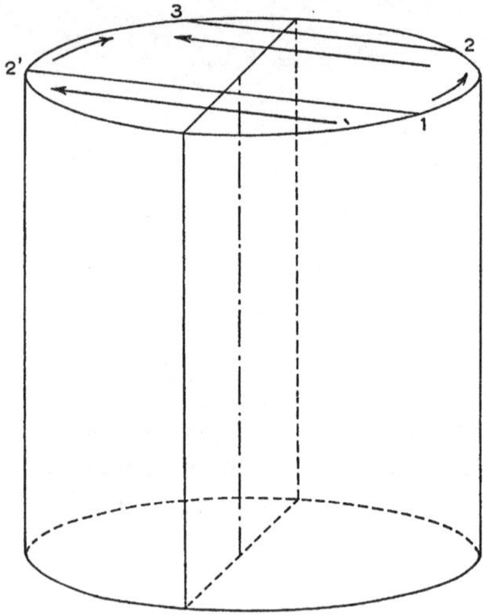

Fig. 4.1. Illustration of the non-commutativity of rotation and reflection operation in a system of cylindrical symmetry.

One can go from the position 1 to the position 3 in two different ways. A rotation S_φ (1 to 2) and a reflection T (2 to 3), or a reflection T (1 to 2′) and then a rotation $\mathrm{S}_{-\varphi}$ (2′ to 3). Thus we obtain

$$\mathrm{T}\mathrm{S}_\varphi = \mathrm{S}_{-\varphi}\mathrm{T}, \tag{4.16}$$

where the succession of the operation must be read from right to left as usual. First let us consider the subgroup $\mathscr{D}_2$ of $\mathscr{D}'_2$. It induces in the function space a representation which we have studied in the preceding section: the invariant subspaces are one-dimensional and are spanned by a set of basis functions ψ_m, such that $\mathrm{S}_\varphi \psi_m = \mathrm{e}^{\mathrm{i}m\varphi}\psi_m$. Let us suppose $m > 0$.

Likewise ψ_{-m} is characterized by:

$$\mathrm{S}_\varphi \psi_{-m} = \mathrm{e}^{-im\varphi}\psi_{-m} \quad \text{or} \quad \mathrm{S}_{-\varphi}\psi_{-m} = \mathrm{e}^{im\varphi}\psi_{-m}.$$

But the equation (4.16) gives

$$\mathrm{TS}_\varphi \psi_m = \mathrm{Te}^{im\varphi}\psi_m = \mathrm{e}^{im\varphi}\mathrm{T}\psi_m = \mathrm{S}_{-\varphi}\mathrm{T}\psi_m$$

or taking into account the last of the preceding equations

$$\mathrm{T}\psi_m = \psi_{-m}. \tag{4.17}$$

As could be expected the operation T leads to a coarser reduction of our representation.

It is necessary to combine every function ψ_m with every function ψ_{-m} in order to build a two-dimensional subspace of the space of the states invariant with respect to every element of the group $\mathscr{D}'_2$. In this subspace the element S_φ is represented by the diagonal matrix

$$\boldsymbol{S}_m(\varphi) = \begin{pmatrix} \mathrm{e}^{-im\varphi} & 0 \\ 0 & \mathrm{e}^{im\varphi} \end{pmatrix} \tag{4.18}$$

and the element T by the non-diagonal matrix which results from (4.17)

$$\boldsymbol{T}_m = \begin{pmatrix} 0 & 1 \\ 1 & 0 \end{pmatrix}. \tag{4.18a}$$

Then the representations and the levels are classified according to the absolute values $|m|$, $+m$ and $-m$ being associated in the same representation. The eigenvalues for which $|m| = 0, 1, 2, \ldots$ are labelled by the symbols $\Sigma, \pi, \Delta, \ldots$. The state $m = 0$ is a particular case:

$\mathrm{S}_\varphi \psi_0 = \psi_0$, $\mathrm{S}_0(\varphi) = 1$, as (4.16) gives us no more information $\boldsymbol{T}_0$ is determined by the condition $\mathrm{T}^2 = \mathrm{E}$, i.e., $\boldsymbol{T}_0^2 = 1$. Then we can take at will

$$\boldsymbol{T}_0 = \pm 1.$$

There remain two one-dimensional representations which are distinguished by their property with respect to reflection 1 and -1 (cf. Chapter 5, § 8). Finally we obtain the following irreducible representations and systems of eigenvalues

$$\mathscr{G}_0^+, \mathscr{G}_0^-, \mathscr{G}_1, \mathscr{G}_2 \ldots \mathscr{G}_m, \ldots$$

$$\Sigma^+, \Sigma^-, \pi, \Delta, \ldots.$$

Appendix 4.I

We will give some examples: Consider the case of a two-dimensional irreducible representation $\mathcal{G}_2$ which occurs m-times in $\mathcal{G}$. This representation is part of a $2m$-dimensional subspace $\mathfrak{R}_{2m}$ which is subdivided into m irreducible 2-dimensional spaces. The large boxes of the matrices $\boldsymbol{H}$ and $\boldsymbol{S}$ which correspond to $\mathfrak{R}_{2m}$ can be represented according to (4.13) and (4.13a) in the form:

$$\boldsymbol{S}_2 = \begin{pmatrix} \begin{array}{cc|} S_{11} & S_{12} \\ S_{21} & S_{22} \\ \hline \end{array} & & \\ & \begin{array}{|cc|} S_{11} & S_{12} \\ S_{21} & S_{22} \\ \hline \end{array} & \\ & & \ddots \end{pmatrix};$$

$$\boldsymbol{H}_2 = \begin{pmatrix} H'_{11} & 0 & H'_{12} & 0 & \dots H'_{1m} & 0 \\ 0 & H'_{11} & 0 & H'_{12} \dots & 0 & H'_{1m} \\ H'_{21} & 0 & \cdot & \cdot & \cdot & \cdot \\ 0 & H'_{21} & \cdot & \cdot & \cdot & \cdot \\ \cdot & \cdot & \cdot & \cdot & \cdot & \cdot \\ H'_{m1} & 0 & \cdot & \cdot & \cdot & \cdot \\ 0 & H'_{m1} & \cdot & \cdot & \cdot & \cdot \end{pmatrix}$$

that is to say that, if we designate the $2m$ basis functions by $\varphi_1, \varphi_1', \varphi_2, \varphi_2' \dots \varphi_m, \varphi_m'$, the matrices $\boldsymbol{S}_2$ transform as:

$$\begin{matrix} \mathrm{s}\varphi_1 = S_{11}\varphi_1 + S_{12}\varphi_1' \\ \mathrm{s}\varphi_2' = S_{21}\varphi_1 + S_{22}\varphi_1' \end{matrix}; \quad \begin{matrix} \mathrm{s}\varphi_2 = S_{11}\varphi_2 + S_{12}\varphi_2' \\ \mathrm{s}\varphi_2' = S_{21}\varphi_2 + S_{22}\varphi_2' \end{matrix}; \dots \text{etc.} \tag{4.A1}$$

$$\left.\begin{aligned} \mathrm{H}_2\varphi_1 &= H'_{11}\varphi_1 + H'_{12}\varphi_2 + \dots H'_{1m}\varphi_m \\ \mathrm{H}_2\varphi_2 &= H'_{21}\varphi_1 + H'_{22}\varphi_2 + \dots H'_{2m}\varphi_m \\ &\dots\dots\dots\dots\dots\dots \\ \mathrm{H}_2\varphi_1' &= H'_{11}\varphi_1' + H'_{12}\varphi_2' + \dots H'_{1m}\varphi_m' \\ \mathrm{H}_2\varphi_2' &= H'_{21}\varphi_1' + \quad \dots \quad H'_{2m}\varphi_m' \\ &\dots\dots\dots\dots\dots\dots \end{aligned}\right\}. \tag{4.A2}$$

We try to find the eigenfunctions of the Hamiltonian H_2 by putting

$$\psi = \alpha_1\varphi_1 + \alpha_2\varphi_2 \dots \alpha_m\varphi_m$$
$$\psi' = \beta_1\varphi_1' + \beta_2\varphi_2' \dots \beta_m\varphi_m'.$$

As a result of the equations (4.A1) and (4.A2), we have

$$\mathrm{s}\psi = S_{11}\psi + S_{12}\psi'; \qquad \mathrm{s}\psi' = S_{21}\psi + S_{22}\psi'.$$

$$\begin{aligned} \mathrm{H}_2\psi &= (\alpha_1 H'_{11} + \alpha_2 H'_{21} + \ldots \alpha_m H'_{m1})\varphi_1 \\ &\quad + (\alpha_1 H'_{12} + \alpha_2 H'_{22} + \ldots \alpha_m H'_{m2})\varphi_2 + \ldots \end{aligned}$$

$$\begin{aligned} \mathrm{H}_2\psi' &= (\beta_1 H'_{11} + \ldots \beta_m H'_{m1})\varphi'_1 + \ldots \\ &\quad + (\beta_1 H'_{12} + \ldots \beta_m H'_{m2})\varphi'_2 + \ldots. \end{aligned}$$

We will reduce H_2 to its diagonal form and we will have $\mathrm{H}_2\psi = E\psi$; $\mathrm{H}_2\psi' = E\psi'$ if we satisfy the equations

$$\begin{aligned} \alpha_1 H'_{11} + \alpha_2 H'_{21} \ldots \alpha_m H'_{m1} &= E\alpha_1 \\ \alpha_1 H'_{12} + \alpha_2 H'_{22} \ldots \alpha_m H'_{m2} &= E\alpha_2 \\ \ldots\ldots\ldots\ldots\ldots\ldots \\ \alpha_1 H'_{1m} + \ldots \qquad\qquad &= E\alpha_m \end{aligned} \tag{4.A3}$$

and a similar set of equations in β, with the same coefficients $H'_{11}, \ldots$. We have to reduce in both cases the same secular problem in m dimensions (and not in $2m$ dimensions). The equation

$$\begin{pmatrix} H'_{11}-E & H'_{21} & \ldots & H'_{m1} \\ H'_{12} & H'_{22}-E & \ldots & H'_{m2} \\ \ldots & \ldots & \ldots & \ldots \\ H'_{1m} & H'_{2m} & \ldots & H'_{mm}-E \end{pmatrix} = 0$$

has in general m distinct roots. As a result we have a set of m constants $\alpha_1 \ldots \alpha_m$ given by (4.A3) and m constants $\beta_1 \ldots \beta_m$ which are equal to the first set.

CHAPTER 5

ROTATIONS IN 3-DIMENSIONAL SPACE: GROUP $\mathscr{D}_3$

1. Spherical Harmonics and Representation of the Rotation Group

We will follow the historical order. Let us consider an atom with only one electron in a central field of spherical symmetry (Bohr-Sommerfeld). We suppose that the origin of the polar coordinates r, ϑ, φ is at the nucleus. The potential energy is an arbitrary function of the distance r. The Schrödinger equation and the Hamiltonian H are invariant with respect to the group $\mathscr{D}_3$ of the 3-dimensional rotations about arbitrary axes passing through the nucleus.

According to the Wigner theorem the eigenfunctions of this Hamiltonian can be classified in a number of different systems. Each of them acts as a coordinate system in function space and constitutes the basis of an irreducible representation of the group $\mathscr{D}_3$. In other words, each of these eigenfunctions becomes after rotation a linear combination of functions *of the same system* and the matrices of these transformations form an irreducible representation of the group.

We shall show that the solutions of the hydrogen atom problem as calculated by Schrödinger have this property. These solutions can be written as follows:

$$\psi_{nl}^{(m)} = f_{nl}(r) Y_l^{(m)}(\varphi, \vartheta), \tag{5.1}$$

with

$$Y_l^{(m)} = \mathrm{e}^{im\varphi}(\sin \vartheta)^{-m} P_l^{(m)}(\cos \vartheta), \tag{5.1a}$$

where $Y_l^{(m)}$ is the usual notation for the Laplace spherical harmonics. The associated Legendre polynomial is defined by

$$P_l^{(m)}(z) = \frac{\mathrm{d}^{(l-m)}}{\mathrm{d}z^{(l-m)}} (1-z^2). \tag{5.1b}$$

The Laplace spherical harmonic is homogeneous in $\cos \vartheta$ and $\sin \vartheta$. Its degree l, *the azimuthal quantum number or the quantum number of angular momentum* is a positive integer: $l = 0, 1, 2 \ldots$. Given a certain value l, after (5.1b), $(l-m)$ can only have integer values between 0 and $2l$. Hence the *magnetic quantum number* m can assume only the $(2l+1)$ values $m = -l, -l+1 \ldots l-1, l$; to a given value l there correspond $(2l+1)$ independent spherical harmonics $Y_l^{(m)}$ all of degree l in $\cos \vartheta$ and $\sin \vartheta$.

The third quantum number n is always a positive integer which can indicate either the *radial quantum number* of the Bohr-Sommerfeld quantum theory (and then assumes the values $n_r = 0, 1, 2 \ldots$) or else the total quantum number $n = l+1, l+2 \ldots l+n_r+1 \ldots$, which is the usual choice.

The energy levels $E(n, l)$ are independent of m. They are $(2l+1)$-fold degenerate and one can choose as corresponding eigenfunctions $(2l+1)$ independent functions of the form

$$f_{nl}(r) Y_l(\varphi, \vartheta)$$

Y_l being a linear combination of the $Y_l^{(m)}$.

Thus the eigenfunctions ψ are products of two factors, the first of which $f_{nl}(r)$ only depends on the particular interaction between the electron and the remaining part of the atom; the second one exhibits the symmetry properties of the operator H with respect to the group $\mathscr{D}_3$.

Since $f_{nl}(r)$ is left invariant under an arbitrary rotation s of the system about the nucleus we have

$$\psi_{nl}^{(m)} \to \mathrm{s}\psi_{nl}^{(m)} = f_{nl}(r) \mathrm{s} Y_l^{(m)}(\varphi, \vartheta),$$

but $\mathrm{s}\psi_{nl}^{(m)}$ *is also an eigenfunction of the energy level* E_{nl}, hence $\mathrm{s}Y_l^{(m)}$ is a linear combination of the $(2l+1)$ functions $Y_l^{(m)} (m = -l, \ldots, +l)$:

$$\mathrm{s} Y_l^{(m)}(\varphi, \vartheta) = \sum_{p=-l}^{p=+l} Y_l^{(p)} S_{pm}^{(l)} \tag{5.2}$$

where $S_l = (S_{pm}^{(l)})$ is a matrix of dimension $(2l+1)$.

The $Y_l^{(m)}$, in the space of functions $f(\varphi, \vartheta)$, span a $(2l+1)$-dimensional subspace which is invariant with respect to the operations of the group $\mathscr{D}_3$.

The equation (5.2) is a known property of spherical harmonics. It follows immediately from the invariance of the degree l of an homogeneous polynomial in (x, y, z)[1] by a rotation of these coordinate axes. Moreover we

[1] As $x+iy = r \sin \vartheta\, e^{i\varphi}$, $z = r \cos \vartheta$ every homogeneous polynomial of degree l in x, y, z has the form: $r^l Y_l(\varphi, \vartheta)$. The $Y_l^{(m)}$ are basis polynomials, homogeneous in x, y, z and defined on the unit sphere by the condition that the difference between the exponents of $(x+iy)$ and $(x-iy)$ be equal to m (cf. (5.1a)). The expansion of an arbitrary polynomial Y_l in a linear combination of $(2l+1)$ polynomials $Y_l^{(m)}$ follows from this definition.

conclude from the well-known orthogonality properties of spherical harmonics that these matrices $\boldsymbol{S}_l$ are unitary.

The function space $\mathfrak{R}$ is decomposed in subspaces $\mathfrak{R}_l$ invariant with respect to the group $\mathscr{D}_3$. Each of these invariant subspaces corresponds to a representation $\mathscr{D}_l$ of the group $\mathscr{D}_3$ consisting of matrices $\boldsymbol{S}_l$ of dimension $(2l+1)$. It can be shown that these representations are irreducible. This is almost obvious as a result of the theory of spherical harmonics. This representation occurs an infinite number of times because the radial or total quantum number n can increase to $+\infty$, l remaining the same. The spaces $\mathfrak{R}_l$ filling the big square matrices of the formulae (4.13) and (4.14) are of infinite dimension. They can be decomposed in an infinity of subspaces $\mathfrak{R}(n, l)$ with dimension $(2l+1)$ each corresponding to an energy level $E(n, l)$ and with one "copy" of the representation $\mathscr{D}_l$ filling each one of the small square matrices.

In particular if the rotation is around the axis Oz through an angle $\omega_z(\varphi' = \mathrm{s}\varphi = \varphi+\omega_z)$, it follows from (4.2b), (5.2) and (5.1a) that the matrix $\boldsymbol{S}(\omega_z)$ takes the diagonal form

$$\boldsymbol{S}_l(\omega_z) = \begin{pmatrix} \exp[-\mathrm{i}l\omega_z] & 0 & \cdots & 0 \\ 0 & \exp[-\mathrm{i}(l-1)\omega_z] & \cdots & 0 \\ \cdot & \cdot & \cdot & \cdot \\ 0 & & & \cdots \exp[\mathrm{i}l\omega_z] \end{pmatrix}. \tag{5.3}$$

All these representations are of odd degree $(2l+1)$. One can wonder why this is so and if no other exists. According to Chapter 4, § 3.2, concerning the plane rotations $\mathscr{D}_2$, the exponents m of the exponentials $\exp(\mathrm{i}m\omega_z)$ have to be integers in order to obtain an unambiguous representation for all values of the parameter ω_z. The matrices $\boldsymbol{S}_l$ are the only ones which give a faithful representation. But if we give up the condition of faithfulness for arbitrary values of the parameters and maintain this restriction only *in the vicinity of the identity* $\omega_z = 0$, it is possible to find other matrices.

Sommerfeld has shown in the old quantum theory that *internal quantum numbers j* which can be *half integers*, must be introduced in order to describe certain spectra. It is known that these half integer numbers come from the spin (cf. § 6.2). By analogy with (5.3), we find for the rotations ω_z

$$\boldsymbol{S}_j(\omega_z) = \begin{pmatrix} \exp[-\mathrm{i}j\omega_z] & 0 & \cdots & 0 \\ 0 & \exp[-\mathrm{i}(j-1)\omega_z] & \cdots & 0 \\ \cdot & \cdot & \cdot & \cdot \\ 0 & & & \cdots \exp[\mathrm{i}j\omega_z] \end{pmatrix} \tag{5.3a}$$

with $j = 0, \frac{1}{2}, 1, \frac{3}{2} \ldots$. All the exponents of the matrix are either integers or half integers. We shall add to the matrices of odd dimension in which j is integer, matrices of even dimension with $j = \frac{1}{2}(2p+1)$. But in the latter there is *not a one-to-one correspondence between the representation of the rotation group* and the abstract elements of this group. When ω_z is increased by 2π the matrix elements in (5.3a) will reverse their signs because the factor

$$j = \tfrac{1}{2}(2p+1) = \tfrac{1}{2}+\text{integer}$$

gives rise to a -1 in the matrix element. To each angle ω_z there are two corresponding matrices $\boldsymbol{S}_j(\omega_z)$ and $-\boldsymbol{S}_j(\omega_z)$ in these representations of even degree. They are two-valued representations. Between the representation and the group $\mathscr{D}_3$ there exists only a homomorphism. Both matrices 1 and -1 correspond to an angle zero or 2π.

2. Rotation Group and Two-Dimensional Unitary Group

2.1. RELATION BETWEEN THE ROTATION GROUP AND THE UNITARY GROUP.

E. Cartan then H. Weyl have shown how to build "a priori" all of these representations[1]. The group $\mathscr{D}_3$ of the rotations around a center is a 3 parameter group. Two of the parameters fix the direction of the axis, the third one the rotation angle. Hence we have the problem of constructing a correspondence between rotations in three dimensions and a unitary three parameter group.

The simplest unitary group is the unitary unimodular group $\mathscr{U}_2$ with two complex variables. The transformations σ of this group can be written

$$\sigma \rightarrow \begin{cases} \xi' = \alpha\xi+\beta\eta \\ \eta' = -\beta^*\xi+\alpha^*\eta. \end{cases} \tag{5.4}$$ [2]

[1] Cf. E. Cartan, thesis [1894]; and his papers: "The projective groups which leave no plane manifold invariant (irreducibles)", [1913], and [1914]. See also Cartan [1938].

Cartan's work has been resumed and completed by another method by H. Weyl [1925] on certain points. These studies are very general and they aim at building all the possible irreducible linear groups "a priori" and at finding their structure (compare the footnote on p. 17 and Appendix III).

[2] A mapping of a two-dimensional complex space on itself can be written as:

$$\sigma \rightarrow \xi' = \alpha\xi+\beta\eta, \quad \eta' = \gamma\xi+\delta\eta.$$

For simplicity we impose the condition that *this* will be a *unimodular transformation*, i.e.

$$\alpha\delta-\beta\gamma = 1.$$

The matrices of the adjoint and inverse transformations are:

$$\tilde{\sigma} = \begin{pmatrix} \alpha^* & \gamma^* \\ \beta^* & \delta^* \end{pmatrix}, \qquad \sigma^{-1} = \begin{pmatrix} \delta & -\beta \\ -\gamma & \alpha \end{pmatrix}.$$

(*Footnote continued on the next page.*)

The determinant of σ is 1; this condition

$$\alpha\alpha^* + \beta\beta^* = 1 \tag{5.4a}$$

which couples the real and imaginary parts of α and β and lowers the number of independent parameters from four to three.

By a stereographic projection we establish a correspondence between every rotation s and a transformation σ of the type (5.4), with certain complex coefficients α and β.

Let x, y, z be the coordinates of a point P on the unit sphere: $x^2+y^2+z^2 = 1$. Let $x = 0$, $y = 0$, $z = -1$ be the coordinates of the south pole S which we use as center of projection. The plane of projection will be the equatorial plane and the coordinates of P′, the projection of P, will be called x' and y' (compare Fig. 5.1).

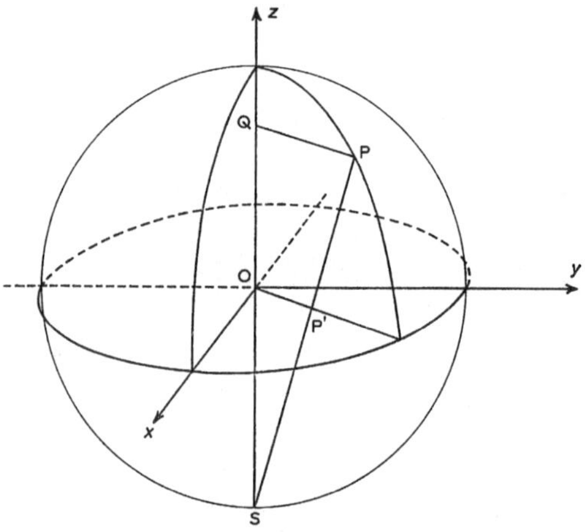

Fig. 5.1. Stereographic projection of the point P into P′.

In order that σ be unitary $\tilde{\sigma} = \sigma^{-1}$ from which follows that:

$$\delta = +\alpha^*, \quad \gamma = -\beta^*$$

i.e.

$$\sigma = \begin{pmatrix} \alpha & \beta \\ -\beta^* & \alpha^* \end{pmatrix}.$$

A unitary but not unimodular transformation would only satisfy:

$$\delta = D\alpha^*, \quad \gamma = -D\beta^*$$
$$\text{mod } D = 1, \quad \alpha\alpha^* + \beta\beta^* = 1$$

where D is the determinant of the coefficients. This would be a 4 parameter transformation.

Let us put $x'+iy' = \zeta$.
We have

$$\rho = \frac{SP'}{SP} = \frac{x'}{x} = \frac{y'}{y} = \frac{1}{1+z} = \frac{\zeta}{x+iy} = \frac{\zeta^*}{x-iy} = \sqrt{\frac{1+\zeta\zeta^*}{2(1+z)}}$$

and

$$\rho = \frac{1+\zeta\zeta^*}{2}, \qquad x+iy = \frac{2\zeta}{1+\zeta\zeta^*}$$

$$x-iy = \frac{2\zeta^*}{1+\zeta\zeta^*}, \qquad z = \frac{1-\zeta\zeta^*}{1+\zeta\zeta^*}$$

or finally using complex homogeneous coordinates ξ and η, such that $\zeta = \eta/\xi$ with the supplementary condition:

$$\xi\xi^*+\eta\eta^* = 1 \tag{5.5}$$

we obtain

$$\begin{aligned} x+iy &= 2\eta\xi^*, \qquad & x-iy &= 2\xi\eta^* \\ x &= \eta\xi^*+\xi\eta^*, \qquad & y &= -i(\eta\xi^*-\xi\eta^*) \\ & & z &= \xi\xi^*-\eta\eta^*. \end{aligned} \qquad (5.6)[1]$$

To each pair of numbers ξ and η there corresponds a point on the unit sphere, since as a result of (5.6), (5.5) is equivalent to $x^2+y^2+z^2 = 1$.

Every unitary transformation leaves (5.5) invariant and consequently transforms a point on the sphere into an other point on the sphere. It is easy to see that such a transformation leaves the angles between two lines from O to two different points C invariant (because the transformation is unitary and unimodular). Hence the transformation is a rotation.

To each transformation σ of the type (5.4) there corresponds a rotation s of $\mathscr{D}_3$. The converse statement is not quite the same for if we change the sign of α and β: i.e. $\alpha' = -\alpha$, $\beta' = -\beta$, the signs of ξ' and η' change, but after (5.6) x', y', z' remain invariant. Consequently to each rotation s of the group $\mathscr{D}_3$ there correspond *two* transformations of the unitary unimodular group $\mathscr{U}_2$: $+\sigma$ and $-\sigma$. Since the transformations are linear there corresponds to the product $s_2 s_1$ of two successive rotations a product $\sigma_2\sigma_1$. *The group $\mathscr{U}_2$ is a two-dimensional representation of the rotation group $\mathscr{D}_3$.*

A rotation about Oz leaves the difference $\xi\xi^*-\eta\eta^* = z$ and the sum

[1] When the point P is on a sphere of radius $r \neq 1$ it is sufficient to multiply these formulae by the real constant r. The conclusions do not change.

$\xi\xi^* + \eta\eta^*$ invariant. Hence we have $\xi'\xi'^* = \xi\xi^*$, $\eta'\eta'^* = \eta\eta^*$ i.e. taking into account (5.4) and (5.4a) we find that $\alpha\alpha^* = 1$ or,

$$\xi' = e^{i\varphi}\xi \quad \text{and} \quad \eta' = e^{-i\varphi}\eta,$$

where φ is an arbitrary quantity. We see, however, from

$$x' + iy' = 2\eta'\xi'^* = e^{-2i\varphi}2\eta\xi^* = e^{-2i\varphi}(x+iy)$$

that $2\varphi = -\omega_z$, the angle of rotation, i.e.

$$\sigma(\omega_z) = \begin{pmatrix} \exp(-\frac{1}{2}i\omega_z) & 0 \\ 0 & \exp(\frac{1}{2}i\omega_z) \end{pmatrix}. \tag{5.7}$$

A similar calculation (compare problem 5.2) gives the matrix which represents a rotation ω_y about the axis Oy

$$\sigma(\omega_y) = \begin{pmatrix} \cos\frac{1}{2}\omega_y & i\sin\frac{1}{2}\omega_y \\ i\sin\frac{1}{2}\omega_y & \cos\frac{1}{2}\omega_y \end{pmatrix} \tag{5.7a}$$

and a matrix of the same type $\sigma(\omega_x)$. An arbitrary rotation s (compare Fig. 5.2) defined by the Euler angles φ, ϑ and ψ is the product of 3 rotations:

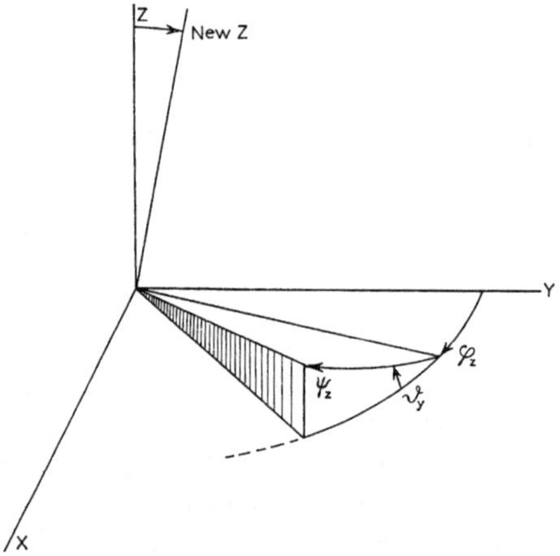

Fig. 5.2. Eulerian angles.

a rotation $s(\varphi_z)$ through an angle φ around the axis Oz, then a rotation $s(\vartheta_y)$ around the new axis Oy, finally $s(\psi_z)$ around the new axis Oz. Then we have $s(\varphi, \vartheta, \psi) = s(\psi_z) \cdot s(\vartheta_y) \cdot s(\varphi_z)$ and the corresponding matrix can be written

$$\sigma(\varphi, \vartheta, \psi) = \sigma(\psi_z) \cdot \sigma(\vartheta_y) \cdot \sigma(\varphi_z)$$

$$= \begin{pmatrix} \exp(-\tfrac{1}{2}i(\varphi+\psi)) \cos \tfrac{1}{2}\vartheta & +i \exp(+\tfrac{1}{2}i(\varphi-\psi)) \sin \tfrac{1}{2}\vartheta \\ i \exp(-\tfrac{1}{2}i(\varphi-\psi)) \sin \tfrac{1}{2}\vartheta & \exp(+\tfrac{1}{2}i(\varphi+\psi)) \cos \tfrac{1}{2}\vartheta \end{pmatrix} \tag{5.7b}$$

it has the form required by (5.4).

An alternate way to introduce the spinor variables and their transformation properties is to use the so-called "null-vector method" used by Kramers. (See KRAMERS [1937] and BRINKMAN [1956].)

2.2. THE REPRESENTATIONS OF THE GROUP $\mathscr{U}_2$ AS REPRESENTATIONS OF THE GROUP $\mathscr{D}_3$

The preceding method gives us at once an infinite number of representations of the rotation group, *because all the representations of the group $\mathscr{U}_2$ are obviously representations of the group $\mathscr{D}_3$ also.* They are easy to construct.

Let us form tensors in the two-dimensional unitary space ξ, η. A symmetric tensor of rank v has $v+1$ components

$$\xi^v, \xi^{v-1}\eta, \ldots, \xi\eta^{v-1}, \eta^v \tag{5.8}$$

(compare Chapter 1, § 1.3).

By a transformation σ (5.4) on ξ and η we obtain

$$\xi'^{v-i}\eta'^i = (\alpha\xi+\beta\eta)^{v-i}(-\beta^*\xi+\alpha^*\eta)^i = \sum_{k=0}^{k=v} S_{ik}^{(v)}\xi^{v-k}\eta^k. \tag{5.8a}$$

The components of a tensor of order v undergo a linear transformation. The transformation matrix is $S^{(v)} = (S_{ik}^{(v)})$. In the space of all the functions of x, y, z, to which they are related by (5.6), they span a subspace of dimension $(v+1)$. This subspace remains invariant under the rotations s. The transformation (5.8a) is not yet unitary. It becomes unitary if one uses the variables (5.8b) instead of (5.8)

$$q_k = \frac{\xi^{v-k}\eta^k}{\sqrt{(v-k)!k!}}, \qquad k = 0, 1, 2 \ldots v \tag{5.8b}$$

because if (5.5) is used one finds

$$\sum_{k=0}^{v} q_k^* q_k = \sum_k \frac{(\xi\xi^*)^{v-k}(\eta\eta^*)^k}{(v-k)!k!} = \frac{1}{v!}(\xi\xi^*+\eta\eta^*)^v = \text{invariant}. \tag{5.9}$$

The matrices $S^{(v)}$ can be multiplied among themselves like the matrices σ. Then we obtain an infinite number of representations of the rotation group, each arising from a tensor of order $v = 0$[1], $1, 2 \dots$.

The customary notations in quantum mechanics lead us to put $v = 2j$ and to label these representations by the symbols $\mathscr{D}_j$, $j = 0, \frac{1}{2}, 1, \frac{3}{2} \dots$. Consequently, their dimension is $2j+1$. It is convenient to call $j-k = m$. The variables q of the equation (5.8b) take then the symmetrical form

$$q_m^{(j)} = \frac{\xi^{j+m}\eta^{j-m}}{\sqrt{(j+m)!(j-m)!}} \qquad m = j, j-1, \dots -j. \tag{5.8c}$$

The group $\mathscr{U}_2$, from which we started, has as a basis the *vectors* of the two-dimensional space ξ, η, with $v = 1$ or $j = \frac{1}{2}$. These vectors are usually called *spinors*. It is the same as the representation $\boldsymbol{D}_{\frac{1}{2}}$ which is double valued as we have seen in § 2.1. This space is usually called spinor-space.

By taking the real and imaginary parts of the variables ξ, η and the coefficients α, β explicitly into account in order to have 4 real equations equivalent to (5.4), it is easily proved that if $\xi, \eta; \xi', \eta'$, satisfy this system, they are satisfied also by

$$\begin{aligned} X &= \lambda\xi + \mu\eta^*, \qquad & Y &= -\mu\xi^* + \lambda\eta \\ X' &= \lambda\xi' + \mu\eta'^*, \qquad & Y' &= -\mu\xi'^* + \lambda\eta' \end{aligned}$$

where λ and μ are arbitrary coefficients. In order to have $XX^*+YY^* = 1$, we must put: $\lambda\lambda^*+\mu\mu^* = 1$.

The above transformation, which is very peculiar since it connects X and Y simultaneously with ξ and η and their conjugates, shows the extent to which the bases of the unitary unimodular group are undetermined.

In particular if we put $\lambda = 0$, $\mu = 1$ we obtain a pair of variables $(\eta^*, -\xi^*)$ which undergo the same transformation as (ξ, η). This can be immediately verified. Consequently after choosing $x+iy = 2\eta\xi^*$, $x-iy = 2\xi\eta^*$, $z = \xi\xi^*-\eta\eta^*$ as the basis of the rotation group $\mathscr{D}_3$ in the 3-dimensional configuration space, these expressions transform respectively as $-\eta^2$, ξ^2 and $-\xi\eta$, i.e. as the three components of the second order tensor in ξ, η *or according to* (5.8c) *as* $q_{-1}^{(1)}$, $-q_1^{(1)}$ *and* $(1/\sqrt{2})q_0^{(1)}$. Thus the group $\mathscr{D}_3$ can be identified with the representation $\boldsymbol{D}_1$.

In the same way, the representation $\boldsymbol{D}_j$ *where* $j = l$ *is an integer* is nothing but the representation $\boldsymbol{D}_l$ (obtained in section 1) the basis of which are the

[1] To the value $v = 0$ corresponds the identical representation; to $v = 1$, the group $\mathscr{U}_2$.

$(2l+1)$ spherical harmonics $Y_l^{(m)}$. For a more rigorous proof of this statement see Appendix I. These functions are homogeneous polynomials of degree l in $(x+iy)$, $(x-iy)$ and z on the unit sphere i.e. they are sums of terms like:

$$A_{\sigma\tau}(\eta\xi^*)^\sigma(\xi\eta^*)^\tau(\xi\xi^*-\eta\eta^*)^{l-\sigma-\tau} \text{ or } A_{\sigma\tau}(\xi^{l-\sigma}\eta^{*\tau}\eta^\sigma\xi^{*l-\tau}-\xi^\tau\eta^{*l-\sigma}\eta^{l-\tau}\xi^{*\sigma}).$$

Let us put $m = \tau-\sigma$. The above remark shows that these terms are transformed like $\xi^{l+m}\eta^{l-m}$ or like $q_m^{(l)}$. Thus the functions $Y_l^{(m)}$ are independent linear combinations of $(2l+1)$ functions of ξ, ξ^*, η, η^* which, though not identical with the $q_m^{(l)}$, transform in the same way. They are the basis of a linear group which is equivalent to $\boldsymbol{D}_j$, $j = l$.

These representations are faithful. The others for which $j = \frac{1}{2}(2l+1)$ are double valued.

They are all irreducible and no others exist (this will be proved in note II which should be read after the following section).

One final remark: in the space $\mathfrak{R}_j$ of the representation $\mathscr{D}_j$ the rotations ω_z around Oz induce the following transformations, which can immediately be deduced from (5.7) and (5.8c).

$$q'_m = \exp(-im\omega_z)q_m.$$

The corresponding matrices $\boldsymbol{S}^{(j)}(\omega_z)$ have the form (5.3a) as was postulated before.

In the following sections we shall assume that the representations $\mathscr{D}_j$ where j is half integer can play a role in physics. This will be justified in § 6.

3. Infinitesimal Transformations and Angular Momentum

The rotation group is a continuous group. It contains operations which differ as little as wanted from the identity. These infinitely small differences were termed by Sophus Lie *infinitesimal transformations*. The *differential operators* which give rise to these transformations and which for the rotations are the components of the angular momentum, are the quantum analogues of classical quantities. This is true for all continuous groups. The quantum theory gives a physical meaning to discontinuous operators, such as reflections and permutations, which are considered as pure abstractions by classical theory.

3.1. INFINITESIMAL TRANSFORMATIONS OF A CONTINUOUS GROUP

Let us consider for example the rotations. *An infinitesimal rotation*, an element of a 3 parameter group, is defined by the three components of an axial vector $\mathbf{d\vartheta} = \boldsymbol{\omega}\mathrm{d}t$ lying along the rotation axis. The length of this vector is equal to the rotation angle $\mathrm{d}\vartheta$, $\mathrm{d}t$ is an auxiliary parameter (the time in kinematics), $\boldsymbol{\omega}$ similar to an angular velocity with components ω_x, ω_y, ω_z. ($\omega_x = \mathrm{d}\vartheta_x/\mathrm{d}t \ldots$). The transformation of the coordinates x, y, z by this rotation is linear and homogeneous in x, y, z according to the known formulae

$$\mathrm{d}x = (\omega_y z - \omega_z y)\mathrm{d}t, \text{ etc.} \ldots, \quad x' = x - \omega_z y \mathrm{d}t + \omega_y z \mathrm{d}t, \text{ etc.} \ldots. \tag{5.10}$$

A finite rotation is the *integral* of a continuous succession of infinitesimal transformations (5.10).

More generally, let us consider a continuous group with r parameters $s_1, s_2, \ldots s_r$, *which are all zero for the identical transformation.* In the vicinity of the identity every element of the group is defined if the values of these parameters are given and conversely. Let n be the dimension of the space in which the group operates, either configuration space or representation space. A transformation S is defined by the values $s_1, s_2, \ldots s_r$ of the parameters and by the transformation formulae

$$x_i' = \varphi_i(x_1, x_2, \ldots x_n; s_1, s_2, \ldots s_r) \tag{5.11}$$

the φ_i are supposed to be continuous and differentiable with respect to the s_λ.

The transformations (5.11) *form a group* when two successive transformations

$$\mathrm{S}: x_i \to x_i' \quad \text{and} \quad \mathrm{T}: x_i' \to x_i'' = \varphi_i(x_1' \ldots x_n'; t_1 \ldots t_r)$$

give rise to a unique transformation

$$\mathrm{U} = \mathrm{TS}: x_i \to x_i'' = \varphi_i(x_1, \ldots x_n; u_1, \ldots u_r),$$

where the u_λ are, in the vicinity of zero, continuous and differentiable functions of the variables s and t:

$$u_\lambda = u_\lambda(s_1, \ldots s_r; t_1, \ldots t_r). \tag{5.12}$$

The nature of the continuous group, that is its "multiplication rules", are contained in these functions.

An infinitesimal transformation is defined by the following equations

$$\mathrm{d}x_i = \sum_{\lambda=1}^{r} \left(\frac{\partial \varphi_i}{\partial s_\lambda}\right)_0 \mathrm{d}s_\lambda$$

if we put $\mathrm{d}s_\lambda = \omega_\lambda \mathrm{d}s$, $\mathrm{d}s$ being an auxiliary infinitesimal small coefficient, as was dt in (5.10),

$$\mathrm{d}x_i = \sum_{\lambda} \left(\frac{\partial \varphi_i}{\partial s_\lambda}\right)_0 \omega_\lambda \mathrm{d}s. \tag{5.13}$$

The symbol $(\ )_0$ means the s_λ must be taken as zero in the derivatives.

It often happens, and this is the case for the rotations (5.10), that the group is not defined by its finite equations (5.11) and (5.12) but by its infinitesimal transformations. The infinitesimal transformations are then defined by the following equations.

$$\mathrm{d}x_i = \sum_{\lambda} \alpha_i^{(\lambda)}(x_1, \ldots x_n)\omega_\lambda \mathrm{d}s \tag{5.13a}$$

where the $\alpha_i^{(\lambda)}$ depend only on the variables x. These functions have of course to fulfil certain integrability conditions.

Two successive infinitesimal transformations are equivalent to a unique transformation described by equations which are obtained in the first order by replacing ω_λ in (5.13a) by *the sum* $(\omega_\lambda + \omega'_\lambda)$. An arbitrary infinitesimal transformation of a given group is then *a linear combination of r basis transformations with arbitrary coefficients* ω_λ; *each of them is defined by n functions*

$$\alpha_i^{(\lambda)}(x_1, \ldots x_n), \qquad i = 1, 2, \ldots n; \qquad \lambda = 1, 2, \ldots r.$$

These r transformations i.e. the rectangular matrix which has as elements the functions $\alpha_i^{(\lambda)}$ determine the nature of the group completely.

The integrability conditions mentioned above are that the dx_i must be exact differentials and hence they contain the second derivatives of the φ_i i.e. infinitesimal elements of the second order.

This can be stated in the following form:

Two transformations, S and T, infinitely near the identity do not generally commute; the difference between ST and TS is then of the second order. There must exist a third transformation *in the same group*, Γ(S, T), which connects these two products in such a way that:

$$\mathrm{ST} = \Gamma \mathrm{TS} \quad \text{or} \quad \Gamma(\mathrm{S}, \mathrm{T}) = \mathrm{STS}^{-1}\mathrm{T}^{-1}. \tag{5.14}$$

Γ is called the *commutator* of S and T; this operation differs only from the identity by an infinitesimal transformation. *It must be a linear combination of the same basis transformations as the infinitesimal transformation of* S *and* T.[1]

3.2. LINEAR SUBSTITUTIONS

This is the only type of substitution we are interested in. The equations (5.10) of the infinitesimal rotations are linear and the same thing is by definition true for any representation of an arbitrary group. Hence we suppose in (5.13a) that the functions $\alpha_i^{(\lambda)}$ are linear and we write

$$dx_i = \sum_{k=1}^{n} \sum_{\lambda=1}^{r} \alpha_{ik}^{(\lambda)} x_k \omega_\lambda ds = \sum_k a_{ik} x_k ds \tag{5.15}$$

with the abridged notation

$$a_{ik} = \sum_{\lambda=1}^{r} \alpha_{ik}^{(\lambda)} \omega_\lambda .$$

We will put: $\mathsf{A} = (a_{ik})$, $\mathsf{A}^{(\lambda)} = (\alpha_{ik}^{(\lambda)})$ and call $\boldsymbol{x}$ the vector with components x_i and use the conventions of Chapter 1.

$$d\boldsymbol{x} = \mathsf{A}\boldsymbol{x}ds; \quad \boldsymbol{x}' = \boldsymbol{x}+d\boldsymbol{x} = (1+\mathsf{A}ds)\boldsymbol{x}. \tag{5.15a}$$

Each infinitesimal transformation s is defined by a matrix A, a linear combination with the coefficients ω_λ of r basis matrices $\mathsf{A}^{(\lambda)}$. Let S and T be two successive transformations. We have

$$\mathrm{T} : \boldsymbol{x}' \to \boldsymbol{x}'' = (1+\mathsf{B}dt)\boldsymbol{x}'$$

and for the resulting transformation

$$\mathrm{TS} : \boldsymbol{x} \to \boldsymbol{x}'' = (1+\mathsf{B}dt)(1+\mathsf{A}ds)\boldsymbol{x} = (1+\mathsf{A}ds+\mathsf{B}dt+\mathsf{BA}ds\,dt)\boldsymbol{x}.$$

In the same way

$$\mathrm{ST} : \boldsymbol{x} \to \boldsymbol{x}''' = (1+\mathsf{A}ds+\mathsf{B}dt+\mathsf{AB}ds\,dt)\boldsymbol{x}$$

(5.14) shows that the commutator Γ has the form

$$\Gamma(\mathrm{S}, \mathrm{T}) = 1+\mathsf{C}ds\,dt, \qquad \mathsf{C} = \mathsf{AB}-\mathsf{BA}, \tag{5.16}$$

[1] As a classical example which illustrates the theory of non-holonomic systems in mechanics we consider a sphere rolling without sliding on an horizontal plane. Its infinitesimal displacements have three degrees of freedom: the two angles which fix the rotation axis and the rotation angle. But these displacements are not integrable. Its finite displacements form a five parameter group: these parameters are the coordinates x and y of its center and the three rotation parameters.

C *must be a linear combination of the same basis matrices as* A *and* B *are.*[1]

Let us verify these results for the rotation group $\mathscr{D}_3$. Let us consider the three infinitesimal basis operations, the elementary rotations ρ_x, ρ_y, ρ_z around the three Cartesian axes. We obtain them by assuming that we have in (5.10) successively:

$$\begin{aligned} &1)\ \mathrm{d}\vartheta_x = \omega_x \mathrm{d}t, && \omega_y = \omega_z = 0 \\ &2)\ \mathrm{d}\vartheta_y = \omega_y \mathrm{d}t, && \omega_z = \omega_x = 0 \\ &3)\ \mathrm{d}\vartheta_z = \omega_z \mathrm{d}t, && \omega_x = \omega_y = 0. \end{aligned}$$

According to (5.15a) they displace a vector $\boldsymbol{v}$ in ordinary space by an amount $\mathrm{d}\boldsymbol{v}$:

$$\mathrm{d}\boldsymbol{v} = \boldsymbol{R}_x \boldsymbol{v}\,\mathrm{d}\vartheta_x \qquad \mathrm{d}\boldsymbol{v} = \boldsymbol{R}_y \boldsymbol{v}\,\mathrm{d}\vartheta_y \qquad \mathrm{d}\boldsymbol{v} = \boldsymbol{R}_z \boldsymbol{v}\,\mathrm{d}\vartheta_z \tag{5.15b}$$

with the transformation matrices (cf. 5.10)

$$\boldsymbol{R}_x = \begin{pmatrix} 0 & 0 & 0 \\ 0 & 0 & -1 \\ 0 & 1 & 0 \end{pmatrix} \quad \boldsymbol{R}_y = \begin{pmatrix} 0 & 0 & 1 \\ 0 & 0 & 0 \\ -1 & 0 & 0 \end{pmatrix} \quad \boldsymbol{R}_z = \begin{pmatrix} 0 & -1 & 0 \\ 1 & 0 & 0 \\ 0 & 0 & 0 \end{pmatrix}. \tag{5.17}$$

The commutators of these three transformations considered in pairs are obtained from (5.16), and the multiplication rules for matrices. If we write

$$\Gamma(\rho_x, \rho_y) = 1 + \mathsf{C}_{xy}\mathrm{d}\vartheta_x \mathrm{d}\vartheta_y, \text{ etc.} \ldots$$

one finds:

$$\begin{aligned} \mathsf{C}_{xy} = \boldsymbol{R}_x\boldsymbol{R}_y - \boldsymbol{R}_y\boldsymbol{R}_x = \boldsymbol{R}_z; \qquad \boldsymbol{R}_y\boldsymbol{R}_z - \boldsymbol{R}_z\boldsymbol{R}_y = \boldsymbol{R}_x \\ \boldsymbol{R}_z\boldsymbol{R}_x - \boldsymbol{R}_x\boldsymbol{R}_z = \boldsymbol{R}_y. \end{aligned} \tag{5.17a}$$

If we put

$$\boldsymbol{L}_x = \mathrm{i}\hbar \boldsymbol{R}_x, \text{ etc.} \tag{5.18}$$

we find again the commutation rules (2.20) for angular momenta. The above calculation is, as a matter of fact only a particular case of the calculation in Chapter II, § 3, because (5.15b) is the expression of (2.19) when the ψ's are linear functions of x, y, z.

[1] If one considers the r infinitesimal basis transformations $\mathsf{A}^{(\lambda)}$, $\mathsf{A}^{(\mu)} \ldots$ (as will be done later for rotations) one must have:

$$\mathsf{C}^{(\lambda\mu)} = \mathsf{A}^{(\lambda)}\mathsf{A}^{(\mu)} - \mathsf{A}^{(\mu)}\mathsf{A}^{(\lambda)} = \sum_{\nu=1}^{r} c_{\lambda\mu,\nu}\mathsf{A}^{(\nu)}$$

the $c_{\lambda\mu,\nu}$ being constants. S. Lie has proved that this condition is sufficient in order that the infinitesimal transformations considered give rise to a group. The structure of this group is determined by the constants $c_{\lambda\mu,\nu}$. These theorems are the basis of Cartan's work.

3.3. REPRESENTATIONS OF THE ROTATION GROUP. MATRICES FOR THE ANGULAR MOMENTA

Let us consider the irreducible representations $\boldsymbol{D}_j$ of the group $\mathscr{D}_3$. Each infinitesimal rotation

$$\rho : \boldsymbol{v}' = (1+\boldsymbol{R}\mathrm{d}\vartheta)\boldsymbol{v}$$

induces in the space of representations $\mathfrak{R}_j$, in which the components of the unitary vectors $\boldsymbol{\varphi}$ are the $q_m^{(j)}$, a transformation

$$\rho_j : \boldsymbol{\varphi}' = (1+\boldsymbol{R}^{(j)}\mathrm{d}\vartheta)\boldsymbol{\varphi} \tag{5.19}$$

or going over to algebraic language (cf. (5.15) and (5.15a))

$$\mathrm{d}q_m^{(j)} = \sum r_{mm'}^{(j)} q_{m'}^{(j)} \mathrm{d}\vartheta, \qquad m, m' = j, j-1, \ldots, -j. \tag{5.19a}$$

Following the definition of the representations we know that *the rules of multiplication of the matrices* $\boldsymbol{R}^{(j)} = (r_{mm'}^{(j)})$ *are similar to those of matrices* $\boldsymbol{R}$. In particular: to the matrices $\boldsymbol{R}_x$, $\boldsymbol{R}_y$, $\boldsymbol{R}_z$ of equations (5.15b) correspond matrices $\boldsymbol{R}_x^{(j)}$, $\boldsymbol{R}_y^{(j)}$, $\boldsymbol{R}_z^{(j)}$, which likewise satisfy the commutation conditions (5.17a). This remark is the basis of the proof in Appendix II.

Let us put as in (5.18)

$$\boldsymbol{L}_x^{(j)} = \mathrm{i}\hbar\boldsymbol{R}_x^{(j)}; \qquad \boldsymbol{L}_y^{(j)} = \mathrm{i}\hbar\boldsymbol{R}_y^{(j)}; \qquad \boldsymbol{L}_z^{(j)} = \mathrm{i}\hbar\boldsymbol{R}_z^{(j)}. \tag{5.18a}$$

We shall assume these three quantities are the three *components of the angular momentum in the quantum state j corresponding to the representation* $\boldsymbol{D}_j$.

We justify this definition by the following more precise statement: A physical system must have a spherical symmetry in order to admit the group $\mathscr{D}_3$. *Hence we are dealing with an atom with a radial field, i.e. a mononuclear system.* As a result of the Wigner theorem, there corresponds to each irreducible representation $\boldsymbol{D}_j$ of the rotation group for such a system a big square matrix inside the matrix $\boldsymbol{H}$ (4.14) i.e. a "system of eigenvalues" labelled by the indices j an integer or a half-integer. We will call this index according to the terminology introduced by Sommerfeld in 1920, the *internal quantum number* or better, *the angular quantum number*, and the states of the atom with such an energy, the quantum states j. Each state E_{nj} is described by *the whole set* of the $(2j+1)$ wave functions corresponding to the representation $\boldsymbol{D}_j$.

As for the quantities $\boldsymbol{L}_x^{(j)}$, $\boldsymbol{L}_y^{(j)}$, $\boldsymbol{L}_z^{(j)}$, we are allowed to consider them as the components of the total angular momentum of the atom: first because they satisfy the commutation relations (2.20) and the definition (2.19), as is shown by (5.18a) and (5.17a); secondly because they are constants of the motion. $\boldsymbol{R}_x^{(j)}\mathrm{d}\vartheta$ is a differential operator of the group $\mathscr{D}_3$ which leaves H invariant by hypothesis. The Hamiltonian commutes with $\boldsymbol{R}_x^{(j)}\mathrm{d}\vartheta$ and also with $\boldsymbol{L}_x^{(j)}$, so these matrices are independent of time (cf. Chapter 4, § 1.4).

For simplicity we will use units $\hbar$:

$$\boldsymbol{M}_x^{(j)} = \mathrm{i}\,\boldsymbol{R}_x^{(j)}, \qquad \boldsymbol{M}_y^{(j)} = \mathrm{i}\,\boldsymbol{R}_y^{(j)}, \qquad \boldsymbol{M}_z^{(j)} = \mathrm{i}\,\boldsymbol{R}_z^{(j)} \tag{5.20}$$

$$\boldsymbol{L}_x^{(j)} = \hbar \boldsymbol{M}_x^{(j)}, \text{ etc.} \tag{5.20a}$$

Henceforth we shall call these quantities $\boldsymbol{M}_x^{(j)}$ *the components of the angular momentum in the state j.* The square of these momenta can be written:

$$(\boldsymbol{M}^{(j)})^2 = (\boldsymbol{M}_x^{(j)})^2 + (\boldsymbol{M}_y^{(j)})^2 + (\boldsymbol{M}_z^{(j)})^2. \tag{5.21}$$

3.4 PAULI MATRICES

It remains now to obtain the explicit expression for these matrices belonging to the different representations $\boldsymbol{D}_j$. Let us begin with $j = \frac{1}{2}$.

The variables of $\boldsymbol{D}_{\frac{1}{2}}$ are the complex numbers ξ and η connected with x, y, and z by

$$\begin{aligned} x+\mathrm{i}y &= 2\eta\xi^* \\ x-\mathrm{i}y &= 2\xi\eta^* \end{aligned} \qquad z = \xi\xi^* - \eta\eta^* \tag{5.6}$$

their transformation formulas are those of the group $\mathscr{U}_2$

$$\begin{cases} \xi' = \alpha\xi + \beta\eta \\ \eta' = -\beta^*\xi + \alpha^*\eta \end{cases} \tag{5.4}$$

$$\alpha\alpha^* + \beta\beta^* = 1. \tag{5.4a}$$

For infinitesimal transformations near the identity we will use a set of small real numbers κ, λ, μ, ν with which we express α and β as follows:

$$\alpha = 1+\kappa+\mathrm{i}\lambda; \quad \beta = \mu + \mathrm{i}\nu.$$

Equation (5.4a) can now be written:

$$1+2\kappa+\kappa^2+\lambda^2+\mu^2+\nu^2 = 1.$$

Hence κ is of the second order and negligible. There remains:

$$\mathrm{d}\xi = \mathrm{i}\lambda\xi + (\mu+\mathrm{i}\nu)\eta, \quad \mathrm{d}\eta = -(\mu-\mathrm{i}\nu)\xi - \mathrm{i}\lambda\eta.$$

The three basis transformations are then:

1^0 $\lambda = 0, \quad \mu = 0$

$$d\xi = i\nu\eta, \quad d\eta = i\nu\xi \tag{5.22}$$

$$d(x+iy) = -d(x-iy) = 2i\nu(\xi\xi^*-\eta\eta^*) = 2i\nu z$$
$$dx = 0, \quad dy = 2\nu z, \quad dz = 2i\nu(\xi^*\eta-\eta^*\xi) = -2\nu y.$$

It is a rotation around the x axis over an angle $d\vartheta_x = -2\nu$.

2^0 $\lambda = 0, \quad \nu = 0$

$$d\xi = -\mu\eta, \quad d\eta = \mu\xi. \tag{5.22a}$$

A similar calculation gives

$$dx = 2\mu z, \quad dy = 0, \quad dz = -2\mu x,$$

i.e. a rotation around the y axis over an angle $d\vartheta_y = 2\mu$.

3^0 $\mu = 0, \quad \nu = 0$

$$d\xi = i\lambda\xi, \quad d\eta = -i\lambda\eta \tag{5.22b}$$

$$dx = 2\lambda y, \quad dy = -2\lambda x, \quad dz = 0,$$

i.e. a rotation around the z axis over an angle $d\vartheta_z = -2\lambda$.

The matrices $R_x^{(\frac{1}{2})}$, $R_y^{(\frac{1}{2})}$, $R_z^{(\frac{1}{2})}$ are defined by (5.19a) with $j = \frac{1}{2}$. The variables q_m are then equal to ξ and η. They are the matrices of the equations (5.22) to (5.22b) where ν, μ, λ, must be replaced by their values $-\frac{1}{2}d\vartheta_x$, $\frac{1}{2}d\vartheta_y$, and $-\frac{1}{2}d\vartheta_z$. Finally, by suppressing the factors $d\vartheta_x$, $d\vartheta_y$, $d\vartheta_z$ and multiplying by i according to (5.20), one obtains:

$$M_x^{(\frac{1}{2})} = \frac{1}{2}\begin{pmatrix} 0 & 1 \\ 1 & 0 \end{pmatrix}, \quad M_y^{(\frac{1}{2})} = \frac{1}{2}\begin{pmatrix} 0 & -i \\ i & 0 \end{pmatrix}, \quad M_z^{(\frac{1}{2})} = \frac{1}{2}\begin{pmatrix} 1 & 0 \\ 0 & -1 \end{pmatrix}. \tag{5.23}$$

The Hermitian matrices which are here multiplied by the factor $\frac{1}{2}$ are the matrices S_x, S_y, S_z used for the first time by Pauli in 1927 in his theory of spin. It is convenient to introduce

$$\begin{aligned} M_+^{(\frac{1}{2})} &= M_x^{(\frac{1}{2})}+iM_y^{(\frac{1}{2})} = \begin{pmatrix} 0 & 1 \\ 0 & 0 \end{pmatrix} \\ M_-^{(\frac{1}{2})} &= M_x^{(\frac{1}{2})}-iM_y^{(\frac{1}{2})} = \begin{pmatrix} 0 & 0 \\ 1 & 0 \end{pmatrix} \end{aligned} \tag{5.24}$$

in which case we deal only with real matrices.

These operators no longer represent infinitesimal rotations because multiplying the equations (5.210) by i completely alters their features.

However, these operators satisfy the following commutation relations

$$\begin{aligned} M_+M_- - M_-M_+ &= 2M_z, \\ M_-M_z - M_zM_- &= M_-, \\ M_zM_+ - M_+M_z &= M_+, \end{aligned} \tag{5.25}$$

where we have suppressed the superscript $\frac{1}{2}$, since they hold in general like (5.17a) and (2.20).

Let us introduce three infinitesimal parameters, $\mathrm{d}s$, $\mathrm{d}t$, $\mathrm{d}u$; then the transformations arising from $M_+^{(\frac{1}{2})}$, $M_-^{(\frac{1}{2})}$ and $M_z^{(\frac{1}{2})}$ are

$$\begin{aligned} M_+^{(\frac{1}{2})} &= \begin{cases} \mathrm{d}\xi = \eta\,\mathrm{d}s \\ \mathrm{d}\eta = 0 \end{cases} \\ M_-^{(\frac{1}{2})} &= \begin{cases} \mathrm{d}\xi = 0 \\ \mathrm{d}\eta = \xi\,\mathrm{d}t \end{cases} \\ M_z^{(\frac{1}{2})} &= \begin{cases} \mathrm{d}\xi = \frac{1}{2}\xi\,\mathrm{d}u \\ \mathrm{d}\eta = -\frac{1}{2}\eta\,\mathrm{d}u. \end{cases} \end{aligned} \tag{5.24a}$$

Before investigating the consequences of these formulae, let us verify that the matrices $M_x^{(j)} = \mathrm{i}R_x^{(j)} \ldots$, which represent the components of the angular momentum are Hermitian. This is necessary in order that these quantities have physical meaning. The matrices $R_x^{(j)}$, $R_y^{(j)}$, $R_z^{(j)}$ give rise to unitary infinitesimal transformations in the state space and this should be the case in order to keep the system of fundamental bases functions orthogonal. This remark suggests the following general theorem, the proof of which is very easy.

The matrices of the infinitesimal transformations of a unitary linear group become Hermitian when they are multiplied by the factor $\mathrm{i}(\mathrm{i}^2 = -1)$.

Let us consider a finite dimensional space in which the coordinates x_i undergo a linear infinitesimal transformation:

$$\mathrm{d}x_i = \sum a_{ik} x_k \mathrm{d}s. \tag{5.15}$$

In order that this transformation be unitary $\left(\sum_i x_i^* x_i = \text{const.}\right)$ we must have the following condition fulfilled:

$$\sum_i x_i^* \frac{\mathrm{d}x_i}{\mathrm{d}s} + \sum_i x_i \frac{\mathrm{d}x_i^*}{\mathrm{d}s} = 0$$

i.e.

$$\sum_{ik} (a_{ik} + a_{ki}^*) x_i^* x_k = 0$$

for every value of x_i and x_k. Hence we have

$$a_{ik} = -a_{ki}^*$$

or, if we put $\mathsf{A}' = \mathrm{i}\mathsf{A}$

$$a'_{ik} = a'^*_{ki}.$$

3.5. ANGULAR MOMENTUM IN THE STATE j.

We will calculate the expressions $\boldsymbol{M}_+^{(j)}$, $\boldsymbol{M}_-^{(j)}$, $\boldsymbol{M}_z^{(j)}$ and $(\boldsymbol{M}^{(j)})^2$. The coordinates of the space $\mathfrak{R}_j$ form the basis for the unitary representation $\boldsymbol{D}_j$. They are the $(2j+1)$ monomials $q_m^{(j)}$ of degree $2j$ in ξ and η given by the formulas (5.8c). We obtain by differentiation

$$\mathrm{d}q_m^{(j)} = \frac{(j+m)\xi^{j+m-1}\eta^{j-m}\mathrm{d}\xi + (j-m)\xi^{j+m}\eta^{j-m-1}\mathrm{d}\eta}{\sqrt{(j+m)!(j-m)!}}.$$

The infinitesimal transformations induced in the space $\mathfrak{R}_j$ by $\boldsymbol{M}_+^{(\frac{1}{2})}$, $\boldsymbol{M}_-^{(\frac{1}{2})}$, $\boldsymbol{M}_z^{(\frac{1}{2})}$ operating in the space $\mathfrak{R}_{\frac{1}{2}}$ are, according to (5.24a)

$$\boldsymbol{S}_+ : \mathrm{d}q_m^{(j)} = \frac{(j+m)\xi^{j+m-1}\eta^{j-m+1}\mathrm{d}s}{\sqrt{(j+m)!(j-m)!}} = \sqrt{(j+m)(j-m+1)}\, q_{m-1}^{(j)}\mathrm{d}s,$$

$$\boldsymbol{S}_- : \mathrm{d}q_m^{(j)} = \frac{(j-m)\xi^{j+m+1}\eta^{j-m-1}\mathrm{d}t}{\sqrt{(j+m)!(j-m)!}} = \sqrt{(j-m)(j+m+1)}\, q_m^{(j)}\mathrm{d}t,$$

$$\boldsymbol{S}_z : \mathrm{d}q_m^{(j)} = \ldots = mq_m^{(j)}\mathrm{d}u.$$

Hence the elements of the corresponding matrices $\boldsymbol{M}^{(j)}$ are (cf. (5.15))

$$\begin{cases} \langle m|\boldsymbol{M}_+^{(j)}|m-1\rangle = \langle m|\boldsymbol{M}_x^{(j)}+\mathrm{i}\boldsymbol{M}_y^{(j)}|m-1\rangle = \sqrt{(j+m)(j-m+1)} \\ \langle m|\boldsymbol{M}_-^{(j)}|m+1\rangle = \langle m|\boldsymbol{M}_x^{(j)}-\mathrm{i}\boldsymbol{M}_y^{(j)}|m+1\rangle = \sqrt{(j-m)(j+m+1)} \\ \langle m|\boldsymbol{M}_z^{(j)}|m\rangle = m. \end{cases} \tag{5.26}$$

These are the fundamental formulas for the components of the angular momentum. Each of these matrices has in each row and in each column only one element different from zero. Only the last one is diagonal. In the state j, the component $\boldsymbol{M}_z^{(j)}$ can assume the values m, $\boldsymbol{L}_z^{(j)}$ the values $m\hbar$ with $m = j, j-1, \ldots, -j$. In order to separate the different states corresponding to these different values it is necessary to destroy the spherical symmetry (Stern and Gerlach experiment).

We have according to (5.21), (5.24) and (5.25)

$$M^2 = \frac{M_+M_- + M_+M_-}{2} + M_z^2 = M_+M_- - \frac{M_+M_- - M_+M_-}{2} + M_z^2$$
$$= M_+M_- - M_z + M_z^2. \tag{5.27a}$$

If one uses this formula for the state j using (5.26) and the known multiplication rules one finds that the matrix $(M^{(j)})^2$ is diagonal and has the value:

$$(M^{(j)})^2_{m,m} = (j+m)(j-m+1) - m + m^2 = j(j+1). \tag{5.27}$$

This matrix is a multiple of the unit matrix. It is *invariant under rotations* and commutes with $M_x^{(j)}$, $M_y^{(j)}$ and $M_z^{(j)}$. Hence in a given quantum state j, the square of the length of the angular momentum has an accurately known value. Its component along the z axis can assume (in units $\hbar$) the discrete values $m = j, j-1, \ldots, -j$ and these values can be measured in a magnetic field $\boldsymbol{H}$ which fixes, in laboratory space, the particular direction of Oz. If we know the value m of $M_z^{(j)}$ it is not possible to say anything accurate about $M_x^{(j)}$ and $M_y^{(j)}$, which are represented by non-diagonal matrices and thus are undetermined. We are very far from the classical picture.

However, (5.27) and the last equation (5.26) can be considered as the foundation and the correct interpretation of the vector model which is still very useful in discussing experiments. It can be seen from these equations how the vector model can be used practically, i.e. how much they are wrong and how one must correct quantitatively the qualitative exact conclusion to which they lead us. For example $M_z^{(j)}$ behaves like the projection of a vector $M^{(j)}$ on the z axis, the angle between these being fixed (space quantization) but the maximum of this projection is, in the vector model, not equal to $\sqrt{(M^{(j)})^2} = \sqrt{j(j+1)}$ but to j. The usual geometrical notions, as for example the theorem of Pythagoras, have to be modified if we want them to apply to the eigenvalues of the quantum vectors.

4. Transition from the Group $\mathscr{D}_3$ to the Subgroup $\mathscr{D}_2$

4.1. ZEEMAN EFFECT

An atom in the state j, with a $(2j+1)$-fold degenerate energy E_{nj}, is subjected to an external magnetic field H. The spherical symmetry of the system is replaced by cylindrical symmetry. The perturbation λW arising from this field is now invariant only with respect to the *Abelian* subgroup $\mathscr{D}_2$ of $\mathscr{D}_3$, the elements of which are the rotations around Oz (Chapter 4, § 3). The invariant subspaces of the group are then *one-dimensional* and, according to the Wigner theorem, each of the levels E_{nj} is split in such a field into $(2j+1)$ discrete levels.

From this the reduction of $\boldsymbol{D}_j$ follows at once. All the matrices of this group representing rotations ω_z around the z axis are in the diagonal form (5.3a). As a result of the perturbation λW they will be divided into $(2j+1)$ one dimensional matrices without any change in the matrix elements and at the same time the eigenfunctions are supplemented by terms of the order of λ. Under a rotation ω_z these eigenfunctions undergo the following transformations:

$$\psi'_m = \psi_m \exp(-im\omega_z) \qquad m = j, j-1, \dots -j. \tag{5.28}$$

To each of them there corresponds a perturbed level E_{njm}. If one counts these levels one obtains the number j which is characteristic of the angular momentum state of the atom. This can be done by a determination of the number of Zeeman components of the lines of the spectrum in a weak magnetic field (λ is small), and by trying to fit this number to a certain multiplicity of the initial and final states of the transition. *By this method Sommerfeld has found it was necessary to use odd numbers* $2j+1$ *and half integers* j *and* m.

The solution of the secular equation (2.47) can be found as easily as the reduction of $\boldsymbol{D}_j$. In the zeroth approximation we only take into account the principal term W_0 of the perturbation matrix, the elements of which connect the $(2j+1)$ states among themselves (cf. 2.39a) so we have

$$W_0 \psi_m = \sum_{m'=-j}^{+j} w_{mm'} \psi_{m'}.$$

If we carry out a rotation ω_z we have, according to the invariance properties of W_0 and the transformation formulas (5.28),

$$W_0 \psi_m \exp(-im\omega_z) = \sum_{m'} w_{mm'} \psi_{m'} \exp(-im'\omega_z).$$

This relation can only be satisfied for all the values of the angle ω_z if $w_{mm'} = w_m \delta_{mm'}$.

Then the matrix W_0 *is diagonal and the perturbed energy levels can be written*

$$E_{njm} = E_{nj} + \lambda w_m.$$

This formula embodies all the information which can be obtained by group theory. The perturbing function and consequently the value of the constants w_m depend on the way in which the field influences the atom. For the first attempt we will try to apply the classical formulas of the Lorentz electromagnetic theory to the Bohr-Sommerfeld atom model: i.e. to the angular momentum L_z of the electrons of the atom around the z axis there corresponds a magnetic moment (i.e. its projection on the z axis)

$$\mathscr{M}_z = \mathbf{L}_z \frac{e}{2m_0 c} = M_z \frac{e}{2m_0 c} \hbar \tag{5.29}$$

and an energy

$$\lambda W = -H\mathscr{M}_z = -H \frac{e}{2m_0 c} \mathbf{L}_z$$

where m_0 is the mass of the electron, e its charge, c the velocity of light (H plays the role of λ). Since $\mathbf{L}_z$ is in diagonal form $\mathscr{M}_z$ and W are also in diagonal form, i.e. to each element $m\hbar$ of the matrix $\mathbf{L}_z$ there corresponds an element of the matrix λW

$$\lambda w_m = -H \frac{e\hbar}{2m_0 c} m = 2\pi\hbar\omega_{\mathrm{L}} m, \qquad \omega_L = \frac{eH}{2m_0 c}, \tag{5.30}$$

ω_{L} is the Larmor precession frequency.

The experiments did not verify this formula. To include the anomalous Zeeman effect, the second member of (5.30) must be multiplied by a certain number g, *the Landé factor* which changes value from one spectrum to another. The simple Bohr model which would hardly have explained the fractional numbers j, must be modified. This statement is the origin of the theory of spin.

Before treating this problem it is necessary to come back to the theory of angular momentum and to establish their quantum addition formulas. In this way the classical vector model of the atom will be completely justified and perfected.

5. Product of Two Representations. Reduction Formula

5.1. KINEMATIC COUPLING OF TWO SYSTEMS WITH SPHERICAL SYMMETRY

Let us first consider the problem from a physical point of view: suppose one adds the last electron to a monovalent positive ion in order to form an atom. The state of the ion is known. First we neglect perturbations, i.e. we suppose that the action of the ion on the electron is described by a central field. The electron forms a second system in a state that is easily defined and calculated by the Schrödinger theory. Then the ion and the electron are further coupled by taking a certain perturbing function into account. The problem is to find the state of the complete atom formed in this way. That is what levels may result from a certain initial level as a result of this coupling. A detailed dynamical discussion will not be undertaken here; this discussion would be difficult and in fact has been done only in some simple cases. We shall only give a preliminary purely kinematical study leading to a classification of the energy levels.

Two groups are essential here: the rotation group and the permutation group. Only the first one is actually of interest to us but the method is general. We couple two systems with a spherical symmetry whose angular momenta are respectively j_1 and j_2 and we look for the angular momentum j of the total system submitted to a coupling with spherical symmetry. A similar problem arises concerning the spin and orbital momentum.

5.2. PRODUCT OF TWO REPRESENTATIONS

Two systems for which the Hamiltonian is invariant under the operations of a group $\mathscr{G}$ are brought together. Their interactions are supposed to allow the same group. They are respectively in the quantum states corresponding to the irreducible representations $\mathscr{G}_1$ and $\mathscr{G}_2$ of $\mathscr{G}$ (cf. (4.13) and (4.14)). How can the resulting states of the total system be classified?

According to our hypothesis on the interactions this classification is *independent of their magnitude*; this classification depends only on the representation of $\boldsymbol{G}$ in the function space and can be found in the limiting case where the dynamical coupling is infinitely small (but different from zero). The gradual strengthening of the coupling alters the positions of the levels themselves without any change in their distribution over level-systems and their degeneracy.

Let $x_1, x_2, \ldots, x_k$, be the coordinates of the particles of the first system, $y_1, y_2, \ldots, y_l$, those of the second system, n_1 and n_2 the dimensions of the representations $\mathscr{G}_1$ and $\mathscr{G}_2$, i.e. the degeneracy of the corresponding levels. The first system is in the energy level E_1, the eigenfunctions of which are arbitrary linear orthogonal combinations of n basis functions, i.e.

$$\psi = \sum_{m=1}^{n_1} q_m \psi_m(x_1, x_2 \ldots x_k).$$

The "components" q_m are the variables of the representation $\mathscr{G}_1$ and the ψ_m the basic vectors of the space $\mathfrak{R}_1$.

In the same way the second system is in the energy level E_2, the eigenfunctions of which are

$$\varphi = \sum_{\mu=1}^{n_2} q'_\mu \varphi_\mu(y_1, y_2 \ldots y_l).$$

The q'_μ are the variables of the representation $\mathscr{G}_2$, the φ_μ are the basis vectors of $\mathfrak{R}_2$.

It is known that if the coupling is loose the Schrödinger equation is nearly separable in two independent equations, the eigenfunctions of which differ little from the zeroth approximation eigenfunctions $\Psi = \psi\varphi$ i.e. linear combinations of the $n_1 n_2$ basis functions $\Psi_{m\mu} = \psi_m \varphi_\mu$. The perturbed levels differ little from the level $E = E_1 + E_2$ the degeneracy of which is $n_1 n_2$ for coupling of strength zero. All these results are well known. We have:

$$\Psi = \psi\varphi = \sum_{m\mu} q^j_m q^{j'}_\mu \psi_m \varphi_\mu = \sum_{m,\,\mu} Q_{m\mu} \Psi_{m\mu}. \tag{5.31}$$

The Ψ are vectors of a $n_1 n_2$-dimensional space which is spanned by the basis vectors $\Psi_{m\mu}$ and which is designated by the symbol

$$\mathfrak{R}_1 \times \mathfrak{R}_2. \tag{5.32}$$

It is *the direct product of the two spaces* $\mathfrak{R}_1$ and $\mathfrak{R}_2$ and its coordinates are

$$Q_{m\mu} = q^j_m q^{j'}_\mu \tag{5.31a}$$

i.e. all possible products of the coordinates of $\mathfrak{R}_1$ and $\mathfrak{R}_2$. *This is equivalent to building a tensor of the second rank from two vectors.* (Compare Chapter 1, § 1.2).

By iteration one can obtain tensors of higher rank.

Under the operations of the group $\mathscr{G}$ the components $Q_{m\mu}$ undergo linear transformations which form a group $\mathscr{G}$, a representation of $\mathscr{G}$. This representation is the *direct product of the two representations* $\mathscr{G}_1$ and $\mathscr{G}_2$.

$$\mathscr{G} = \mathscr{G}_1 \times \mathscr{G}_2. \tag{5.32a}$$

From

$$\psi_m \to \mathrm{S}\psi_m = \sum_{r=1}^{n_1} \psi_r c_{rm}$$

and

$$\varphi_m \to \mathrm{S}\varphi_m = \sum_{\rho=1}^{n_2} \varphi_\rho \gamma_{\rho\mu}$$

one gets:

$$\Psi \to s\Psi = \sum Q_{m\mu} s\Psi_{m\mu}$$
$$= \sum_{r\rho} Q'_{r\rho} \Psi_{r\rho}$$

with

$$Q'_{r\rho} = \sum_{m,\,\mu=1}^{n_1 n_2} c_{rm} \gamma_{\rho\mu} Q_{m\mu} = \sum c_{rm,\,\rho\mu} Q_{m\mu} \tag{5.33}$$

a transformation whose matrix can be written in a symbolic way

$$C = C_1 \times C_2, \quad C = (C_{rm,\,\rho\mu}), \quad C_1 = (c_{rm}), \quad C_2 = (\gamma_{\rho\mu}). \tag{5.33a}$$

If $\mathcal{G}_1$ and $\mathcal{G}_2$ are two representations of $\mathcal{G}$, $\mathcal{G}$ is also a representation of $\mathcal{G}$. This can be easily verified. *But generally this representation is not irreducible even if $\mathcal{G}_1$ and $\mathcal{G}_2$ are.*

Consequently, if we want to use the Wigner theorem and determine in how many distinct levels E will be separated under the influence of the interaction between the two systems and what will be the degeneracy of each of these levels, the reduction has to be performed further. Our physical problem is then translated into a purely mathematical problem: *How to reduce the product $\mathcal{G} = \mathcal{G}_1 \times \mathcal{G}_2$ in its irreducible components.* To each irreducible component, that is to each invariant subspace of $\mathfrak{R}_1 \times \mathfrak{R}_2$, there corresponds a level whose degeneracy is equal to the dimensions of this invariant subspace.

5.3. REDUCTION OF THE DIRECT PRODUCT OF TWO REPRESENTATIONS. GROUP $\mathcal{U}_2$ AS AN EXAMPLE. CLEBSCH-GORDAN FORMULA

The problem which has just been formulated is easily solved in the case of the unitary unimodular group $\mathcal{U}_2$, or what amounts to the same in the case of the rotation group $\mathcal{D}_1$. From these one can obtain the rules for representations of higher dimensions. The reduction formula finally arrived at is called by mathematicians the Clebsch-Gordan formula:

$$D_j \times D_{j'} = D_{j+j'} + D_{j+j'-1} + \ldots D_{|j-j'|}. \tag{5.34}$$

First let us consider rotations ω_z around Oz. The matrices $S_j(\omega_z)$ and $S_{j'}(\omega_z)$ are diagonal and have the form (5.3a): one of them contains the elements $\exp(-im\,\omega_z)$ with $m = j, j-1, \ldots, -j$; the other one the elements $\exp(-im'\,\omega_z)$ with $m' = j', \ldots, -j'$. To the operation ω_z there corresponds in the product $D_j \times D_{j'}$ a matrix which remains diagonal after (5.33) and the elements of which are the $(2j+1)(2j'+1)$ exponentials: $\exp -i(m+m')\omega_z = \varepsilon^{m+m'}(\varepsilon = \exp(-i\omega_z))$ among these $[2(j+j')+1]$ only are distinct.

They can be classified symmetrically with respect to a horizontal line which contains the terms of exponent zero, the positive exponents are above the line, the negative ones below; if we suppose that $j > j'$ one obtains the following table in which we have written all the exponents which have the same value on a given line. It is sufficient to read this table through the vertical columns in order to verify the above formula

					$m+m'=$
$j+j'$					$j+j'$
$j-1+j'$	$j+j'-1$				$j+j'-1$
$j-2+j'$	$j-1+j'-1$	$j+j'-2$			$j+j'-2$
$\cdots$	$\cdots$	$\cdots$			$\cdots$
$j-2j'+j'$	$\cdots$	$\cdots$	$\cdots$	$j-j'$	$j-j'$
$\cdots$	$\cdots$	$\cdots$	$\cdots$	$\cdots$	$\cdots$
$-j'+j'$	$\cdots$	$\cdots$	$\cdots$	$j'-j'$	0
$\cdots$	$\cdots$	$\cdots$	$\cdots$	$\cdots$	$\cdots$
$-(j-2j'+j')$	$\cdots$	$\cdots$	$\cdots$	$-(j-j')$	$-(j-j')$
$\cdots$	$\cdots$	$\cdots$	$\cdots$		$\cdots$
$-(j-2+j')$	$-(j-1+j'-1)$	$-(j+j'-2)$	$\cdots$		$-(j+j'-2)$
$-(j-1+j')$	$-(j+j'-1)$	$\cdots$	$\cdots$		$-(j+j'-1)$
$-(j+j')$	$\cdots$	$\cdots$	$\cdots$		$-(j+j')$

We will be satisfied here with this hint which is not a proof. It mainly serves to illustrate that the number of basis functions is sufficient to obtain (5.34).[1]

The formula (5.34) is equivalent to

$$\boldsymbol{D}_j \times \boldsymbol{D}_{j'} = \boldsymbol{D}_{j+j'} + \boldsymbol{D}_{j-\frac{1}{2}} \times \boldsymbol{D}_{j'-\frac{1}{2}} \tag{5.34a}$$

since (5.34) can be obtained by successive applications of (5.34a).

To establish this equation rigorously the space $\mathfrak{R}_j \times \mathfrak{R}_{j'}$ must be decomposed in an irreducible subspace $\mathfrak{R}_{j+j'}$ which corresponds to the representation $\boldsymbol{D}_{j+j'}$ and one other reducible subspace $\mathfrak{R}_{j-\frac{1}{2}} \times \mathfrak{R}_{j'-\frac{1}{2}}$. This decomposition can be obtained by looking for the basis vectors of $\mathfrak{R}_{j+j'}$ (cf. Appendix III).

The formula (5.34) is the symbolic translation of the addition rules of angular momenta into representation theory language. According to H. Weyl it is the fundamental formula of the classification of atomic spectra as well as of the theory of chemical valency.

5.4. TOTAL ANGULAR MOMENTUM

Let us come back to our example: we bring an electron and an ion together, two systems both of which have spherical symmetry. One of them is in state

[1] This can be completed using the theorems about the characters (Chapter 3, § 11.2).

j corresponding to the representation $\boldsymbol{D}_j$ of $\mathscr{D}_3$, the other is in state j'. We want to calculate the components of the total angular momentum.

The representation of $\mathscr{D}_3$ which determines the states of the total system is $\boldsymbol{D}_j \times \boldsymbol{D}_{j'}$ in the space $\mathfrak{R}_j \times \mathfrak{R}_{j'}$ with the variables $Q(mm') = q^j_m q^{j'}_{m'}$, q^j_m and $q^{j'}_{m'}$ having the form (5.8c).

As a result of (5.19a) and (5.20) the component of the total angular momentum along a given axis, which we will call M, is obtained by writing the equations of the infinitesimal transformations of the variables $Q(mm')$ induced by a rotation $\mathrm{d}\vartheta$ around this axis

$$\mathrm{d}Q(mm') = -\mathrm{i} \sum_{m_1 m'_1} M_{mm';\, m_1 m'_1} Q(m_1, m'_1) \mathrm{d}\vartheta.$$

But

$$\mathrm{d}Q(mm') = \mathrm{d}q^j_m q^{j'}_{m'} + q^j_m \mathrm{d}q^{j'}_{m'} = -\mathrm{i} \left[\sum_{m_1} M^{(j)}_{mm_1} q^j_{m_1} q^{j'}_{m'} + q^j_m \sum_{m'_1} M^{(j')}_{m'm'_1} q^{j'}_{m'_1} \right] \mathrm{d}\vartheta$$

i.e.

$$M_{mm';\, m_1 m'_1} = M^{(j)}_{mm_1} \delta_{m'm'_1} + \delta_{mm_1} M^{(j')}_{m'm'_1}. \tag{5.35}$$

Or using the notation defined by (5.33a)

$$M = (M^{(j)} \times \mathrm{I}) + (\mathrm{I} \times M^{(j')}), \tag{5.35a}$$

an equation which expresses the additivity of momenta.

The properties of the angular momentum as well as the vector model of the atom will be considered more rigorously in the following sections. To do this it is necessary to take the interactions between the two partial systems into account and to reduce the representation $\boldsymbol{D}_j \times \boldsymbol{D}_{j'}$ according to (5.34) in order to separate the levels which were mixed before.

5.5. HELIUM ATOM WITHOUT SPIN

Let us consider first two electrons in the Coulomb field of a nucleus. If we neglect their interactions and if we do not take into account their spin then their states are described according to the Schrödinger theory by wave functions of the type (5.1):

$$\psi^{(m)}_{nl} \quad \text{and} \quad \psi^{(m')}_{n'l'}.$$

Here the integer l plays the role of j and the values $l = 0, 1, 2 \ldots$ correspond to the so-called s, p, d, ... states of the electron. In the zeroth approximation, i.e. when one neglects the interactions between the electrons, the wave functions of the total system are linear combinations of the products $\psi_{nl}^{(m)} \cdot \psi_{n'l'}^{(m')}$ the coefficients of which are the variables of the representation $\boldsymbol{D}_l \times \boldsymbol{D}_{l'}$. The levels $E = E_{nl} + E_{n'l'}$ depend neither on m nor on m'.

If we take into account the Coulomb interaction between the electrons the representation $\boldsymbol{D}_l \times \boldsymbol{D}_{l'}$ must be reduced in its irreducible components as in the formula (5.34). To each of its components there corresponds a level $E_{nn'L}$ and an angular momentum operator $M^{(L)}$ (where $L = l+l', l+l'-1 \ldots |l-l'|$) i.e. a well-defined state of the atom, *L is the total azimuthal or orbital quantum number.*

To give a geometrical picture of this decomposition into $2l'+1$ distinct states ($l' < l$) attribute an angular momentum equal to $\hbar l$ to each electron, i.e. a vector $M^{(l)} = \boldsymbol{l}$, and add the two vectors $\boldsymbol{l}$ and $\boldsymbol{l}'$ provided that they can take only such relative orientations which give a resulting vector of integral length. When $L = l+l'$ the two vectors are parallel, when $L = l-l'$ they are antiparallel.

It can be seen from this simple example that the equation (5.34) is really the exact basis of the vector model of the atom. All the formulas from (5.20) to (5.27) remain true provided that j be replaced by the integer L; in particular $(M^{(L)})^2 = L(L+1)$ and not L^2. One labels the states L of the total atom ($L = 0, 1, 2, \ldots$) by the symbols S, P, D, ...

Let us give some examples:

1⁰ Two s electrons; $l = l' = 0$; $\boldsymbol{D}_0 \times \boldsymbol{D}_0 = \boldsymbol{D}_0$; the atom is in a state S (ground state of helium).

2⁰ One s electron and one p or d . . . electron; $\boldsymbol{D}_0 \times \boldsymbol{D}_1 = \boldsymbol{D}_1$; $\boldsymbol{D}_0 \times \boldsymbol{D}_2 = \boldsymbol{D}_2 \ldots$ the atom is in a P or D . . . state.

3⁰ Two p electrons; $l = l' = 1$; $\boldsymbol{D}_1 \times \boldsymbol{D}_1 = \boldsymbol{D}_2 + \boldsymbol{D}_1 + \boldsymbol{D}_0$. As a result of their interaction a single level is divided in three levels: S, P and D. (Compare also problem 5.3.)

These rules are general, more complex atoms can be built up in the same way step by step. The use of half-integers j does not change anything essential except that the letter L must be replaced by J (§ 6).

We must mention an important consequence: If two systems are coupled, the first of which is in a $j = 0$ state i.e. a closed shell of any degree of complexity and the second one in a state j', one always obtains following (5.34) *only one resulting state $J = j'$*. The simplicity of the spectra of the alkali metals which consist of a closed shell plus one electron is thus explained at least if one neglects the spin.

6. The Electron Spin

6.1. UHLENBECK AND GOUDSMIT HYPOTHESIS

For each electron the Schrödinger theory uses only three quantum numbers n, l, and m (cf. § 1). On the other hand, the study of the anomalous Zeeman effect has led us to consider half-integers j which play a role similar to the number l and which correspond to the representations of the group $\mathscr{D}_3$ of even dimension. Then it is necessary to consider a fourth quantum number in order to complete the hypotheses of wave mechanics and to set up a connection between l and j.

Let us consider as Sommerfeld did the experimental results concerning the spectra of alkali metals. We have just seen at the end of the preceding section that these atoms have completely filled shells for which j is zero, plus a single external electron the state of which is defined by the total quantum number n and the orbital quantum number l playing here the role of j'. In this case the formula (5.34) shows that the levels resulting from the coupling must be simple. The experiments, however, show that they are double. The two states of this doublet are distinguished by a fourth quantum number and this number *can only assume two distinct values.* We foresee that these values will probably not be integers because the Zeeman effect for the alkali metals is anomalous in a magnetic field; the spectral terms are divided in an even number of components $(2j+1)$ so that j is a half-integer. The numbers l are known from the study of spectral series and selection rules. The experiments show that j is always equal to $l \pm \frac{1}{2}$.

Consequently the two states of the electron which form the doublet must be distinguished by the quantum number $s = \pm\frac{1}{2}$ with $j = l+s$.

In the more complex spectra involving many electrons such as the alkaline earths, these hypotheses and the vector model were able to give an explanation of the experimental facts. This work was done by Sommerfeld (1920–1923).

As a result of a precise discussion of empirical results Landé discovered in 1923 a remarkable relation between the splitting factor g and the numbers J, L, and S which replace j, l and s in atoms with many electrons. This relation can be written, in the case of alkali metals

$$g = \frac{2j+1}{2l+1}. \tag{5.36}$$

The theoretical explanation of these results stayed somewhat confused until 1925. However, that time Uhlenbeck and Goudsmit had the idea to connect these facts to some phenomena of a very different kind: the *gyromagnetic effects* and thereby found the key to the problem.

Einstein and De Haas have measured the change in angular momentum of a ferromagnetic substance when the magnitization is suddenly reversed.[1] Barnett studied the inverse effect where a certain magnetization is created by rotation. These experiments which were improved more and more showed that the formula (5.29) connecting the angular momentum L_z of the atom to its magnetic moment $\mathcal{M}_z$ (which comes theoretically from the existence of the electron *orbits*) is not true for the elementary moments of the ferromagnetic materials. For these substances the right-hand side of this equation must be multiplied by the factor $g = 2$

$$\mathcal{M}_z = L_z \frac{e}{m_0 c} \,. \tag{5.29a}$$

It is sufficient to substitute in equation (5.36) $l = 0, j = \frac{1}{2}$ in order to obtain this factor.[2]

All these facts suggest the following hypotheses which we can express as follows by using the vector model:

The quantum number s is related to a fourth and last degree of freedom of the electron which can only be a rotational degree of freedom. The electron then has an angular momentum or *spin*, the projection of which on a given axis Oz can only assume the values

$$L_z = \pm \tfrac{1}{2}\hbar; \qquad M_z = s = \pm \tfrac{1}{2}. \tag{5.37}$$

The corresponding magnetic moment is given by (5.29a). Hence we have:

$$\mathcal{M}_z = \pm \frac{e\hbar}{2m_0 c} \tag{5.37a}$$

according to (5.29) and (5.26) this moment is equal to the magnetic moment of a p-orbit ($l = m = 1$) i.e. a Bohr magneton.[3] The ratio of the magnetic

[1] A similar effect can easily be observed when a gyroscope is suddenly turned over.

[2] In experiments SUCKSMITH [1930] succeeded in measuring the Landé factor g for certain paramagnetic ions directly by gyromagnetic experiments and he found a value equal to the spectroscopic number.

[3] According to the correspondence principle one finds in classical electromagnetic theory the formula (5.29) for an electric charge moving along an orbit. The formula (5.29a) for a spinning electrically charged sphere corresponds to considerations in the theory of Relativity (compare for instance MØLLER [1952]).

moment to the angular momentum is twice as large for the spin as for the orbital motion.

As the energy difference between the components of doublets and multiplets is always small the secular equation of Schrödinger is a good first approximation and the dynamical interactions of the spin (interaction with the orbit or with the spin of an other electron) can be considered as perturbations.[1]

6.2. TRANSLATION IN QUANTUM THEORY (PAULI)

Let us consider a system with only one electron: the Schrödinger wave function $\psi(x, y, z)$ must be replaced by a function of four variables $\psi(x, y, z, s)$ where s can only assume the values $\pm\frac{1}{2}$.

If we know exactly how the spin is oriented (for this we must use an external magnetic field to define a certain direction Oz of the space as in the Stern and Gerlach experiments) we can fix the value of s. If spherical symmetry is preserved, which is usually the case, two values of s are possible, each one with a certain probability. It is necessary to use two functions at the same time,

$$\psi_1(x) = \psi(x, y, z, +\tfrac{1}{2}) \quad \text{and} \quad \psi_2(x) = \psi(x, y, z, -\tfrac{1}{2})$$

where $\psi_1^* \psi_1 \, d\tau$ and $\psi_2^* \psi_2 \, d\tau$ represent the respective probabilities for the two values of s in the elementary volume $d\tau = dx\,dy\,dz$. These two functions can be considered as the two components of a vector in a two-dimensional space $\mathfrak{R}_s$ or *spinspace*. This vector is sometimes called a *spinor*. The state of an atomic system with one electron is then represented by a two-dimensional vector or spinor:

$$\psi = \psi_1(x, y, z)u_1 + \psi_2(x, y, z)u_2. \tag{5.38}$$

u_1 and u_2 are two orthogonal unit vectors; to each of them there corresponds a well-defined state of spin orientation: *they are pure spin functions.* Their orthogonality means that an electron cannot at the same time be in the two states of spin $+\frac{1}{2}$ and $-\frac{1}{2}$:

$$u_1(-\tfrac{1}{2}) = 0, \qquad u_2(+\tfrac{1}{2}) = 0;$$
$$u_1(-\tfrac{1}{2})u_2(-\tfrac{1}{2}) + u_1(+\tfrac{1}{2})u_2(+\tfrac{1}{2}) = 0.$$

If we go back to the definitions (5.31) to (5.32a) of the direct product of

[1] This last hypothesis is justified by classical pictures. The magnetic forces coming from the rotation are small with regard to electrostatic forces.

two spaces and of the direct product of two representations we see that *the space of the functions* ψ *i.e. the total function including the spin is the product space* $\mathfrak{R}_s \times \mathfrak{R}$. This is exactly expressed by the formula (5.38) and this is the starting point for the theory of the spin.

It now remains to establish by some hypothesis how the vectors of the space $\mathfrak{R}_s$ behave under a rotation of the system. We shall assume that rotations in ordinary space induce in the space $\mathfrak{R}_s$ transformations which form an *irreducible* representation $\boldsymbol{D}_j$ of the group $\mathscr{D}_3$.

This assumption seems so natural that it is difficult to make a different one. In fact since $\mathfrak{R}_s$ is a two-dimensional space and if the representation $\boldsymbol{D}_j$ were reducible, it could be decomposed in two one-dimensional representations which would not tell us anything.

Since the space $\mathfrak{R}_j$ which corresponds to $\boldsymbol{D}_j$ is $(2j+1)$-dimensional, j must be equal to $\frac{1}{2}$ and the group $\mathscr{D}_3$ induces in the space $\mathfrak{R}_s \times \mathfrak{R}$ the representation $\boldsymbol{D}_{\frac{1}{2}} \times \boldsymbol{D}$.

6.3. APPLICATIONS

It now remains to consider some consequences of this result:

1⁰ Consider an alkali atom. Let us neglect the spin and hence the simple Schrödinger theory can be applied. The energy level E_{nl} and the angular momentum state are well defined by the integer l and the irreducible representation $\boldsymbol{D}_l$ of $\mathscr{D}_3$.

We now take into account the spin and the perturbations which arise from this. To the energy level E_n there corresponds now the representation $\boldsymbol{D}_{\frac{1}{2}} \times \boldsymbol{D}_l$ which can be reduced according to (5.34)

$$\boldsymbol{D}_{\frac{1}{2}} \times \boldsymbol{D}_l = \sum \boldsymbol{D}_j = \boldsymbol{D}_{l+\frac{1}{2}} + \boldsymbol{D}_{l-\frac{1}{2}}. \tag{5.39}$$

We obtain a splitting of the levels corresponding to the two internal quantum numbers $j = l+\frac{1}{2}$, $j' = l-\frac{1}{2}$.

2⁰ Let us submit this alkali atom to an infinitesimal rotation ρ. In the space $\mathfrak{R}_{\frac{1}{2}} \times \mathfrak{R}$ this notation induces an infinitesimal linear transformation the matrix M of which represents the component along the rotation axis of the total angular momentum. After (5.35a) we have

$$M = (M^{(\frac{1}{2})} \times \mathrm{I}) + (\mathrm{I} \times M^{(l)}). \tag{5.39b}$$

The momentum M is the sum of the orbital momentum $M^{(l)}$ and the spin momentum $M^{(\frac{1}{2})}$. The first can be defined by (2.18) (apart from a factor $\hbar$) but the definition of the second one can be given only with the more or less explicit help of the theory of the rotation group.

3^0 The components of the spin momentum $M^{(\frac{1}{2})}$ along the three axes are given by the Pauli formulas. Only $M_z^{(\frac{1}{2})}$ is in a diagonal form with eigenvalues $\pm\frac{1}{2}$. These are the two observable values of the spin momentum projected on a fixed direction of the space Oz. Whenever the last one is determined then $M_x^{(\frac{1}{2})}$ and $M_y^{(\frac{1}{2})}$ cannot be determined simultaneously because they are non-diagonal matrices.

6.4. COMPLEX ATOMS

These results can be generalized to an atom where r electrons participate in the emission of light. It is sufficient to build this atom by attaching these different electrons one by one. We neglect first the mutual interactions and suppose that the first electron is in the orbital state l_1, the second one in the state $l_2 \ldots$, and that the wave functions have the form:

$$\psi(x_1, y_1, z_1, x_2, y_2, z_2 \ldots x_r, y_r, z_r;\ s_1, s_2, \ldots s_r).^1$$

To each representation of the group $\mathscr{D}_3$:

$$D = D_{l_1} \times D_{l_2} \times \ldots \times D_{l_r} \times D_{\frac{1}{2}} \times D_{\frac{1}{2}} \ldots$$

there corresponds a state i.e. a system of unperturbed levels with a $(2l_1+1) \ldots (2l_r+1) \cdot 2^r$-fold degeneracy.

If we now take the electrostatic interactions and the spin interactions into account each of these levels is divided in a set of distinct terms the number of which is equal to the number of irreducible representations D_j in D.

The reduction is done step by step. But in practice one will take into

[1] The wave functions ψ_1 and ψ_2 of (5.38) can be calculated in the zeroth approximation by scalar Schrödinger theory and they will have the form (5.1) after a convenient normalization. Consequently in an atom with r electrons ψ can be written in the same approximation as

(a)
$$\psi(x_1, \ldots, s_r) = \sum_{i, k \ldots l} \psi_i^{(1)} \psi_k^{(2)} \ldots \psi_l^{(r)} u_i^{(1)} u_k^{(2)} \ldots u_l^{(r)},$$

where the upper indices label the electrons. The indices $i, k, l \ldots$ can only assume the values 1 and 2 corresponding respectively to

$$s = +\tfrac{1}{2} \quad \text{and} \quad s = -\tfrac{1}{2}.$$

If there are interactions between the electrons the product of individual space functions is replaced by a single function and we have

(b)
$$\psi(x_1, \ldots, s_r) = \sum_{i, k \ldots l} \psi_{ik \ldots l}(x_1 \ldots z_r) u_i^{(1)} u_k^{(2)} \ldots u_l^{(r)}.$$

If the perturbations from the spin are small $\psi_{ik \ldots l}$ can in principle be calculated by the scalar theory (with only the Coulomb interactions).

account the order of magnitude of the different perturbations which may change from one atom to another.

Generally *when the order of magnitude of these terms of the perturbation is normal* the Coulomb interactions play the main role, then come the interactions among spins and finally the interactions of the total spins with the total orbital angular momentum. This corresponds *to the Russell-Saunders coupling*.

First one reduces the representation:

$$\boldsymbol{D}_{l_1} \times \boldsymbol{D}_{l_2} \times \ldots \times \boldsymbol{D}_{l_r} = \sum \boldsymbol{D}_L .$$

This operation gives one term for each possible L-value: L is the total orbital quantum number; to the values $L = 0, 1, 2, \ldots$ correspond the states S, P, D ... of the atom.

Then one reduces

$$\boldsymbol{D}_{\frac{1}{2}} \times \boldsymbol{D}_{\frac{1}{2}} \times \ldots \times \boldsymbol{D}_{\frac{1}{2}} = \sum \boldsymbol{D}_s .$$

To each number S there corresponds a state with a given value of the total spin; for example:

$$\begin{aligned} \boldsymbol{D}_{\frac{1}{2}} \times \boldsymbol{D}_{\frac{1}{2}} &= \boldsymbol{D}_1 + \boldsymbol{D}_0 \\ \boldsymbol{D}_{\frac{1}{2}} \times \boldsymbol{D}_{\frac{1}{2}} \times \boldsymbol{D}_{\frac{1}{2}} &= \boldsymbol{D}_1 \times \boldsymbol{D}_{\frac{1}{2}} + \boldsymbol{D}_0 \times \boldsymbol{D}_{\frac{1}{2}} = \boldsymbol{D}_{\frac{3}{2}} + \boldsymbol{D}_{\frac{1}{2}} + \boldsymbol{D}_{\frac{1}{2}} \end{aligned} \qquad (5.40)$$

S is the quantum number of total spin.

Finally when S and L are known, one reduces

$$\boldsymbol{D}_S \times \boldsymbol{D}_L = \sum \boldsymbol{D}_J \qquad J = L+S, L+S-1 \ldots |L-S|, \qquad (5.41)$$

J is the total angular momentum quantum number.

It is easy to translate these results in the vector language: first addition of orbital momenta into a single total orbital momentum, then addition of the spin vectors and finally the coupling of $\boldsymbol{S}$ and $\boldsymbol{L}$ into a total angular momentum $\boldsymbol{J}$.

This last coupling gives rise to *the multiplets* because it is the weakest. The multiplicity of a level is, following (5.41), $2S+1$ if $L \geqq S$, $2L+1$ if $L < S$ and 1 if $L = 0$ (a singlet or S state).

The examples (5.40) show that the numbers S are integers or half-integers, the multiplicities $2S+1$ are odd or even according to the number r of electrons being even or odd: the alkali metals have doublets, the alkaline earth metals have singlets and triplets and so on.

In certain atoms there exist other kinds of coupling, particularly the $\boldsymbol{j} \cdot \boldsymbol{j}$

coupling in which the coupling of the spin of each electron with its own orbit is more important and gives a resulting angular momentum $\boldsymbol{j}$; the momenta $\boldsymbol{j}$ of the different orbits are coupled according to the equation

$$\boldsymbol{D}_{j_1} \times \boldsymbol{D}_{j_2} \times \ldots \times \boldsymbol{D}_{j_r} = \sum \boldsymbol{D}_J .$$

7. Selection Rules

These well-established rules can easily be confirmed by using group theory. We will consider only the case of atoms in this section.

Generally [1] the radiation is determined by the electric moment $\boldsymbol{\mu}$ which is a vector in the ordinary three-dimensional space. Its vector nature shows itself by two characteristic properties:

1⁰ it has three components μ_x, μ_y, μ_z or more conveniently:

$$\mu_+ = \mu_x + i\mu_y, \qquad \mu_- = \mu_x - i\mu_y \quad \text{and} \quad \mu_z .$$

2⁰ If the system undergoes a rotation s these components undergo the same linear transformations as the coordinates of a point, $x+iy$, $x-iy$, z, i.e. one of the transformations of the representation $\boldsymbol{D}_1$ (cf. §§ 2.1 and 2.2).

In quantum mechanics the components of the momentum are considered as operators applied to the wave functions. Let us suppose that the function space $\mathfrak{R}$ decomposed in certain $(2j+1)$-dimensional subspaces $\mathfrak{R}_j$ which are invariant and irreducible with respect to the rotation group. Each of these subspaces is spanned by a set of "orthogonal axes" ψ_{jm} which constitute a complete system with $m = j, j-1, \ldots -j$ (cf. § 2.2 and eq. (5.3a)). With these axes each component of $\boldsymbol{\mu}$ is represented by a matrix defined by the equations (1.25) which can be written as

$$\begin{aligned} \mu_+ \psi_{jm} &= \sum_{j'm'} \psi_{j'm'} (\mu_+)_{j'm',\, jm} \\ \mu_- \psi_m &= \ldots; \qquad \mu_z \psi_m = \ldots . \end{aligned} \tag{5.42}$$

For example each of the constants, $(\mu_z)_{jm,\, j'm'}$, raised to the second power is proportional to the transition probability from the state jm to the state $j'm'$ and to the intensity of the corresponding spectral line (with a polarization along Oz). Without magnetic fields the levels are independent of m hence this index can be suppressed and one can investigate the selection rules for transitions $j \to j'$ only.

[1] We will neglect here, as in Chapter 2, § 6, the radiation of higher order multipoles (quadrupoles, etc.)

If we submit the system to a rotation s, the left-hand sides of the three equations (5.42) are the products of a component of a vector $\boldsymbol{\mu}$ belonging to the three-dimensional space $\mathfrak{R}_1$ and a component of a vector belonging to the $(2j+1)$-dimensional function space $\mathfrak{R}_j$. Hence they are the components of a vector belonging to the product space $\mathfrak{R}_1 \times \mathfrak{R}_j$. This vector will transform under the rotation s according to $\boldsymbol{D}_1 \times \boldsymbol{D}_j$ which are reduced with the formula (5.34)

$$\boldsymbol{D}_1 \times \boldsymbol{D}_j = \boldsymbol{D}_{j+1} + \boldsymbol{D}_j + \boldsymbol{D}_{j-1}\,.$$

The right-hand side of the equations (5.42) in which the matrix elements are constants, is a sum of terms which transform like the components of the vectors belonging to the spaces $\mathfrak{R}_{j'}, \ldots$ i.e. according to $\boldsymbol{D}_{j'}, \ldots$, with $j' = \frac{1}{2}$, $1, \frac{3}{2} \ldots$.

Since both sides must transform in the same way under the rotation s, we find that all the terms of the right-hand side must be zero except those for which $j' = j+1,\ j$, or $j-1$.

Hence we have the following selection rule (the arrow shows the possible transitions)

$$j \rightarrow j-1, \quad j, \quad j+1. \tag{5.43}$$

In case $j = 0$ we have, however, $\boldsymbol{D}_1 \times \boldsymbol{D}_0 = \boldsymbol{D}_1$; the only possible transition is $j = 0 \rightarrow j = 1$, i.e. $0 \rightarrow 0$ is forbidden.

One finds by the same method the selection rules concerning the transitions of the magnetic number m *in a magnetic field.* The operations of the group $\mathscr{D}_2$ i.e. the rotations around Oz through an arbitrary angle ω are the only allowed operations: μ_+ is then multiplied by $\exp(\mathrm{i}\omega)$, ψ_{jm} by $\exp(-\mathrm{i}m\omega)$, $\psi_{j'm'}$ by $\exp(-\mathrm{i}m'\omega)$.[1]

For a rotation with an arbitrary ω one has

$$\mu_+ \psi_{jm} \exp(-\mathrm{i}\omega(m-1)) = \sum_{j'm'} \psi_{j'm'} \exp(-\mathrm{i}m'\omega)(\mu_+)_{jm,\, j'm'}\,.$$

All the terms of the sum in the right-hand side are zero except those for which $m' = m-1$. By a similar argument for μ_- and μ_z one obtains finally the following selection rule: the only allowed transitions are:

$$m \rightarrow m-1, \quad m, \quad m+1. \tag{5.44}$$

The first and the last one give circularly polarized light in the xy plane with two opposite directions of rotation. The transition $m \rightarrow m$ gives linearly polarized light with a polarization plane parallel to the z-axis.

[1] For the minus sign, see § 1, equation (5.3).

8. Parity or Reflection Character. Approximate Selection Rules

8.1. PARITY; THE RULE OF LAPORTE

The group of pure rotations is not the only one that is implied by the spherical symmetry of atoms. This symmetry is also preserved under reflections, which are all the products of rotations and the only operation connected with "symmetry with respect to a center", i.e. inversion of the axes:

$$x' = -x \quad y' = -y \quad z' = -z.$$

We shall label this operation by the symbol κ; it satisfies the equation:

$$\kappa^2 = \mathrm{E}.$$

The group $\mathscr{D}_3$ is extended by this operation, which commutes with all the rotations, to a group $\mathscr{D}_3'$. As a result of the preceding equation this inversion operation will have representations in the form of a diagonal matrix containing either $+1$ or -1. The eigenfunctions or basis vectors are multiplied by the factor $\delta = \pm 1$ as a result of this operation. This factor δ is called the *parity*[1], *signature* or *reflection character* of the representation.

The parity of a wave function $\psi(x_1, y_1, z_1, x_2, \ldots; s_1, s_2, \ldots)$ depends only on the space coordinates of the electrons and not on their spin s. The spin can be considered as an *axial vector* (angular momentum, magnetic moment) the components of which remain invariant under inversion of the coordinate axes. The operation κ does not operate on the vectors of the space $\mathfrak{R}_s$, but does on the vectors of the space $\mathfrak{R}$.

In the one electron problems the orbital wave functions (ψ_1 and ψ_2 in (5.38)) are as a result of (5.1) homogeneous polynomials of degree l in x, y and z, l being the orbital quantum number, hence they are multiplied by $(-1)^l$ by inversion of the axes. The spectral terms have then alternatively the reflection characters $\delta = +1$ and $\delta = -1$, i.e. they are *positive* or *even*, *negative* or *odd* in the following order: $s_+, p_-, d_+, f_- \ldots$ which is the most frequent order even in very complex atoms. This order is characteristic of those terms which are called *normal* by the spectroscopists.

The parity of the states of an atom with f electrons can be calculated *a priori* whenever one can give to each electron a well-determined orbital quantum number l, particularly in the case of Russell-Saunders coupling.

[1] Recent developments in the theory of elementary particle interactions again focus attention on the subject of parity. For this development we refer to the literature quoted by LEE [1960]. A group theoretical classification was given by MELVIN [1960].

The zeroth approximation eigenfunctions are then the products $\psi_1 \psi_2 \ldots \psi_f$ of the individual wavefunctions and their parity is

$$\delta = (-1)^{l_1+l_2+\ldots l_f}. \tag{5.45}$$

A perturbation will arise from the coupling of the electrons; this perturbation can be important, nevertheless it always preserves the spherical symmetry of the atom and changes neither the representations of the group $\mathscr{D}'_3$ nor the parity δ. The value of δ is unchanged according to (5.45).

In the helium atom the first electron is generally in the state $s(l = 0)$ and δ is completely determined by the quantum number $l = L$ of the second electron: the terms are normal.

This is not generally true. For example let us consider the atoms containing two electrons in their non-filled shell [1] and let us suppose that both are in a p state ($l = 1$). The formula $\boldsymbol{D}_1 \times \boldsymbol{D}_1 = \boldsymbol{D}_0 + \boldsymbol{D}_1 + \boldsymbol{D}_2$ shows that three kinds of states may occur: the S states ($L = 0$), P states ($L = 1$) or D states ($L = 2$). These three kinds of states are known in Mg; they are the so-called primed terms and they all have the same parity

$$\delta = (-1)^{l_1+l_2} = (-1)^2 = 1.$$

The *experimental* importance of the parity δ is shown by a selection rule discovered empirically by Laporte, Russell and Saunders.

The sign of the components of the electric moment $\boldsymbol{\mu}$ and those of every polar vector are changed under inversion of axes: the rotations induce in ordinary space the transformations of the group $\mathscr{D}_1$ with the parity $\delta = -1$.

Let us come back to the equations 5.42 and replace the indices j and m by $l_1, l_2, \ldots l_f$; if the system undergoes the operation κ: the sign of μ_p is changed, $\psi_{l_1 l_2 \ldots l_f}$ is multiplied by $\delta = (-1)^{l_1+l_2+\ldots+l_f}$ and $\psi_{l'_1 l'_2 \ldots l'_f}$ by $\delta = (-1)^{l'_1+l'_2+\ldots+l'_f}$. Since the matrix elements $(\mu_+)_{l_1 \ldots l_f, l'_1 \ldots l'_f}$ are constants it is necessary that

$$\delta' = -\delta. \tag{5.46}$$

Consequently: *in an allowed transition the sum of the orbital quantum numbers $l_1 + l_2 + \ldots + l_f$ can only be changed by an odd number* (selection rule of Laporte).

[1] The completely filled shells do not contribute to the total orbital number L. This is a consequence of the Pauli principle.

8.2 APPROXIMATE SELECTION RULES

The selection rules for the numbers L and S as well as those of j can be obtained by considering the rotation group, but they are only approximate.

The operator of the electric moment μ_+, μ_-, μ_z changes only the orbital part of the wave functions ($\psi_{ik\ldots l}$ in the formula (a) page 160 footnote) without any action on the pure spin functions ($u_i^{(1)} u_k^{(2)} \ldots$). As long as the perturbations resulting from the spin are weak and the distance between the components of the multiplets is small, these functions ψ can be calculated by the scalar Schrödinger theory and they have an exact total quantum number L and the spin functions have an exact total quantum number S. Hence when one writes the expansions (5.42) of the components of the electric moment *the pure spin functions are the same in both members of each equation and the expansions of the second members can only be done in terms of space eigenfunctions* $\psi_{ik\ldots l}$. An argument identical to that of the above section leads to the rules:

$$L \to L+1, \quad L, \quad L-1 \quad \text{or} \quad \Delta L = 0, \quad \pm 1 \quad S \to S \quad \text{or} \quad \Delta S = 0. \tag{5.47}$$

The inversion of axes forbids the transition $L \to L$.

These rules are the basis of the classification of lines in the series. For one electron atoms they can be obtained directly by the theory of spherical harmonics.

But, while the rules (5.43), (5.44) and (5.46) are rigorous the last ones are only approximate and cease to be valuable as soon as the perturbations due to the spin "blur" the sharp values of the vectors $\boldsymbol{L}$ and $\boldsymbol{S}$. In fact, there are many exceptions for heavy atoms, where the series disappear almost completely.

Since the parity rule is violated for *strong* interactions (large deviation from the Russell-Saunders scheme), at first it seems slightly contradictory that in elementary particle theory the rule is violated for *weak* interactions. The term "weak interactions" is, however, purely generic; it indicates interactions that are weak compared to the nucleon-nucleon interaction.

9. Stark Effect. Anomalous Zeeman Effect. Line Components Intensity. Landé Splitting Factor. Paschen Back Effect

The results obtained at the ènd of Appendix III allow us to complete at certain points the theory of the Zeeman effect (§ 4) and to say something about the Stark effect.

9.1. GENERAL THEORY

From the group theoretical point of view the essential difference between these two phenomena comes from the different symmetries of the magnetic and electric fields. The first one has the symmetry of a rotating cylinder, it admits only the *Abelian* group formed by the rotations $\mathscr{D}_2$ around the field and a reflection with respect to a plane perpendicular to the field. The second one has the symmetry of a cone of revolution which allows the *non-Abelian* group $\mathscr{D}'_2$ of rotations and reflections. Obviously $\mathscr{D}_2$ is a subgroup of $\mathscr{D}'_2$. If there is a degenerate level corresponding to the irreducible representation $\mathscr{D}_j$ of the group $\mathscr{D}_3$, a magnetic field will split the level into $2j+1$ components ($m = j, j-1, \ldots, -j$) according to Chapter 4, §§ 3.2 and 4. The splitting in an electric field will be less complete. The two values $\pm m$ of the magnetic quantum number (or we could say the electric quantum number $|m|$) correspond to a unique level which cannot be split by the electric perturbation (essential degeneracy). Only the level $m = 0$ contains two terms, one positive, the other negative with the parity $+1$ and -1 (cf. § 4). Altogether the splitting by Stark effect will consist of $j+1$ separate levels.

Generally this decomposition is of second order except for an accidental degeneracy (hydrogen case), the expansion of the perturbation begins by terms proportional to the square of the field. The actual perturbing function is

$$\lambda W = -E_z \mu_z$$

where E_z is the field and $\mu_z = \sum e_i z_i$ is the projection of the electric moment on this field. The diagonal elements of the perturbation matrix are the scalar products $E_z(\psi_M^{(J)}, \mu_z \psi_{M'}^{(J)})$ which are all zero because all the $\psi_M^{(J)}$ have the same parity for a given J and the total result is anti-symmetric due to the odd parity of the operator μ_z.[1]

The selection rules (5.43) and (5.44) remain exact but (5.47) are violated as soon as the electric field is strong enough to create eigenfunctions differing from their zeroth approximation by a significant amount, i.e. they start to show the symmetry of the new problem. The rule of Laporte is unmodified by a magnetic field which is represented by an axial vector and hence gives no change in parity. This rule is, however, not verified in a strong electric field (a polar vector) since the perturbed wave function does not have the parity of the unperturbated wave function.

[1] Any integrand anti-symmetric in the coordinates will give zero when integrated over the total space.

9.2. INTENSITY OF COMPONENTS

Let us come back to the expansions (5.42) of the components of the electric moment. The left-hand sides of these formulas are the products of the components of two vectors. The first are μ_+, μ_- and μ_z which transform under a rotation as $x+iy$, $x-iy$, z, i.e. as the basis variables of the representation $\mathscr{D}_1$ or more precisely (cf. § 2) as $q_{-1}^{(1)}$, $-q_1^{(1)}$ and $2^{-\frac{1}{2}}q_0^{(1)}$. The second are the functions ψ_{jm} which transform as the set $q_m^{(j)}$ with $m = j, \ldots, -j$. At the right-hand side the constants $\mu_{j'm', jm}$, which will be written with a change of the indices as $(\mu)_{JM, jm}$, are multiplied by the functions ψ_{JM} which transform as the basis variables of the representations $\boldsymbol{D}_J$.

The equations (5.42) are essentially the same as relation (5.A26) in Appendix III, provided we substitute in this last one $j' = 1$, $m' = 1, 0, -1$, $J = j+1, j, j-1$, $m+m' = M = m+1, m, m-1$. Referring to the remark following the equation (5.A26) *the matrix elements* $(\mu_+)_{jm, JM}$, $-(\mu_-)_{jm, JM}$ and $2^{\frac{1}{2}}(\mu_z)_{jm, JM}$ have the form $\rho_J C_{mM}^J$ where the coefficients C are given by Table 5.A.1 in Appendix III, the ρ_J remaining arbitrary. This is the first illustration of the Wigner-Eckhart theorem (compare Chapter 4 and 6).

Thus we obtain, apart from the factors ρ_J, the components of the electric moment i.e. the "amplitudes" of the different components of a given spectral line in the Zeeman or Stark effect, the squares of which represent the radiation intensities. These components are, neglecting some numerical factors,

$$\begin{aligned}
&\text{for } j \to J = j+1, & (\mu_+)_{m, m-1} &= \rho_{j+1}\sqrt{(j-m+1)(j-m+2)}\\
&& -(\mu_-)_{m, m+1} &= \rho_{j+1}\sqrt{(j+m+1)(j+m+2)}\\
&& (\mu_z)_{m, m} &= \rho_{j+1}\sqrt{(j+m+1)(j+m+1)}\\
&\text{for } j \to J = j, & (\mu_+)_{m, m-1} &= \rho_j\sqrt{(j+m)(j-m+1)}\\
&& (\mu_-)_{m, m+1} &= \rho_j\sqrt{(j-m)(j+m+1)}\\
&& (\mu_z)_{m, m} &= \rho_j m\\
&\text{for } j \to J = j-1, & (\mu_+)_{m, m-1} &= \rho_{j-1}\sqrt{(j+m)(j+m-1)}\\
&& -(\mu_-)_{m, m+1} &= \rho_{j-1}\sqrt{(j-m)(j-m-1)}\\
&& -(\mu_z)_{m, m} &= \rho_{j-1}\sqrt{(j+m)(j-m)}.
\end{aligned} \tag{5.48}$$

All the other matrix elements are zero. Hence we found more precisely the selection rules concerning j and m.

These relations have been checked experimentally in weak fields. We notice that the components of the $j \to j$ transition are of exactly the same form as the components $M_+^{(j)}$, $M_-^{(j)}$ and $M_z^{(j)}$ of the angular momentum in the state j (section 3.4). This is not surprising. In both cases the axial or polar character of *the vector* does not make any difference in the pure rotations. For the same reason the formulas (5.48) are valid in Stark effect as well as in the Zeeman effect.

9.3. LANDÉ-FACTOR

In the same way as we have explicitly developed the electric moment matrix i.e. by using the formula (5.A26) in Appendix III, we shall give some further attention to the Zeeman effect and complete it with the help of the preparations of § 4.

Let us remember that in the perturbation matrix we must in the first approximation take into account only the matrix elements that determine the mutual relations among the $(2j+1)$ states which form the degenerated unperturbed level E_{nj}. These elements are related to the representation $\boldsymbol{D}_j$ of the rotation group. We have found that these elements form a diagonal submatrix $\lambda(w_m)$ $m = j, j-1, \ldots -j$. We have now to calculate the terms λw_m after completing the form of the perturbing function with the hypothesis of the spin (§§ 6.1, 6.3 and 6.4).

Let $\boldsymbol{H}$ be the magnetic field. The angular momentum $\boldsymbol{L}$ of the atomic orbits and the angular momentum of spin $\boldsymbol{S}$ (infinitesimal rotations) operate first on the space $\mathfrak{R}_L$ of the orbital wave functions and secondly on the space $\mathfrak{R}_S$ of the spin functions. Their sum operates on the vectors belonging to the space $\mathfrak{R} = \mathfrak{R}_S \times \mathfrak{R}_L$ and can be written as a result of (5.39b)

$$\boldsymbol{M} = [\boldsymbol{M}^{(S)} \times (\mathrm{I})_L + (\mathrm{I})_S \times \boldsymbol{M}^{(L)}]. \tag{5.49}$$

The unit matrices $(\mathrm{I})_S$ and $(\mathrm{I})_L$ complete the matrices $\boldsymbol{M}^{(S)}$ and $\boldsymbol{M}^{(L)}$ in the parts of the space $\mathfrak{R}$ where these do not operate. The above equation can be decomposed in three equations of the same form if we use $\boldsymbol{M}_+$, $\boldsymbol{M}_-$, and $\boldsymbol{M}_z$.

As we saw in section 6 the angular momentum due to the spin must be counted twice in the calculation of the total magnetic moment. To the vector $\boldsymbol{M}$ corresponds a magnetic moment $\mathcal{M}$ and its projection on the field $\boldsymbol{H}$ can be written as a result of (5.29) and (5.29a) as

$$\mathcal{M}_z = \beta[M_z^{(S)} \times \mathrm{I} + M_z^{(S)} \times \mathrm{I} + \mathrm{I} \times M_z^{(L)}] = \beta[M_z^{(S)} \times (\mathrm{I}) + M_z] \tag{5.50}$$

where β is the Bohr magneton ($\beta = e\hbar/2m_0 c$). This formula is equivalent to the assumption that the total magnetic moment is the sum of two vectors $\beta M^{(L)}$ and $2\beta M^{(S)}$.

1°. At first let us suppose that the field is weak with regard to the spin orbit coupling $(L \cdot S)$ which we suppose to be of the Russell-Saunders type. The energy differences of the Zeeman components are small with respect to the splitting among the different components of the multiplet, which will be split separately. The spherical symmetry of the atom is hardly modified. If one makes a suitable choice for axes, the space $\mathfrak{R}_S \times \mathfrak{R}_L$ will be reduced to subspaces $\mathfrak{R}_j$ irreducible with regard to $\mathscr{D}_3$. The operator M of (5.49) is decomposed into a sum of operators $M^{(J)}$, each operating on one of the subspaces $\mathfrak{R}_J$ and each being an infinitesimal operator of the different representations $\mathscr{D}_J$. Thus

$$M = \sum_J M^{(J)} = [M^{(S)} \times \mathrm{I} + \mathrm{I} \times M^{(L)}].$$

In this coordinate system the perturbation function can be written as a result of (5.50) and the preceding equation

$$\lambda W = -H\mathscr{M}_z = -\beta H[M_z^{(S)} \times \mathrm{I} + \sum_J M_z^{(J)}]. \tag{5.51}$$

The splitting of a term J is, in the first order, obtained by considering only that part of the operator, or that part of the corresponding matrix, which is related to the space $\mathfrak{R}_J$. That is only the part related to the $(2J+1)$ basis functions of the representation D_J is considered:

$$\lambda W_J = -\beta H[(M_z^{(S)} \times \mathrm{I})^{(J)} + M_z^{(J)}]. \tag{5.52}$$

We know the operator M_z which has been calculated in § 4. It is defined in the space of total angular momentum by the equation

$$M_z \psi_{Jm} = \sum_{J'} \sum_{m'=-J'}^{+J'} \psi_{J'm'} (M_z)_{m'm}^{J'J}$$

and we know that this matrix is diagonal:

$$M_z = (m\delta_{J'J}\delta_{m'm}), \qquad m = J, J-1, \ldots -J.$$

The matrix $(M_z^{(S)} \times \mathrm{I})$ which we shall designate for simplification by S_z is defined by a similar equation:

$$S_z \psi_{Jm} = \sum_{J'} \sum_{m'=-J'}^{+J'} \psi_{J'm'} (S_z)_{m'm}^{J'J}$$

and these two equations, together with those related to the components

$\boldsymbol{M}_+$ and $\boldsymbol{M}_-$, have the form (5.A26) in Appendix III (where one has to take $j = J, j' = 1, J = J', m+m' = m'$).[1] Then $\boldsymbol{S}_z = \sum \boldsymbol{S}_z^{(J)}$. The matrix elements $(\boldsymbol{M}_z^{(J)})$ and $(\boldsymbol{S}_z^{(J)})$ are proportional to the constants $C^J_{m'm}$, i.e. *proportional to each other* where the constant of proportionality ρ_J is independent of m and is the same for $\boldsymbol{S}_p^{(J)}$ and $\boldsymbol{S}_q^{(J)}$:

$$\boldsymbol{S}_z^{(J)} = \rho_J \boldsymbol{M}_z^{(J)} = \rho_J(m). \tag{5.53}$$

Finally we obtain

$$\lambda W_J = -\beta H(1+\rho_J)(m) = -g\beta H(m) \tag{5.53a}$$

a similar formula to that of section 4 but with a coefficient $g = 1+\rho_J$ which is called *the Landé splitting factor*: *the energy difference between the Zeeman components is g times the normal differences.*

This factor can be very simply calculated as follows: Let us consider the part of the operators $\boldsymbol{M}$ operating in the space $\mathfrak{R}^{(J)}$ and let us put

$$(\mathbf{I} \times \boldsymbol{M}^L)^{(J)} = \boldsymbol{L}^{(J)}.$$

Then (5.49) takes the form

$$\boldsymbol{M}^{(J)} = \boldsymbol{S}^{(J)} + \boldsymbol{L}^{(J)} \quad \text{or} \quad \boldsymbol{L}^{(J)} = \boldsymbol{M}^{(J)} - \boldsymbol{S}^{(J)}$$

which expresses the addition rules of the vector model. This equation shows that $\boldsymbol{M}^{(J)}$, $\boldsymbol{S}^{(J)}$ and $\boldsymbol{L}^{(J)}$ commute among themselves since $\boldsymbol{S}^{(J)}$ and $\boldsymbol{L}^{(J)}$ operate on different spaces. (Spin space and the space of the functions $\psi(\boldsymbol{x})$.) Hence we can write

$$(\boldsymbol{L}^{(J)})^2 = (\boldsymbol{M}^{(J)})^2 + (\boldsymbol{S}^{(J)})^2 - 2\boldsymbol{M}^{(J)} \cdot \boldsymbol{S}^{(J)},$$

or according to (5.53)

$$\begin{aligned} 2\boldsymbol{M}^{(J)} \cdot \boldsymbol{S}^{(J)} &= 2\rho_J(\boldsymbol{M}^{(J)})^2 = 2\rho_J J(J+1) \\ &= J(J+1) + S(S+1) - L(L+1), \end{aligned}$$

and finally

$$g = 1+\rho_J = 1 + \frac{J(J+1)+S(S+1)-L(L+1)}{2J(J+1)}. \tag{5.54}$$

This is the formula discovered empirically by Landé.[2]

[1] The summation $\sum_{m'}$ contains for each component only one term according to the selection rules and formulas (5.26).

[2] For the geometrical meaning of Landé g-factor see Mayer and Mayer [1940] p. 344 fig. 15.2.

2^0 In a strong magnetic field the coupling $(\boldsymbol{L} \cdot \boldsymbol{S})$ disappears. The system allows only the rotation group $\mathscr{D}_2$ around the field direction. The space $\mathfrak{R}_S \times \mathfrak{R}_L$ cannot be split in subspaces $\mathfrak{R}_J$. As $\mathscr{D}_2$ is Abelian, the matrices $M_z^{(L)}$ and $M_z^{(S)}$ can be reduced entirely to the diagonal form and we obtain according to (5.50)

$$\lambda W = -\beta H(m_L + 2m_S) = -\beta H(m), \quad \begin{aligned} m_L &= L, L-1, \dots -L \\ m_S &= S, S-1, \dots -S, \end{aligned}$$

the energy differences between the split levels become normal again: as they are large with regard to the energy difference between the multiplet components and as the selection rules remain valid, the field gives rise to a normal triplet. It is the Paschen-Back effect.

The intermediate cases can be studied without difficulty.

Appendix I

THE CONNECTION BETWEEN A FORMAL SET OF BASIS FUNCTIONS AND THE SPHERICAL HARMONICS

Because we have used a formal choice of basis functions, the q_m^j, we must now consider how to deal with actual physical quantities. It was mentioned, particularly in § 2 and Appendix III, that the results obtained for the representations are independent of the basis. Specifically some of the ξ, η could be replaced by η^* and $-\xi^*$ thus simplifying the basis functions to monomials. As a result we do not have a one-to-one correspondence between ξ, η and x, y, z anymore. It is necessary to consider the inverse process. That is if there is a set of normalized monomials $\xi^{j-m}\eta^{j+m}$ as in (5.8c) and if j is an integer, what is the corresponding description in x, y and z?

The connection between the ordinary variables x, y and z and the formal variables ξ, η are given by the following set of formulas:

$$\begin{aligned} x+iy &= 2\eta\xi^* \sim \mp\eta^2 \\ x-iy &= 2\xi\eta^* \sim \pm\xi^2 \\ z &= \xi\xi^* - \eta\eta^* \sim \mp\xi\eta. \end{aligned} \tag{5.A1}$$

The first part of the equations express the original results obtained in § 2. The second part indicated by "$\sim$" shows the way these quantities transform. Solving this for the separate variables the equations become,

$$x \sim \frac{\xi^2 - \eta^2}{2}; \qquad y \sim \frac{-\eta^2 - \xi^2}{2}; \qquad z \sim -\xi\eta. \tag{5.A2}$$

Using the fact that these three quantities transform in the same way consider a scalar defined by the internal product

$$K = x\frac{\xi^2-\eta^2}{2} + y\frac{-\eta^2-\xi^2}{2} - z\xi\eta. \tag{5.A3}$$

This invariant can be rewritten as follows

$$K = -(x+iy)\eta^2 + (x-iy)\xi^2 - z\xi\eta. \tag{5.A4}$$

Nothing new is obtained since this expression says essentially the same thing as formula (5.A1) or (5.A2). However new results can be obtained if we raise this invariant to an integer power, l,

$$K^l = \sum_m \xi^{l-m}\eta^{l+m}\varphi_l^m(x, y, z). \tag{5.A5}$$

In the right-hand side terms of like powers in ξ and η have been grouped. Each coefficient is a function only of x, y and z. Both the *monomials* in ξ and η and the *polynomials* in x, y, z transform according to a certain representation. Hence knowing that $\xi^{l-m}\eta^{l+m}$ form a basis for an *irreducible* representation we conclude that the $\varphi_l^m(x, y, z)$ do the same. It turns out moreover that the functions constructed in this way are actually the solutions of the Laplace equation. This is not surprising since one of the properties of the Laplace equation is that it does not change under a rotation of coordinate axes. Its set of solutions, forms a representation of the rotation group and it is possible to choose the solutions in such a way that they form an irreducible representation. Hence we conclude that the solutions of the Laplace equation should be a linear combination of the polynomials we have constructed above.

If we went into more detail we could show that the polynomials are actually, up to a mulplicative constant, the well-known spherical harmonics. (See KRAMERS [1937], page 170.)

Appendix II

CONSTRUCTION OF THE IRREDUCIBLE REPRESENTATION OF THE GROUP $\mathscr{D}_3$

The following argument is of a type that is commonly used in quantum mechanics. It does not differ essentially from those which one finds, for example, in Dirac.[1] It goes actually back to Lie and Cartan.

[1] DIRAC [1958] Section 30. See also BORN and JORDAN [1930] Ch. 4, section 27, which we follow closely. This book is full of group theoretical arguments; the authors did not find it necessary, however, to put much emphasis on this fact.

Since the rotation group is continuous, the representations consist of matrices that are continuous and differentiable with respect to three parameters of the group. Hence they are completely determined by three infinitesimal basis transformations, that is by three matrices $\boldsymbol{R}_x, \boldsymbol{R}_y, \boldsymbol{R}_z$ or which comes down to the same $\boldsymbol{M}_{+1}$, $\boldsymbol{M}_{-1}$, $\boldsymbol{M}_0$ (compare (5.19), (5.20) and (5.24)). We demand only from these representations that they are unambiguous *in the neighborhood of the identity*. We know that they have to satisfy the commutation relations (5.24)

$$\boldsymbol{M}_{+1}\boldsymbol{M}_{-1}-\boldsymbol{M}_{-1}\boldsymbol{M}_{+1} = 2\boldsymbol{M}_z \tag{5.A7}$$

$$\boldsymbol{M}_{-1}\boldsymbol{M}_z-\boldsymbol{M}_z\boldsymbol{M}_{-1} = \boldsymbol{M}_{-1} \tag{5.A8}$$

$$\boldsymbol{M}_z\boldsymbol{M}_{+1}-\boldsymbol{M}_{+1}\boldsymbol{M}_z = \boldsymbol{M}_{+1} \tag{5.A9}$$

to which we add (compare (5.27a))

$$\begin{aligned}\boldsymbol{M}^2 &= \boldsymbol{M}_x^2+\boldsymbol{M}_y^2+\boldsymbol{M}_z^2 = \frac{\boldsymbol{M}_{+1}\boldsymbol{M}_{-1}+\boldsymbol{M}_{-1}\boldsymbol{M}_{+1}}{2}+\boldsymbol{M}_z^2 \\ &= \boldsymbol{M}_{+1}\boldsymbol{M}_{-1}-\boldsymbol{M}_z+\boldsymbol{M}_z^2.\end{aligned} \tag{5.A10}$$

As we have seen, $\boldsymbol{M}^2$ represents the square of the total angular momentum. This is an invariant of the rotation group, which consequently commutes with all the operators of this group; in particular with $\boldsymbol{M}_x$, $\boldsymbol{M}_y$ and $\boldsymbol{M}_z$.[1] We have now to find all the possible systems of three matrices $\boldsymbol{M}_{+1}$, $\boldsymbol{M}_{-1}$ and $\boldsymbol{M}_0$ that satisfy the preceding conditions and form an *irreducible* linear group.

Let $\mathscr{D}$ be an arbitrary representation of $\mathscr{D}_3$ and $\mathfrak{R}_0$ the corresponding representation space. $\mathfrak{R}_0$ is a subspace of the function space $\mathfrak{R}$, invariant with respect to the rotation group. We can choose in this subspace a system of basis functions ψ_{mn}, finite or infinite, (we explain in a moment why we use two indices) for which the abelian subgroup $\mathscr{D}_2$ of rotations around the z-axis is completely reduced. In other words the basis functions are chosen in such a way that a rotation θ_z, considered as a change of coordinate axes (compare Chapter 4, § 2.3) induces the transformation

$$\psi_{mn} \rightarrow \psi'_{mn} = \mathrm{e}^{-im\theta_z}\psi_{mn}$$

therein. The infinitesimal rotation is

$$\mathrm{d}\psi_{mn} = -im\psi_{mn}\mathrm{d}\theta_z$$

[1] Independent of all physical interpretation this commutability can be verified by a direct calculation with the help of equations (5.A7), (5.A8) and (5.A9).

where m is an integer of the representation and is unambiguous for all values of θ_z. We have seen that this condition is not necessary if we are content with unambiguity in the neighborhood of the identity (compare Sec. 1). Hence it is sufficient to allow m to be a real or imaginary number. From (5.19) and (5.20) we see that the corresponding matrix $\boldsymbol{M}_z$ is diagonal and contains a sequence of eigenvalues $m, m', m'', \ldots$ which may each occur with a certain degree of multiplicity hence a second index n. The ψ_{mn} are the eigenvectors of $\boldsymbol{M}_z$.

THEOREM I. *In each irreducible representation of $\mathscr{D}_3$ the numbers $m, m' \ldots$ are all smaller in absolute value than a fixed number K.*

Indeed, from Schur's lemma we know that $\boldsymbol{M}^2$, which commutes with all matrices of this irreducible representation should be a multiple of the unit matrix

$$\boldsymbol{M}^2 = K^2 \mathbb{I}$$

where K^2 is a constant. On the other hand, $\boldsymbol{M}_x^2 + \boldsymbol{M}_y^2 = \boldsymbol{M}^2 - \boldsymbol{M}_z^2$ is, in our system of axes, a semi-definite diagonal matrix, whose eigenvalues cannot be negative. Consequently we have

$$K^2 - m^2 \geqq 0 \qquad |m| \leqq |K|. \tag{5.A11}$$

The result is that in our search for irreducible representations of $\mathscr{D}_3$ we can restrict ourselves to representations $\mathscr{D}$ which satisfy the condition (5.A11).

If we let the operators $\boldsymbol{M}_+ \mathrm{d}s$ and $\boldsymbol{M}_- \mathrm{d}t$ act on the basis vectors ψ_{mn}; the transformed vectors $\boldsymbol{M}_+ \psi_{mn} \mathrm{d}s$ and $\boldsymbol{M}_- \psi_{mn} \mathrm{d}t$ stay in $\mathfrak{R}_0$, which is invariant. We may suppress the infinitesimal *numerical* factors ds and dt.

THEOREM II. *$\boldsymbol{M}_{+1} \psi_{mn}$ and $\boldsymbol{M}_{-1} \psi_{mn}$ are eigenvectors of $\boldsymbol{M}_z$ and correspond respectively to the eigenvalues $(m+1)$ and $(m-1)$.*

The verification is done easily: (5.A8) and (5.A9) give:

$$\begin{aligned} \boldsymbol{M}_z \boldsymbol{M}_{-1} \psi_{mn} &= \boldsymbol{M}_{-1}(\boldsymbol{M}_z \psi_{mn} - \psi_{mn}) = (m-1)\boldsymbol{M}_{-1} \psi_{mn} \\ \boldsymbol{M}_z \boldsymbol{M}_{+1} \psi_{mn} &= \boldsymbol{M}_{+1}(\boldsymbol{M}_z \psi_{mn} + \psi_{mn}) = (m+1)\boldsymbol{M}_{+1} \psi_{mn}. \end{aligned} \tag{5.A12}$$

$\boldsymbol{M}_{+1} \psi_{mn}$ is hence a vector in the eigenspace of $\boldsymbol{M}_z$ which corresponds to the eigenvalues $(m+1)$: it is a linear combination of the basis vectors $\psi_{m+1,n}$, where $(m+1)$ is constant and n takes all possible values. As the last are only defined except for a unitary transformation, we can take $\boldsymbol{M}_+ \psi_{mn}$ itself as basis vector under the condition that we multiply it with a convenient constant to normalize it. The same considerations hold for $\boldsymbol{M}_{-1} \psi_{mn}$ in the eigenspace $(m-1)$ of the matrix M_z. Hence, if we now suppress the index n which became superfluous, we can give to our theorem the following interpretation:

If we begin with an arbitrary eigenfunction ψ_m of M_z corresponding to the eigenvalue m, we obtain, by iteration with the operations M_{+1} and M_{-1} a sequence of eigenfunctions of the same operator: $\psi_{m\pm1}\,\psi_{m\pm2}\,\psi_{m\pm3}\,\ldots$ corresponding to a set of eigenvalues which differ by a factor one.

This sequence is obviously limited in both directions since the space $\mathfrak{R}_0$ satisfies the condition (5.A11); hence there should exist a maximum value j and a minimum value k for which one has

$$M_{+1}\psi_j = 0; \quad M_{-1}\psi_k = 0. \tag{5.A13}$$

For the other values of m we write

$$M_{-1}\psi_m = \alpha_m\psi_{m-1}; \quad M_{+1}\psi_m = \rho_m\psi_{m+1}; \quad M_z\psi_m = m\psi_m$$

the α_m and ρ_m are numerical factors. This, together with (5.A7) and (5.A13) gives:

$$M_{+1}\psi_{j-1} = \frac{1}{\alpha_j}M_{+1}M_{-1}\psi_j = \frac{1}{\alpha_j}(M_{-1}M_{+1}+2M_z)\psi_j = \frac{2j}{\alpha_j}\psi_j = \rho_{j-1}\psi_j$$

$$M_{+1}\psi_{j-2} = \frac{1}{\alpha_{j-1}}(M_{-1}M_{+1}+2M_z)\psi_{j-1} = \frac{1}{\alpha_{j-1}}[\rho_{j-1}\alpha_j+2(j-1)]\psi_{j-1}$$
$$= \rho_{j-2}\psi_{j-1}$$

$$M_{+1}\psi_{m-1} = \ldots = \frac{1}{\alpha_m}[\rho_m\alpha_{m+1}+2m]\psi_m = \rho_{m-1}\psi_m.$$

From which, by putting $\beta_m = \rho_m\alpha_{m+1}$, we find the recurrent formula

$$\beta_{m-1} = \beta_m+2m \tag{5.A14}$$

$$\beta_j = 0, \quad \beta_{j-1} = 2j, \quad \beta_{j-2} = 2[j+(j-1)], \ldots, \beta_m = j(j+1)-m(m+1).$$

But for $m = k$ the minimum value of m we know from (5.A13) that $M_{+1}M_{-1}\psi_k = 0$. Hence $\beta_k+2k = 0$ and finally

$$j(j+1)-k(k+1) = 0 \tag{5.A15}$$

an equation that has two solutions. One is impossible: $k = j+1$, the other $k = -j$ is allowed. Hence, the sequence of eigenvalues of M_z, which we have extracted in this way with the help of the operators M_+ and M_-, can be written as $j, j-1, \ldots -(j-1), -j$. They are symmetrically arranged around the value $m = 0$. *The sequence contains* $2j+1$ *terms and since this number should be an integer,* j *is either an integer or a half-integer.*

The corresponding eigenfunctions $\psi_j, \psi_{j-1}, \ldots \psi_{-j}$ span a $2j+1$-dimensional subspace $\mathfrak{R}_j$ *invariant with respect to the rotation group*, since the functions transform among each other under the infinitesimal operators $\boldsymbol{M}_{+1}, \boldsymbol{M}_{-1}$, and $\boldsymbol{M}_z$ of the group. The space is *irreducible* since the operators $\boldsymbol{M}_{+1}$ and $\boldsymbol{M}_{-1}$ iterated in a proper way, will transform each arbitrarily chosen axis ψ_m into any other arbitrarily chosen axis ψ_m apart from a multiplicative constant.

Hence the representation $\boldsymbol{D}_j$ which we have extracted from $\boldsymbol{D}$ and which is determined by the infinitesimal basis transformations $\boldsymbol{M}_{+1}, \boldsymbol{M}_{-1}, \boldsymbol{M}_z$ is an irreducible representation of $\mathscr{D}_3$. The representation turns out to be identical to the one that is constructed in § 2 and which was designated by the same symbol.

It is sufficient to put in (5.A14)

$$\rho_m = \alpha_{m+1} = \sqrt{j(j+1)-m(m+1)}$$

that is

$$\alpha_m = \sqrt{j(j+1)-m(m-1)}.$$

We again have equation (5.26) and we are sure that the basis functions are normalized in the same way.

There is no other irreducible representation $\mathscr{D}_3$ since for each representation $\boldsymbol{D}$ one can, by the preceding method, extract at least one representation $\boldsymbol{D}_j$ and if $\boldsymbol{D}$ is irreducible, it should be identical with $\boldsymbol{D}_j$.

Appendix III

PROOF OF THE FORMULA (5.34a)

$$\boldsymbol{D}_{j_1} \times \boldsymbol{D}_{j_2} = \boldsymbol{D}_{j_1+j_2} + \boldsymbol{D}_{j_1-\frac{1}{2}} \times \boldsymbol{D}_{j_2-\frac{1}{2}}. \tag{5.34a}$$

We have the following problem: The vectors in two unitary spaces are subject simultaneously to the *same* transformation of the unimodular unitary group.

$$\begin{aligned} \xi_1' &= \alpha\xi_1 + \beta\eta_1 & \xi_2' &= \alpha\xi_2 + \beta\eta_2 \\ \eta_1' &= -\beta^*\xi_1 + \alpha^*\eta_1 & \eta_2' &= -\beta^*\xi_2 + \alpha^*\eta_2 \end{aligned} \qquad \alpha\alpha^* + \beta\beta^* = 1.$$

We consider the representations $\boldsymbol{D}_{j_1}$, $\boldsymbol{D}_{j_2}$ and $\boldsymbol{D}_{j_1} \times \boldsymbol{D}_{j_2}$ of this group in the three spaces $\mathfrak{R}_{j_1}$, $\mathfrak{R}_{j_2}$ and $\mathfrak{R}_{j_1}\times\mathfrak{R}_{j_2}$. In these three spaces we take as coordinates respectively: the $(2j_1+1)$ monomials

$$q_{1,m_1}^{(j_1)} = \frac{\xi_1^{j_1+m_1}\eta_1^{j_1-m_1}}{\sqrt{(j_1+m_1)!(j_1-m_1)!}}$$

the $(2j_2+1)$ monomials

$$q_{2,m_2}^{(j_2)} = \frac{\xi_2^{j_2+m_2}\eta_2^{j_2-m_2}}{\sqrt{(j_2+m_2)!(j_2-m_2)!}}$$

and the $(2j_1+1)(2j_2+1)$ products

$$Q_{m_1m_2} = q_{1,m_1}^{(j_1)}q_{2,m_2}^{(j_2)} = \frac{\xi_1^{j_1+m_1}\eta_1^{j_1-m_1}\xi_2^{j_2+m_2}\eta_2^{j_2-m_2}}{\sqrt{(j_1+m_1)!(j_1-m_1)!(j_2+m_2)!(j_2-m_2)!}}.$$

We have to decompose the space $\mathfrak{R}_{j_1}\times\mathfrak{R}_{j_2}$ in two subspaces invariant under the transformations (5.4), such that the first: $\mathfrak{R}_{j_1+j_2}$ is irreducible. To obtain this result we will make a change of coordinates which replaces the $Q_{m_1m_2}$ by linear combinations of these quantities. Let us call $j_1+j_2 = J$. We have to find $2J+1$ independent linear combinations, which transform among each other, without intervention of any other linear combination, under the transformation (5.4). They form the basis for the representation $\boldsymbol{D}_{j_1+j_2}$.

If we take $\xi_1 = \xi_2 = \xi$; $\eta_1 = \eta_2 = \eta$ the last system should reduce to the functions

$$\varphi_P = \frac{\xi^{J+P}\eta^{J-P}}{\sqrt{(J+P)!(J-P)!}} \qquad (P = J, J-1, \ldots, -J).$$

This is what we will call the *condition A*. From this we find that

$$\varphi_J = \frac{\xi_1^{2j_1}\xi_2^{2j_2}}{\sqrt{2J}}$$

should necessarily make a part of the basis of the new irreducible subspace, since this is the only one which reduces itself to

$$\frac{\xi^{2J}}{\sqrt{2J}} \quad \text{if } \xi_1 = \xi_2 \text{ and } \eta_1 = \eta_2.$$

Let us suppress the normalization factor in the denominator in order to simplify the formulas. We can insert it easily at the end of the calculation. Since we know one of the variables of $\boldsymbol{D}_{j_1+j_2}$ let us subject it to a transformation (5.4) and then we must find a linear function of all the variables of this representation. Thus in the usual notation

$$\xi_1^{2j_1}\xi_2^{2j_2} \to \xi_1'^{2j_1}\xi_2'^{2j_2} = (\alpha\xi_1+\beta\eta_1)^{2j_1}(\alpha\xi_2+\beta\eta_2)^{2j_2}$$

$$=(2j_1)!(2j_2)!\sum_{M=-J}^{+J}\alpha^{J+M}\beta^{J-M}\sum_{m_1+m_2=M}\frac{\xi_1^{j_1+m_1}\eta_1^{j_1-m_1}\xi_2^{j_2+m_2}\eta_2^{j_2-m_2}}{(j_1+m_1)!(j_1-m_1)!(j_2+m_2)!(j_2-m_2)!}$$

$$= (2J)!\sum_M \frac{\alpha^{J+M}\beta^{J-M}}{\sqrt{(J+M)!(J-M)!}}\varphi_M \qquad (5.A16)$$

with

$$\varphi_M = \frac{(2j_1)!(2j_2)!}{(2J)!}\sum_{m_1+m_2=M}\sqrt{\frac{(J+M)!(J-M)!}{(j_1+m_1)!(j_1-m_1)!(j_2+m_2)!(j_2-m_2)!}}\,\mathfrak{q}_{1m_1}^{j_1}\mathfrak{q}_{2m_2}^{j_2}. \qquad (5.A17)$$

We state that the polynomials φ_J, $\varphi_{J-1} \ldots \varphi_{-J}$ are the variables or basis functions of the representation $\boldsymbol{D}_{j_1+j_2}$. It is evident from (5.A16) that they satisfy the condition A and are consequently linearly independent.[1] Let us now verify that these basis functions transform among each other under an arbitrary transformation of the group $\mathscr{U}_2$

$$\begin{aligned}\xi_1'' &= a\xi_1+b\eta_1 & \xi_2'' &= a\xi_2+b\eta_2\\ \eta_1'' &= -b^*\xi_1+a^*\eta_1 & \eta_2'' &= -b^*\xi_2+a^*\eta_2\end{aligned} \qquad a^*a+b^*b = 1. \qquad (5.A18)$$

We have actually

$$\alpha\xi''+\beta\eta'' = A\xi+B\eta$$

with

$$A = a\alpha-b^*\beta; \quad B = b\alpha+a^*\beta. \qquad (5.A19)$$

Hence

$$(\alpha\xi_1''+\beta\eta_1'')^{2j_1}(\alpha\xi_2''+\beta\eta_2'')^{2j_2} = (A\xi_1+B\eta_1)^{2j_1}(A\xi_2+B\eta_2)^{2j_2}$$

or as a result of (5.A16) and (5.A17),

$$\sum_M \frac{\alpha^{J+M}\beta^{J-M}}{\sqrt{(J+M)!(J-M)!}}\chi_M = \sum_M \frac{A^{J+M}B^{J-M}}{\sqrt{(J+M)!(J-M)!}}\varphi_M$$

where χ_M is the transformed of φ_M by the operation (5.A18).

If we take (5.A19) into account $A^{J+M}B^{J-M}$ becomes a polynomial of the degree $2J$ in α and β. These two numbers are arbitrary. Hence it is sufficient

[1] This condition made us choose the particular grouping of the coefficients in the equations (5.A16) and (5.A17).

to identify in the last equation the two sides term by term in order to obtain χ_M as a linear combination of φ_M. This verifies that the φ_M form the variables that are the basis of the representation $\boldsymbol{D}_{j_1+j_2}$.

As for the variables of $\boldsymbol{D}_{j_1-\frac{1}{2}} \times \boldsymbol{D}_{j_2-\frac{1}{2}}$, these are obviously expressions $P_{m_1m_2}$ analogous to $Q_{m_1m_2}$, but of degree $2j_1-1$ in $\xi_1\eta_1$ and $2j_2-1$ in $\xi_2\eta_2$. Hence these are not linear combinations of $Q_{m_1m_2}$ since these are of higher degree. In order to reduce the first to the second, it is sufficient that

$$Q'_{m_1m_2} = (\xi_1\eta_2-\eta_1\xi_2)P_{m_1m_2}. \tag{5.A20}$$

These functions are of the degree wanted and transform under a substitution (5.4) exactly like the quantities $P_{m_1m_2}$ since the first factor of the right-hand side is an invariant. We have indeed

$$\xi_1'\eta_2'-\eta_1'\xi_2' = (\xi_1\eta_2-\eta_1\xi_2)(\alpha\alpha^*+\beta\beta^*) = \xi_1\eta_2-\eta_1\xi_2.$$

In this way we have found the basis functions of $\boldsymbol{D}_{j_1+j_2}$ and $\boldsymbol{D}_{j_1-\frac{1}{2}} \times \boldsymbol{D}_{j_2-\frac{1}{2}}$.

We still have to show that each linear combination Φ of expressions $Q_{m_1m_2}$ can be expressed in an unambiguous way in a linear combination of φ_M and Q':

$$\Phi = (a_J\varphi_J+ \ldots a_{-J}\varphi_{-J})+(\xi_1\eta_2-\eta_1\xi_2)\psi \tag{5.A21}$$

where ψ is a linear combination of $P_{m_1m_2}$, that is a homogeneous set of polynomials of degree $(2j_1-1)$ in ξ_1, η_1 and $(2j_2-1)$ in $\xi_2\eta_2$.

The necessary number of arbitrary coefficients, are provided easily since we have,

$$(2j_1+1)(2j_2+1) = [2(j_1+j_2)+1]+2j_1 \cdot 2j_2.$$

Hence it is sufficient to show that the different terms of (5.A21) are independent, that is that Φ cannot be identically zero unless all coefficients in the right-hand side are equal to zero. Let us first take $\xi_1 = \xi_2$, $\eta_1 = \eta_2$; the last term in (5.A21) is zero $\Phi \equiv 0$ gives $a_J = \ldots = a_{-J} = 0$ as the φ_M are independent. The a_M being zero, let us make $\xi_1 \neq \xi_2$, $\eta_1 \neq \eta_2$; $\Phi \equiv 0$ implies necessarily $\psi = 0$; hence the decomposition is unambiguous.

The representation $\boldsymbol{D}_{j_1-\frac{1}{2}} \times \boldsymbol{D}_{j_2-\frac{1}{2}}$ can be reduced in the same way as $\boldsymbol{D}_{j_1} \times \boldsymbol{D}_{j_2}$. We argue as before; taking (5.A20) and (5.A21) into account we see that it is sufficient to look for the coefficients in the development of

$$(\xi_1\eta_2-\eta_1\xi_2)(\alpha\xi_1+\beta\eta_1)^{2j_1-1}(\alpha\xi_2+\beta\eta_2)^{2j_2-1} \tag{5.A22}$$

in order to find the linear combinations of $Q_{m_1m_2}$ which serve as basis of the representation $\boldsymbol{D}_{j_1} \times \boldsymbol{D}_{j_2}$.

If we operate this way, step by step, we succeed in expressing the decomposition (5.34) of the representation $\boldsymbol{D}_{j_1} \times \boldsymbol{D}_{j_2}$ in an explicit form: (we suppose that $j_1 < j_2$)

$$\Phi = \psi_{j_1+j_2} + (\xi_1\eta_2 - \eta_1\xi_2)\psi_{j_1+j_2-1} + \dots$$
$$\dots + (\xi_1\eta_2 - \eta_1\xi_2)^\lambda \psi_{j_1+j_2-\lambda} \dots \quad (5.A21a)$$

which is immediately clear by comparing it with (5.A21) Let us put $J = j_1 + j_2 = \lambda$. The linear combinations φ_M^J of $Q_{m_1 m_2}$ which serve as a basis for the representation $\boldsymbol{D}_{j_1+j_2-\lambda}$ can be obtained by developing the function

$$(\xi_1\eta_2 - \eta_1\xi_2)^\lambda (\alpha\xi_1 + \beta\eta_2)^{2j_1-\lambda} (\alpha\xi_2 + \beta\eta_2)^{2j_2-\lambda} \quad (5.A22a)$$

and since we have

$$(\xi_1\eta_2 - \eta_1\xi_2)^\lambda = \sum_{\nu=0}^{\lambda} (-1)^\nu \frac{\lambda!}{\nu!(\lambda-\nu)!} \xi_1^{\lambda-\nu} \eta_1^\nu \xi_2^\nu \eta_2^{\lambda-\nu}$$

we find, by a calculation similar [1] to the one by which we arrived at (5.A17)

$$\varphi_M^J = \rho_J \sum_{m_1+m_2=M} \sum_\nu (-1)^\nu$$
$$\times \frac{\sqrt{(J+M)!(J-M)!(j_1+m_1)!(j_1-m_1)!(j_2+m_2)!(j_2-m_2)!}}{\nu!(\lambda-\nu)!(j_1-m_1-\nu)!(j_1+m_1-\lambda+\nu)!(j_2+m_2-\nu)!(j_2-m_2-\lambda+\nu)!} Q_{m_1 m_2}$$
$$= \rho_J \sum_{m_1+m_2=M} C_{m_1 m_2}^J Q_{m_1 m_2} \quad (M = J, J-1, \dots, -J). \quad (5.A23)$$

ρ_J is a normalization constant which depends only on J, that is on j_1, j_2 and λ. The index ν varies in principle from 0 to λ, *but one has to put all terms that contain a negative factorial equal to zero.*

The coefficient $C_{m_1 m_2}^J$ is called the *Clebsch-Gordan coefficient*, the *Wigner coefficient*, or the *3-j symbol*. To calculate its numerical value is a tedious task; fortunately tables are available (see references in Chapter 6).

The formula (5.A23) solves completely the problem of the decomposition of the product $\boldsymbol{D}_{j_1} \times \boldsymbol{D}_{j_2}$. It takes the simple form (5.A17) if $\lambda = 0$ and it similarly simplifies in the opposite case: $\lambda = 2j_2$, $J = j_1 - j_2$ $(j_1 > j_2)$

[1] The binomial coefficient of $(\alpha\xi_1 + \beta\eta_1)^{2j-\lambda}$ contains a summation index, say k

$$(\alpha\xi_1 + \beta\eta_1)^{2j-\lambda} = \sum_k \binom{2j-\lambda}{k} (\alpha\xi_1)^{2j-\lambda-k} (\beta\eta_1)^k$$

in order to have the total power of ξ_1 equal to $j+m$ we take $k = j-\nu-m$ and restrict the summation to positive or zero values of this quantity. Similarly for ξ_2, η_2.

because v can only take the value $v = j_2+m_2$ and one has

$$C^{j_1-j_2}_{m_1m_2} = (-1)^{j_1+m_1}\sqrt{\frac{(j_1+m_1)!(j_1-m_1)!}{(j_2+m_2)!(j_2-m_2)!(J+M)!(J-M)!}}\,.$$

We can now give the equations (5.A21) and (5.A21a) a definite form, displaying clearly the decomposition $\boldsymbol{D}_{j_1}\times\boldsymbol{D}_{j_2}$ in its irreducible elements. This is the so-called Clebsch-Gordan series

$$\Phi = \sum_{J=j_1+j_2}^{|j_1-j_2|}\sum_{M=-J}^{J} a^J_M \sum_{m_1+m_2=M} \rho_J C^J_{m_1m_2} Q_{m_1m_2} = \sum_J\sum_M a^J_M \varphi^J_M\,. \tag{5.A21b}$$

The vector Φ is an arbitrary vector of the space $\mathfrak{R}_{j_1}\times\mathfrak{R}_{j_2}$, that is a linear combination of $Q_{m_1m_2}$. The a^J_M depend on the form of this combination. The ρ_J are normalization factors and the $C^J_{m_1m_2}$ are the constants determined by equation (5.A23). The φ^J_M are the orthogonal vectors that span the space $\mathfrak{R}_{j_1}\times\mathfrak{R}_{j_2}$, those which have the same index span a separate invariant subspace which transforms according to the irreducible representation $\boldsymbol{D}_j$. Their mutual orthogonality results from the orthogonality of the different irreducible representation of the same group [1] ($\mathscr{D}_3$ for the indices J, $\mathscr{D}_2$ for the indices M). We can write

$$\varphi^J = \rho_J \sum_{m_1+m_2=M} C^J_{m_1m_2} Q_{m_1m_2}\,. \tag{5.A25}$$

But the $Q_{m_1m_2} = q^{j_1}_{m_1} q^{j_2}_{m_2}$ are also orthogonal, since the factors $q^{j_1}_{m_1}$ and $q^{j_2}_{m_2}$ are. Hence one can, keeping $M = m_1+m_2$ constant, consider (5.A25) as a unitary transformation connecting $\varphi_{m_1-m_2}$ with $Q_{m_1m_2}$. The transformation matrix is $B = (\rho_J C^J_{m_1m_2})$, where J labels the rows and m_1 the columns. We can reverse this transformation by expressing $Q_{m_1m_2}$ as a linear combination of φ^J_M. Since B is unitary and real the inverse matrix is equal to the transposed matrix, hence we have

$$Q_{m_1m_2} = q^{j_1}_{m_1} q^{j_2}_{m_2} = \sum \rho_J C^J_{m_1m_2} \varphi^J_{m_1+m_2}\,. \tag{5.A26}$$

The preceding calculations have been made with the help of a particular choice of basis functions $q^{j_1}_{m_1}$, $q^{j_2}_{m_2}$ and $Q_{m_1m_2}$ of the representations $\boldsymbol{D}_{j_1}$, $\boldsymbol{D}_{j_2}$ and $\boldsymbol{D}_{j_1}\times\boldsymbol{D}_{j_2}$. But the results obtained are all linear *hence they depend only on the representations* and all stay true no matter which functions we use as a basis. In particular, we could replace certain ξ by η^* and certain η by

[1] This statement is only true in case every representation occurs only once, a condition fulfilled for the full rotation group, but not always for a finite group.

$-\xi^*$, which do transform in the same way (compare § 2.2). We can give in this way a simple geometrical or physical meaning to the basis functions. They could be for example the Laplace functions, or more generally the eigenfunctions of an atomic problem. The formulas (5.A17), (5.A21b), (5.A23), (5.A24), (5.A25) and (5.A26) always apply. This statement forms the basis of the proof in Sec. 9 about the selection rules and the calculation of the intensities of the Zeeman effect. In the majority of the applications $j_2 = 1$ and

$$\boldsymbol{D}_j \times \boldsymbol{D}_1 = \boldsymbol{D}_{j+1} + \boldsymbol{D}_j + \boldsymbol{D}_{j-1}.$$

The possible values for m_2 are $m_2 = -1, 0, +1$, hence $M = m_1+1, m_1, m_1-1$. Eq. (5.A17) (5.A23) and (5.A24) give the constants in the following table:

TABLE 5.A.1

J	$m_2 = 1 \quad M = m_1+1$	$m_2 = 0 \quad M = m_1$	$m_2 = -1 \quad M = m_1-1$
$j+1$	$\sqrt{\frac{(j+m+2)(j+m+1)}{2}}$	$\sqrt{(j+m+1)(j-m+1)}$	$\sqrt{\frac{(j-m+2)(j-m+1)}{2}}$
j	$-\sqrt{2(j+m+1)(j-m)}$	$2m$	$\sqrt{2(j+m)(j-m+1)}$
$j-1$	$\sqrt{\frac{(j-m)(j-m+1)}{2}}$	$-\sqrt{(j+m)(j-m)}$	$\sqrt{\frac{(j+m)(j+m-1)}{2}}$

For further tables see the Systematic Bibliography § 7.1.

Having once established the Clebsch-Gordan formula for the decomposition of the *basis* into its irreducible components it is very easy to perform the decomposition of the *representation* into the irreducible representations it contains. We pointed out that the coefficients C^J, for fixed $m_1 m_2$, could be looked upon as a unitary matrix. Hence if we want to transform the matrices with respect to the new coordinates we have to apply the similarity transformation equations (1.7) or (1.27a). Explicitly we have

$$\boldsymbol{D}_{j_1 \times j_2} = \sum_{j_3 = j_1+j_2}^{|j_1-j_2|} C^{j_3} \boldsymbol{D}_{j_3} C^{j_3}. \tag{5.A27}$$

This formula forms the basis for the derivation of the Wigner-Eckhart theorem and will be used in Chapter 6, § 3, equation (6.17).

CHAPTER 6

CONTINUATION OF THE THEORY OF THE ROTATION GROUP

1. Irreducible Tensor Operators

At the end of Chapter 2, § 3 it was stated that if angular momentum is a good quantum number, then polynomials of angular momentum operators can be constructed. These polynomials will undergo certain transformations among themselves if the system undergoes a finite rotation. The angular momentum components form a vector, so the polynomials are *tensors*, since they are products of vector components (or linear combinations of such products).

The reason that we prefer linear combinations of products, instead of the products themselves, is obvious. These tensors must be defined in such a way that they transform like irreducible representations in configuration space. For example, the nine products $\boldsymbol{L}_x^2$, $\boldsymbol{L}_y^2$, $\boldsymbol{L}_z^2$, $\boldsymbol{L}_x\boldsymbol{L}_y$, $\boldsymbol{L}_y\boldsymbol{L}_x \ldots$ contain one invariant $\boldsymbol{L}_x^2+\boldsymbol{L}_y^2+\boldsymbol{L}_z^2$, as well as three linear combinations that transform like a vector $(\boldsymbol{L}_x\boldsymbol{L}_y-\boldsymbol{L}_y\boldsymbol{L}_x) \rightarrow \boldsymbol{L}_z, \ldots$, hence there are actually only five operator polynomials of rank 2.

After this preparation we introduce a set of angular momentum operator polynomials which are usually called *the irreducible tensor operators*. They were originally introduced by RACAH [1942, 1943] (see also FANO and RACAH [1959]), with the following properties:

$$\mathrm{D}_{\mathrm{op}}\, T_\mu^\lambda = \sum_{\mu'=-\lambda}^{\lambda} D_{\mu\mu'}\, T_{\mu'}^\lambda \tag{6.1}$$

that is under a rotation of configuration space, the component T_μ^λ transforms like a spherical harmonic Y_μ^λ. The "quantum number" λ labelling the irreducible representation is always an integer. This is because the operators T are related to physical observables which do not tolerate the ambiguity in sign inherent in half-integer representations.

It is not necessary that the operators T be Hermitian. The conventional choice for the T_μ^λ gives a basis analogous to the spherical harmonic basis functions. The spherical harmonics are of course a specific choice out of many possible equivalent bases. They happen to be complex, although equivalent to real ones. The practical advantages of complex exponentials over separate sines and cosines are well known in mathematics, physics and electrical engineering. The counterpart of non-real numbers are non-Hermitian operators. For instance, $x \pm iy$ will correspond to $J_x \pm iJ_y$, a so-called escalator matrix, which has only zero elements on one side of the diagonal.

Explicit construction of the irreducible tensor sets on the basis of the definition (3.1) can be done in different ways. The simplest would be the Schmidt procedure used for the construction of orthogonal polynomials. Start with a constant and the three tensor operators of rank 1, J_0 and $J_{\pm 1}$, and successively construct polynomials that are orthogonal to these four. The orthogonality is determined by the condition

$$\mathrm{Tr}\, \mathrm{P}_i \mathrm{P}_j = \delta_{ij}$$

where P_i and P_j are any two operator polynomials. The orthogonality provides independency. But if the polynomials are independent, that does not insure that they form a basis that is irreducible. To avoid this uncertainty, start the construction in the same way as Legendre polynomials are constructed and generalize to spherical harmonics, i.e. first take the $\mu = 0$ case. They do not contain $J_{\pm 1}$, only J^2 and J_z. (The non-operator case contains r^2 and z, but not $x \pm iy$.) For successive λ-values we find

$$\begin{aligned}
\lambda &= 0 : \text{constant} \\
\lambda &= 1 : J_z \\
\lambda &= 2 : J_z^2 - \tfrac{1}{3}J^2 \\
\lambda &= 3 : 5J_z^3 - 3J^2 J_z + J_z .
\end{aligned}$$

We see that for $\lambda \leqq 3$ the polynomials are different from the spherical harmonics. The spherical harmonics are homogeneous in x, y, z of degree λ. The operator polynomials contain terms of the degree λ and lower. Actually, the concept of "degree" does not have meaning in this case, since with the insertion of a commutator $J_z = (J_x J_y - J_y J_x)$ the degree can be increased in a rather arbitrary manner.

Instead of pursuing the Schmidt procedure for $\mu \neq 0$ a second, heuristic, but sometimes more effective way of constructing the polynomials will be indicated. Take a spherical harmonic and replace a term like

$$2xy \quad \text{by} \quad (J_x J_y + J_y J_x)$$

i.e., take all possible arrangements and divide by the number of arrangements. This process leads to lengthy expressions which can be shortened and systematized through repeated application of the commutation relation.

The third method (compare EDMONDS [1957]) for the introduction of irreducible tensor sets that transform like spherical harmonics, is based directly on their transformation properties. It does not actually give the operator polynomials but their commutation relation, as well as their matrix representation. Since we are, for all practical purposes, interested in the latter, this method is the one that is most widely used.

The derivation is based on equation (2.19), which can be written as

$$\boldsymbol{J} \cdot \delta\omega\psi = i\hbar\delta\psi. \tag{6.2}$$

Its formal integration leads to an expression for the results of a finite rotation $\boldsymbol{\omega}$ on the wave function $\psi(0)$:

$$e^{-i\hbar^{-1}\boldsymbol{J}\cdot\boldsymbol{\omega}}\psi(0) = \psi(\omega). \tag{6.3}$$

Using this unitary operator for the transformation of the tensor-operator under a finite rotation, we have:

$$\mathrm{sTs}^{-1} = e^{-i\hbar^{-1}\boldsymbol{J}\cdot\boldsymbol{\omega}}\,\mathrm{T}\,e^{i\hbar^{-1}\boldsymbol{J}\cdot\boldsymbol{\omega}} = \mathrm{T} - i\hbar^{-1}[\boldsymbol{J}, \mathrm{T}] \ldots. \tag{6.4}$$

On the other hand, the right-hand side of (6.1) can be series developed in a similar way (compare Chapter 2, § 3)

$$\sum_{\mu'} D^{\lambda}_{\mu\mu'} T^{\lambda}_{\mu'} = T^{\lambda}_{\mu} - i\hbar^{-1} \sum (\boldsymbol{J}^{\lambda})_{\mu\mu'} T^{\lambda}_{\mu'} + \ldots. \tag{6.5}$$

Comparing coefficients gives us the relation

$$[\boldsymbol{J}_i, T^{\lambda}_{\mu}] = \sum (\boldsymbol{J}^{\lambda}_i)_{\mu\mu'} T^{\lambda}_{\mu'} \quad (i = 1, 0, -1). \tag{6.6}$$

From the explicit expressions for $(\boldsymbol{J})_{\mu\mu'}$, well known from the Zeeman effect (eq. 5.48), we find a set of commutation relations:

$$[\boldsymbol{J}_{\pm 1}, T_{\mu}] = T^{\lambda}_{\mu\pm 1}\hbar\sqrt{(\mu\mp\lambda)(\lambda\pm\mu+1)} \tag{6.7a}$$

$$[\boldsymbol{J}_0, T_{\mu}] = T^{\lambda}_{\mu}\hbar\mu \tag{6.7b}$$

which are equivalent to the set of equations originally introduced by Racah in 1942 to define the tensor operators.

In the case of finite groups, the tensor operators introduced above are in many cases no longer irreducible. There will be, of course, only a finite set of representations. A general formula cannot be indicated. For the different finite groups, the irreducible operators were determined by KORRINGA [1954] and KOSTER [1958].

Since the irreducible operators T are determined by their rotational behaviour only, it is possible to indicate their matrix elements with respect to an irreducible set of wave functions without any further specification, except for a part, usually referred to as the "strength" of the interaction, which does not depend on the geometry. This part will stay undetermined in the form of a number of constants, which do not depend on the axial quantum numbers m or μ.

2. Representation of Tensor Operators

The use of group-theoretical methods has been most beautifully demonstrated by the so-called Wigner-Eckardt theorem (WIGNER [1931, 1959], ECKARDT [1930]). The theorem states that if there is an invariant coupling by a bilinear form, such as spin orbit coupling:

$$\boldsymbol{L} \cdot \boldsymbol{S} = L_x S_x + L_y S_y + L_z S_z \tag{6.8}$$

the matrix elements can be calculated by group theoretical methods, except for proportionality factors. In other words the ratio between the matrix elements of such an invariant operator can be determined.

This separation of matrix elements into angular or geometrical versus radial or physical parts, is found in many different forms and the Wigner-Eckardt theorem is actually an example of many similar applications of the following idea:

Suppose an operator O transforms according to a given (irreducible) representation of the rotation group, say λ and we want to determine its matrix elements with respect to two sets of basis functions. One set transforms according to the irreducible representation j_1, the other according to j_2. This idea is not quite new, since in Chapter 5 we treated this problem for the special case of O being a vector (i.e. $\lambda = 1$). Given these irreducible representations the matrix elements will of course transform according to the reprentations contained in the direct product of the three representations

$$\Gamma_{j_1} \times \Gamma_\lambda \times \Gamma_{j_2} = \sum_i \Gamma_i . \tag{6.9}$$

Usually some of the Γ_i's will occur more than once.[1] The general idea is that the unitary transformation which performs this reduction is given by group theory (actually the theory of group representations) except for a

[1] In finite groups a multiple occurring representation may already show up in the direct product of *two* representations, in the rotation group they are singly occurring (Clebsch-Gordan series). With the direct product of *three* irreducible representations of the rotation group, there are always multiple occurring representations, except in trivial cases like $j_1 = 0$. Compare § 4 on Racah coefficients.

certain ambiguity due to the multiple occurring irreducible representations.

In order to determine the matrix element, we take the integral over the left-hand side. The integral can be written in polar coordinates and hence consists of an integral over the unit sphere and a radial integral. If the integrand undergoes a rotation in configuration space, the integral will have the same value (provided of course that we transform both operator *and* the wave functions; a rotation in configuration space "induces" a rotation in function space — compare Chapter 4, § 1) since such a rotation only amounts to shifting the intervals of φ and ϑ on the unit sphere. We conclude that the only representation on the right-hand side which will give a non-zero result is the unit representation Γ_1. All the others will average out under spherical integration, and the only one left, the unit representation, needs no angular integration. The integral for this term is 4π (the surface of the unit sphere) times the radial integral.

Let us consider the left-hand side of the symbolic formula (6.9). It stands for a collection of products of wave functions and tensor components. There are $(2m+1)(2\mu+1)(2m'+1)$ possible products. Instead of integrating, we write them first as linear combinations:

$$\sum_{\mu m m'} a^{(i)}_{\mu m m'} \psi_m \mathrm{O}_\mu \psi_{m'} = b_i \tag{6.10}$$

such that they transform like irreducible representations. That is, b_i *is* a basis (in case the representation is one-dimensional), or a *certain number of* b_i's *form* (in case the representation has more than one dimension) a basis of an irreducible representation. If we integrate now, we find that all integrals over b_i are zero, except b_1 which corresponds to the unit representation.

$$\sum_{\mu m m'} a^{(i)}_{\mu m m'} \langle m|\mathrm{O}_\mu|m'\rangle = 4\pi\delta_{1i}\int b_1 r^2 \mathrm{d}r = \delta_{1i} B_1 . \tag{6.11}$$

If we take the inverse of the matrix $a^{(i)}_{\mu m m'}$ (i labels the columns and the triple index labels the rows) we have the matrix elements expressed as a product

$$\langle m|\mathrm{O}_\mu|m'\rangle = (a^1_{\mu m m'})^{-1} B_1 \tag{6.12}$$

of the radial integral B_1 and a factor which depends only on the reduction procedure. Hence this factor is entirely determined by representation theory.

We went through this argument in order to stress two points. The first is the question how many constants B_1 are involved. This is now easily answered. The proof above was given for fixed values of j_1, j_2 and λ. We have to repeat it for every other pair of values j_1 and j_2. How many possibilities does one have for a given λ? That answer is given by the triangular rule, which is of course the same condition as demanding that (6.9) will contain the unit representation. For instance, for $\lambda = 1$ the vectors j_1 and j_2 may differ by ± 1, or 0. Hence there are three different constants B_1 for every given value of j_1 (or j_2). These three, or in general $2\mu+1$, constants depend on the radial shape of the wave functions and on the radial dependence of the operator.

Since it is possible that there is more than one wave function with the same j value, another quantum number α may be added in order to distinguish between the different (radial) wave functions. Hence B depends on j_1, j_2, α_1 and α_2.

The second point is the remark often overlooked in importance that for given α_1 and α_2 the unit representation occurs *once* (or not at all). The possibility of multiple occurring representations has already been mentioned in a footnote. This is not the case for the unit representation. It occurs only once, since the direct product of *two* representations contains each irreducible representation only once.

$$\Gamma_\lambda \times \Gamma_{j_1} = \sum \Gamma_i \qquad (i = \lambda+j_1, \ldots, |\lambda-j_1|) \tag{6.13}$$

and the only way one can form Γ_1 out of the direct product of $\Gamma_{j_2} \times \Gamma_i$ is by taking $i = j_2$.[1]

This is not true for finite groups.

3. Wigner-Eckhardt Theorem, Reduced Matrix Elements

If the wave function and the operator all refer to one type of function, the theorem in the last paragraph is rather trivial, since the only invariant operator is the operator with $\lambda = 0$. In physical examples it is nearly always the Hamiltonian or parts thereof. The theorems tel lus that the energy states are characterized by certain constants (the eigenvalues of H) which do not depend on m or m'. From the triangular rule one concludes that $j = j'$. The only thing that is of any interest is the statement that for $j = j' \neq 0$ several matrix elements have the same value: a $(2j+1)$-fold degenerate level. This case is already adequately covered in Chapter 4. The Hamiltonian was supposed to consist of two parts, the kinetic energy operator and a potential energy, a static field essentially. It is, however, well known that the

[1] This last step has to be more carefully formulated if the representations are of the second kind (j is half-integer). One has to take the antisymmetrical product of Γ_i and the spin conjugate of Γ_{j_2}.

Hamiltonian many contain other terms, which could be called dynamic. For instance terms like spin-orbit coupling or similar vector couplings well known in the theory of the atomic spectra (CONDON and SHORTLEY [1935]).

These products are scalars and hence invariants, but they are made up of vectors, that is three operators that transform as the coordinates of a point. The complication arises from the fact that one vector operates on one set of functions (for instance the angular momentum operator components act on the orbital wave function) and the other vector on a different set of functions (the spin functions). Hence the wave functions have to be the product of two different functions, say the space and the spin functions. Generally speaking, this does not have to be literally true, the total wave function has only to *transform like the direct product of two sets of wave functions.* In case the total wave function is n-fold degenerate we again find that n levels are degenerate, according to the Wigner theorem. One can, however, be more demanding since it turns out to be possible to calculate the matrix elements of each of the six components of the scalar product in the n-dimensional function space without any knowledge beyond the transformation properties. This procedure will be described below and is based on the repeated use of Clebsch-Gordan coefficients. Instead of considering the vector coupling first and generalizing this later to tensor coupling we may just as well give the description in terms of tensors right from the beginning.

The tensor coupling expression is very similar to the scalar product of two vectors. For practical reasons we write the scalar product

$$a_x b_x + a_y b_y + a_z b_z = a_{+1} b_{-1} + a_0 b_0 + a_{-1} b_{+1} \tag{6.8'}$$

where $a_{\pm 1} = 2^{-\frac{1}{2}}(a_x \pm i a_y)$. In the same way we can construct invariants out of tensor components provided we take irreducible tensors and label the components in a proper way. The construction of irreducible tensor operators a priori has been considered in the previous section. It is only necessary to define them here as *quantities that have the same transformation properties as the spherical harmonics.* The general form of an invariant coupling will be

$$\mathrm{O}_{\lambda=0} = \mathrm{H}_{\text{coupl.}} = \sum_{\mu=-\lambda}^{\lambda} T_{\mu}^{\lambda}(1) T_{-\mu}^{\lambda}(2), \tag{6.14}$$

where the first operator acts on one type of wave function, the other on another type of wave function. These were originally the spin and orbital wave functions; we call them 1 and 2. They may have irreducible representations characterized by j_1 and j_2 (and m_1 and m_2). These representations usually are reduced, i.e. written as a set of wave functions with total j ($j = j_1 + j_2, \ldots, |j_1 - j_2|$). Hence the matrix elements of (6.14) depend on many labels: $j_1 m_1, j_2 m_2, j_1', m_1', j_2' m_2'$ and $j = j'$, $m = m'$ as well as μ and λ. All unprimed quantities refer to the initial state, all primed to the final state. Again, we can show that these matrix elements are a product of "constants", that is, factors which depend on the different j's and λ, times expressions containing the m's. This m-dependence is completely determined by the representation theory.

We will give a proof slightly different from § 2 as follows: A transformation of configuration space of an irreducible tensor operator gives, according to Chapter 1, eq. (1.7):

$$\langle|\mathrm{s}^{-1} T_\mu^\lambda|\rangle = \langle|\mathrm{S} T_\mu^\lambda \mathrm{S}^{-1}|\rangle. \tag{6.15}$$

This equation states that a transformation in configuration space is equivalent with another transformation in function space indicated by the operators S. The second one is the inverse of the transformation induced in function space by the first (compare Chapter 4, § 1.2). The proof of equations (6.7) was also based on the same idea.

If we write this in components we have

$$\begin{aligned} \langle jk|T_\mu|j'l\rangle &= \langle k|\mathrm{S}^{-1}|m\rangle \langle m|\mathrm{S}T_\mu \mathrm{S}^{-1}|n\rangle \langle n|\mathrm{S}|l\rangle \\ &= \langle k|\mathrm{S}^{-1}|m\rangle \langle m|\mathrm{s}^{-1}T_\mu|n\rangle \langle n|\mathrm{S}|l\rangle = \sum_{vmn} {}^\lambda D_{\mu v}^{-1}\, {}^j D_{km}^{-1}\, {}^{j'} D_{nl} \langle m|T_v|n\rangle. \end{aligned} \tag{6.16}$$

So far we have done nothing but explicitly expressed the fact that the bases and the tensor components transform according to irreducible representations. We can now, using again the fact that all rotations are determined by the same set of Euler angles, apply the Clebsch-Gordan reduction formula as expressed in Chapter 5, Appendix III, eq. (5.A27), that is we can *reduce* the product $\mathscr{D}^\lambda \mathscr{D}^j$, by choosing a linear combination, such that each of these transform according to the representations $j_2 = \lambda + j, \lambda + j - 1, \ldots |\lambda - j|$:

$${}^\lambda D_{\mu v}^{-1}\, {}^j D_{km}^{-1} = \sum_{j''} C_{\mu k}^{j'', j\lambda}\; {}^{j''} D_{\mu+k,\, v+m}^{-1}\, C_{vm}^{j'', j\lambda} \tag{6.17}$$

and after this substitution we let the parameters (Euler angles) contained in S run through all possible values. We integrate over these variables and apply the orthogonality relations between the matrix elements (see Chapter 3). As a result (6.16) reduces to:

$$\langle jk|T_\mu|j'l\rangle = \sum_{j''=j-\lambda}^{j+\lambda} C^{j'',\,j\lambda}_{\mu k} \sum_{mn\nu} C^{j'',\,j\lambda}_{\nu m} \frac{\delta_{j''j'}\delta_{\nu+m,\,n}\delta_{k+\mu,\,l}}{2j'+1} \langle jm|T_\nu|j'n\rangle$$

$$= C^{j',\,j\lambda}_{\mu k}\delta_{\mu+k,\,l} \sum_{mn\nu} C^{j',\,j\lambda}_{\nu m} \frac{\delta_{\nu+m,\,n}}{2j'+1} \langle jm|T_\nu|j'n\rangle. \qquad (6.18)$$

The last sum is independent of m, n and ν and we denote it by

$$\langle j||T||j'\rangle = \sum_{m,\,n\nu} C^{j',\,j\lambda}_{\nu m} \frac{\delta_{\nu+m,\,n}}{2j'+1} \langle jm|T_\nu|j'n\rangle. \qquad (6.19)$$

The result is that the ratio between the matrix elements $\langle jk|T_\mu|j'l\rangle$ for different k and l values is completely determined by group theory. The proportionality constant depends only on j, j' and whatever other quantum numbers are used to describe the system. The double bar matrix element is called the *reduced matrix element*. It is at this moment that we see the full importance of the Clebsch-Gordan coefficients in physics.

Introducing the notation [1] used by CONDON and SHORTLEY [1935]:

$$\langle j_1 m_1|T^\lambda_\mu|jm\rangle = \frac{(-1)^{\lambda-j_1+j}}{\sqrt{2j_1+1}} (\lambda\mu jm|\lambda j j_1 m_1)\langle j_1|T^\lambda|j\rangle. \qquad (6.19a)$$

This is the Wigner-Eckardt theorem in its original form and by using equations (6.14) and (6.19a) we have for the scalar product of two irreducible tensor operators:

$$\langle\alpha j_1 j_2 JM|\sum_{\mu=-\lambda}^{\lambda} T^\lambda_\mu(1)\cdot T^\lambda_{-\mu}|\alpha' j'_1 j'_2 JM\rangle$$

$$= \sum C^{J,\,j_1j_2}C^{J,\,j'_1j'_2}\langle j_1 m_1|T^\lambda_\mu(1)|j'_1 m'_1\rangle\langle j_2 m_2|T^\lambda_{-\mu}(2)|j'_2 m'_2\rangle$$

$$= \sum C^{J,\,j_1j_2}C^{J,\,j'_1j'_2}C^{j'_1,\,j_1\lambda}C^{j'_2,\,j_2\lambda}\langle j_1||T(1)||j'_1\rangle\langle j_2||T(2)||j'_2\rangle. \qquad (6.20)$$

We will come back to eq. (6.20) later in § 4. Eq. (6.19) has a wide range of applicabilities; for instance transitional probabilities, that is, matrix elements in which $j' \neq j$. One can, of course, also deduce transition probabilities from (6.20) by taking as one of the tensors the field that causes the transition.

Besides the application of (6.19) to the calculation of the dipole moments,

[1] Compare EDMONDS [1957]. This monograph contains a very practical table that indicates the connection with all other notations. Our Clebsch-Gordan coefficient is related to Condon and Shortley's by:

$$C^J_{m_1m_2} = (j_1j_2JM|j_1m_1j_2m_2).$$

which will lead us again to the expressions (5.48) it is also easy to determine the corresponding equations for the higher multipole moments.

Another application is made in crystalline field theory where the potential energy is replaced by a linear combination of irreducible tensors that have the same symmetry. The matrix elements of the energy and, after diagonalisation, the eigenvalues can also be determined with the Wigner-Eckhardt theorem. The result contains a number of undetermined proportionality constants. This is not surprising since the replacement Hamiltonian fulfilled the necessary condition of symmetry, but is not sufficient to describe the physical situation adequately.

4. Racah Coefficients

The Racah coefficients were originally introduced in polyelectronic atomic spectroscopy in 1943 and subsequently found a widespread use in nuclear spectroscopy. They are also useful in solid state work. If the crystal symmetry is introduced through an operator-Hamiltonian, sooner or later in a physical calculation the coefficients in front of the different terms of the operator expression have to be calculated. It was realized that these coefficients were similar to the reduced matrix elements and hence connected with the Racah coefficients. In the following pages we will follow Rose's treatment.

The idea of a Racah coefficient is as follows: If two electrons are coupled the Clebsch-Gordan formula states which irreducible representations are contained in the product representation, and that each of these representations is contained only once. However, if three electrons are coupled the irreducible representation of a certain final j-value (total angular momentum) may occur several times. This indeterminacy is shown by coupling j_1 with j_2 into j' and coupling j' with j_3 into j. As a result there is a set of *multiple occurring* irreducible representations. However, if j_2 and j_3 were coupled into, say, j'' and subsequently j'' and j_1 were coupled, then the resulting bases would have been different. Of course the product representation will contain the same set of representations each with the same multiplicity. Since any linear combination of equivalent irreducible representations is again an irreducible representation of the same type, the two different coupling schemes may give different results for the bases with respect to these multiple occurring representations. The Racah coefficient gives the relation between these:

$$\psi_{j,m}(\text{via } j') = \sum_{j'} R_{j'j''}\psi_j(\text{via } j''). \tag{6.21}$$

For instance take $j_1 = 1$ and $j_2 = 2$; then $j' = 3, 2, 1$ and combining this with $j_3 = 3$ we have $j = 6, 0$ (once) 5, 1 (twice) 4, 3, 2 (three times). However $j_2+j_3 = j'' = 5, 4, 3, 2, 1$ and combining this with j_2 gives the same set. A certain possible value of j say $j = 1$ comes in the first case from $j' = 3$ and 2 and in the second from $j'' = 2$ and 1. If we also specify the value of j' from which $j = 1$ is derived, the set of functions thus obtained will be a linear combination of the functions coming from $j'' = 2$ and from $j'' = 1$.

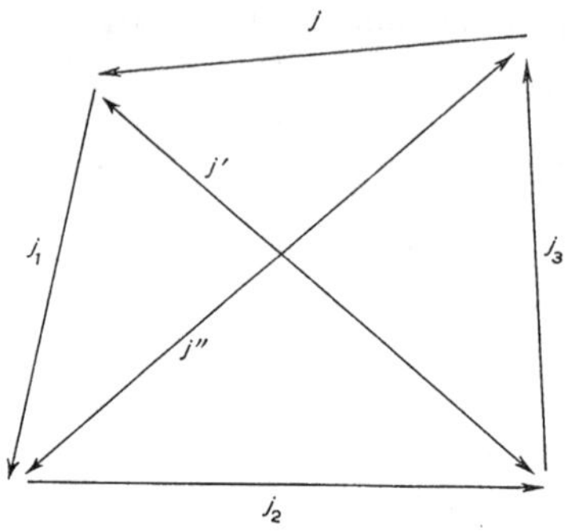

Fig. 6.1. The coupling of four angular momenta.

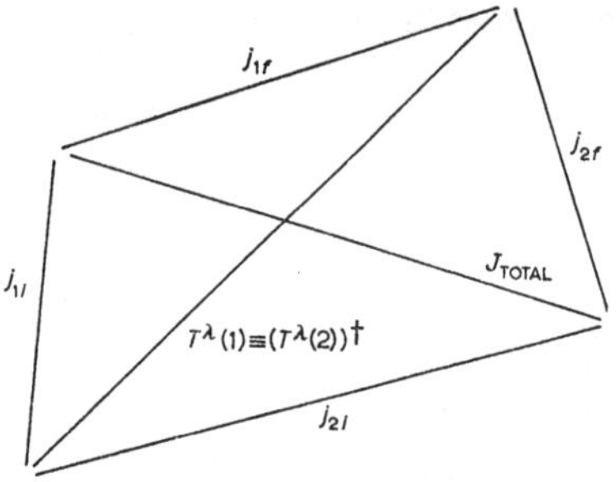

Fig. 6.2. The matrix elements of a scalar bilinear operator compared with the previous scheme.

By working out both coupling schemes the coefficient R can be expressed in terms of a product of four vector coupling coefficients. Some of the summations can be worked out explicitly; but one still remains in the final expression for the Racah coefficient. It is convenient, although to a certain extent misleading, to use a vector diagram. Fig. 6.1 expresses the fact that: $j' = j_1+j_2$; $j'' = j_2+j_3$ and $j = j_1+j_2+j_3$ (the third possibility $j''' = j_1+j_3$ does not lead to anything new).

The coefficient can also be used in describing the coefficient in front of an invariant coupling operator (6.14). Suppose we have a wave function characterized by j and suppose we break this j-value up into j_{1i} and j_{2i}. The tensor operator will take j_{1i} into j_{1f}, or in other words, the tensor operator has non-zero matrix elements between these states. The same holds for that part of the bilinear invariant that acts on j_2. The final result is a set of states j_{1f} and j_{2f} which will be brought together to j_{final}. Since the tensor coupling chosen was invariant, i.e. $T^\lambda(1)$ transforms contragradient to $T^\lambda(2)$, the resulting j_{final} will be the same as the initial one. However, the way in which j

was decomposed into j_{1i} and j_{2i} is not uniquely determined, and all the different values of j_{1i} (to each of which belongs a certain j_{2i}) will give a contribution to matrix elements of the bilinear operator. Each of these contributions has a coefficient which is completely determined by the rotational transformation properties. They all have the same proportionality constant: the "strength" of the coupling. The result of the decoupling and recoupling is equal to the product of this "strength" constant, denoted by $\langle\|T^\lambda(1)\|\rangle\langle\|T^\lambda(2)\|\rangle$, and the Racah coefficient. Hence, if we take the contracted product of two contragradient tensor operators, like (6.14), the dependence of this invariant on the quantum numbers j is determined by a Racah coefficient.

The Racah coefficient can be expressed as a sum of products of four Clebsch-Gordan coefficients, following the recoupling diagram of Fig. 6.2

The first coupling $j_1+j_2 = j'$ is expressed by

$$\psi_{j'm'} = \sum_{\substack{m_1 m_2 \\ (m_1+m_2=m')}} C^{j'}_{m_1 m_2} \psi_{j_1 m_1} \psi_{j_2 m_2} \tag{6.22}$$

and coupling the resultant with j_3 gives:

$$\psi_{jm}(j') = \sum_{\substack{m' m_3 \\ (m'+m_3=m)}} C^{j}_{m' m_3} \psi_{j'm'} \psi_{j_3 m_3}. \tag{6.23}$$

Substituting (6.22) into (6.23) gives,

$$\psi_{jm}(j') = \sum_{m_1}\sum_{m_2} C^{j}_{m', m-m'} C^{j'}_{m_1, m'-m_1} \psi_{j_1 m_1} \psi_{j_2 m'-m} \psi_{j_3 m-m'}. \tag{6.24}$$

In the same way $\psi_{jm}(j'')$ can be found and (6.21) will take the following form

$$\sum_{m_1}\sum_{m'} C^{j}_{m', m-m'} C^{j'}_{m_1, m'-m_1} \psi_{j_1 m_1} \psi_{j_2 m'-m_1} \psi_{j_3 m-m'}$$
$$= \sum_{j''} R_{j'j''} \sum C^{j}_{m'', m-m''} C^{j''}_{m_2, m''-m_2} \psi_{j_2 m_2} \psi_{j_3 m''-m_2} \psi_{j_1 m-m''}. \tag{6.25}$$

Multiplying by $\psi_{j_1\mu_1}\psi_{j_2\mu_2}\psi_{j_3\mu_3}$ will give a set of δ functions at the left-hand side which make $m_1 = \mu_1$; $m'-m_1 = \mu_2$ and $m-m' = \mu_3$, i.e. there are only two summations over the μ's since the third is determined by $\mu_1+\mu_2+\mu_3 = m$ and similarly on the right-hand side. The result is:

$$C^{j_1}_{\mu_1\mu_2} C^{j}_{\mu_1+\mu_2, \mu_3} = \sum_{j''} R_{jj''} C^{j''}_{\mu_2\mu_3} C^{j}_{\mu_1, \mu_2+\mu_3}.$$

Using the orthogonality of the C.-G. coefficients (Chapter 5, Appendix III; note that orthogonality is with respect to the superscript and one of the subscripts) we find

$$R_{jj''} = \sum_{\mu_1}\sum_{\mu_2}\sum_{\mu_3}_{(\mu_1+\mu_2+\mu_3=m)} C^{j'}_{\mu_1\mu_2} C^{j}_{\mu_1+\mu_2,\,\mu_3} C^{j}_{\mu_1,\,\mu_2+\mu_3} C^{j''}_{\mu_2\mu_3}. \tag{6.26}$$

The resulting coefficient depends on j_1, j_2, j_3, j, j' and j'', but not on m, since the position of the tetraeder of Fig. 6.1 in space is irrelevant. The Racah coefficient is usually defined as

$$W(j_1j_2jj_3;j'j'') = [(2j+1)(2j''+1)]^{-\frac{1}{2}}R_{j'j''}. \tag{6.27}$$

The equation (6.26) and the two summations can be reduced to one by some complicated algebra. For this we refer to the original papers of Racah. It has become customary to replace the W-coefficient by the so-called Wigner 6-j symbol:

$$\begin{Bmatrix} j_1 & j_2 & j' \\ j_3 & j & j'' \end{Bmatrix} = (-1)^{j_1+j_2+j_3+j}W(j_1j_2jj_3;j'j'').$$

This notation is convenient because it can be easily generalized. The 3-j symbol or symmetrized Clebsch-Gordan coefficient and the 6-j symbol are the first two members of a family of $3n$-symbols. They refer to the coupling

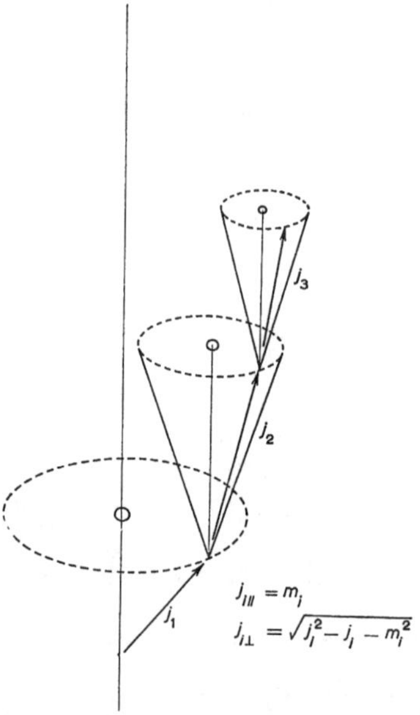

Fig. 6.3. Addition of vectors of given length and given z-component.

of 2 angular momenta ($n = 1$), 3 angular momenta ($n = 2$), 4 angular momenta ($n = 3$), etc.

The W-coefficients have a number of symmetry properties, which reduce the number of entries, but their calculation is nevertheless rather tedious particularly for higher angular momenta. For very high angular momenta they approach the classical vector formula. Fortunately a large number of tables is available.

Finally we would like to make a comment on Figure 6.1. Instead of thinking of angular momenta as classical vectors, an inheritance from pre-quantum mechanical days, we can imagine them as cones produced by rotating vectors. This picture helps the imagination since it stresses the fact that $\boldsymbol{J}_x$ and $\boldsymbol{J}_y$ are not determined, but that $\boldsymbol{J}$ and $\boldsymbol{J}_z$ are. It is only necessary to add the "numerology" that is the fact that $\boldsymbol{J}_z$ is an integer and $\boldsymbol{J}^2 = j(j+1)$ to complete the situation to a correct description. It is obvious from the picture that for given total j and m the orientation of the vectors $j_1 j_2$ and j_3 is not completely fixed. So that for a particular $j' = j_1 + j_2$, a number of different values for j'' can be assigned. The main conclusion from Fig. 6.3 is that Fig. 6.1 has to be regarded as an tetraeder instead of a two-dimensional diagram.

In finite groups the Racah-coefficient becomes quite complicated. The first coupling may already contain multiple occurring representations, that is the Clebsch-Gordan coefficient needs more indices than the rotational case. The main practical difficulty, however, is the lack of a closed formula for the basis function of the different irreducible representations.

CHAPTER 7

SPACE GROUPS

1. Outline

In this and the following chapter we will display those groups that are used in non-relativistic physics in general and in quantum mechanics in particular.

The most important group is the *full rotation group*. The extensive treatment of the representations of this group was given in two of the previous chapters. The full rotation group is used in the description of isolated atoms and ions. It also plays a role in the basic description of the wave function in general, irregardless of the symmetry, since the representations of the full rotation group serve as a starting point for almost every calculation in quantum mechanics.

In dealing with crystalline solids, however, another type of group has to be considered. The most striking property of a crystal is the regularity of its structure. If this structure is perfect, i.e. no errors, infinite in size and free of temperature agitation, this property can be expressed by saying that certain translations are allowed. It is to be expected that a number of physical properties are invariant under the operations of this group and so it would be interesting to study the group properties.

Besides these translations, rotations of the crystal that will bring the lattice into complete covering with itself are possible and the group of translations may be extended to a larger group, the so-called *space group*, which contains both. The meaning of this synthesis in group theoretical terms will be discussed in § 4. First the space group will be considered as a totality and then analyzed in rotational and translational parts.

The space groups contain elements which represent pure translations, elements which represent rotations around a center and elements that represent both. The elements representing rotations will be shown to form a subgroup. These subgroups are called the *crystalline point groups*.

The third set of groups of interest in quantum mechanics are the *point groups*, i.e. the groups representing a finite set of rotations around a center. These are used in molecular problems. The crystalline point groups are a special case of the general point groups. For instance, the five rotations over $\frac{2}{5}\pi$ around a certain axis form a point called C_5. This is not a crystalline point group since it is impossible to construct a periodic structure that contains a five-fold axis, since one cannot cover a plane completely with pentagons.

Most molecules happen to belong to the same point groups that form the crystalline point groups. There are exceptions and nature seems not even to "abhor" five-fold symmetry. There is a substance called Ferrocine that contains a 5-carbon ring. The actual shape of the molecule is somewhat like a sandwich of two of these rings with an iron atom in the middle such that the symmetry around an axis perpendicular to the rings is ten-fold.

The relations between the sets of groups are as follows:

```
1. Full rotation group          2. Space group
         ↓                          /      ↓
3. Point groups                    /   4. Pure translation
              ↘                   ↙
           2′ Crystallographic point groups
```

where each arrow indicates a subgroup.

§§ 1 to 4 deal with the classical space group, § 5 and on with quantum mechanics.

2. Crystallographic Point Groups versus General Point Groups

In order to find out which point groups are crystallographic and which are not it is necessary to translate the restrictions imposed on the crystallographic point groups into mathematical terms. There are two different ways to do this. The first is to construct geometrical bodies, of a certain symmetry and to see whether these "building blocks" can be stacked in such that they fill infinite space. These objects are, of course, the unit cells which fill the space of the total lattice.

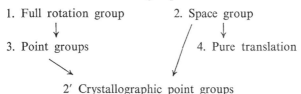

A different and much simpler approach is to make use of the properties of representations. This will be demonstrated for the simple case of a two-dimensional lattice and will serve merely as an illustration for the three-dimensional case. As mentioned before the definition of a point group is: a set of transformations such that the lattice goes over into itself, under the restriction that one point is kept fixed. This group contains a finite number of elements. Consider a certain point P (not identical with O) under all operations of the group (rotations, reflections). This point will remain the same distance from O: OP = OP′. If the group would be infinite all the points P′ would cover a circle. From consideration of the space group it will be seen that any point in the lattice can be described by a set of integers. Hence all operations can be described by a representation which uses only integers. This is again true for the point group contained in the space group. Any representation of the crystallographic point group is equivalent to a finite group of linear substitutions among integers. This important conclusion, which is fully described in the next section, leads in a very simple way to the restrictions on the order of the group (compare BURCKHARDT [1947]).

An arbitrary orthogonal rotation in two dimensions

$$\begin{aligned} x' &= x\cos\varphi - y\sin\varphi \\ y' &= x\sin\varphi - y\cos\varphi \end{aligned} \tag{7.1}$$

can be conveniently expressed with the linear combinations $x+\mathrm{i}y$ and $x-\mathrm{i}y$, which gives rise to a transformation matrix

$$\mathsf{D} = \begin{pmatrix} \mathrm{e}^{\mathrm{i}\varphi} & 0 \\ 0 & \mathrm{e}^{-\mathrm{i}\varphi} \end{pmatrix}.$$

As stated above, every representation is equivalent to a representation with integers. Hence the trace of the above matrix, which is an invariant, should always be an integer,

$$\mathrm{e}^{\mathrm{i}\varphi} + \mathrm{e}^{-\mathrm{i}\varphi} = n. \tag{7.2}$$

Since

$$|n| \leqq |\mathrm{e}^{\mathrm{i}\varphi}| + |\mathrm{e}^{-\mathrm{i}\varphi}| = 2 \tag{7.3}$$

the possible values of n are $n = -2, -1, 0, 1, 2$. Let $\mathrm{e}^{\mathrm{i}\varphi} = \varepsilon$ then multiplying (7.2) with ε gives,

$$\varepsilon^2 - n\varepsilon + 1 = 0, \tag{7.4}$$

an equation which can be solved for the five different values of n:

$$\begin{aligned} n &= -2 \text{ gives } \varepsilon = -1 \\ n &= -1 \text{ gives } \varepsilon = \tfrac{1}{2}(-1 \pm \mathrm{i}\sqrt{3}) \\ n &= 0 \text{ gives } \varepsilon = \pm \mathrm{i} \\ n &= 1 \text{ gives } \varepsilon = \tfrac{1}{2}(1 \pm \mathrm{i}\sqrt{3}) \\ n &= 2 \text{ gives } \varepsilon = 1. \end{aligned} \tag{7.5}$$

Thus the cyclic groups can only be of order 1, 2, 3, 4, and 6.

It is possible to explore by similar considerations the number of possible crystallographic point groups in three dimensions. For this analysis we refer to the literature (compare BURCKHARDT [1947] §§ 11 and 12).

There are two conventions in use for the indication of (crystallographic) point groups. The first one was originally the Herman-Mauguin notation, now called the International notation; the second is the Schoenfliess notation. The last one consists of a capital and one (or more) subscripts. The capital letter is either C, D, S, T or O, standing for Central, Dihedral, Screw, Tetrahedral and Octohedral. The subscript n refers to the order of the axis; for instance C_5 refers to an abelian group of order five, the five rotations over $\frac{2}{5}\pi$. Dihedral is the symmetry of a "two-plane" that is a body from which we consider back and front side equivalent; for instance D_3 is the group of 6 elements consisting of rotations around the 3-fold axis and the three flip-over operations around the bisectrices which we considered in Chapter 3, § 7.4. It is the custom to consider the main axis placed in the vertical positions and hence the two-fold axes are called horizontal axes.

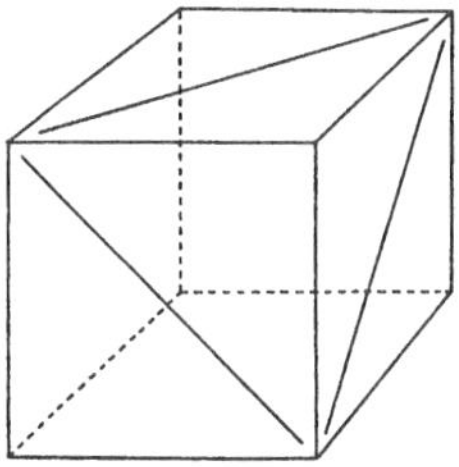

Fig. 7.1. A tetrahedron can be imbedded in a cube showing that its rotation group has to be a subgroup of the cubical group.

Additional indices are attached to C and D if we include the improper rotations: i for the inversion, v for the reflection with respect to a vertical plane and h for the reflection with respect to the horizontal plane. S indicates a screw axis, a combination of translation and rotational motion. This possibility is only mentioned for completeness; it does not belong to the point group operations. The groups T and O are the only two point groups in which the axis of highest symmetry occurs more than once, that is, in all other crystallographic point groups there is always one dominant axis which we took vertical. If we do not restrict ourselves to the crystallographic point groups there is still one more group. The dodecaeder group, consisting of 10 pentagons. This group has 60 elements.[1]

[1] The octahedron is obtained by connecting all midpoints of the faces of the cube, hence has the same group. A similar reciprocal relation exists for the icosahedron with respect to the dodecahedron. This, together with the tetrahedron (which is self reciprocal) completes all five possible regular bodies in three dimensions.

As a memory devise geometrical objects can be constructed which have the proper symmetry. Some pictures are indicated in this section. In particular Fig. 7.1 serves to illustrat· that T is a subgroup of O. Another method is the stereographic projection. Table 8.2 gives all the crystallographic point groups. The extension to non-crystallographic point groups is easy to establish and can be found in the chemical literature.[1] A third method using pieces of cardboard has been described by SCHIFF [1954].

3. Space Groups

After the introductory remarks in § 1 a notation must be established for the elements of the space group. These elements consist of a combination of a rotation and a translation. By rotation we mean a pure rotation, a reflection, an inversion or a combination of these. It should also be pointed out that an element may be a product of a rotation without translation and a translation without rotation, but that this is not necessarily so.

In certain space groups there are screw elements, that is operations consisting of a translation and a rotation which is *not* a product of a translation without rotation and a rotation without translation where both are elements of the group themselves. The screw elements do occur in nature, for instance, we know of the existence of optical effects, such as the rotation of the polarization plane. This occurs if the index of refraction for left circular polarized light is different from the index for right polarized light.

The most general translation allowed in a three-dimensional crystalline structure is described with help of three *primitive translation vectors* $\boldsymbol{a}$, $\boldsymbol{b}$, $\boldsymbol{c}$.

These are three non-coplanar or independent vectors, that is

$$\boldsymbol{a} \cdot (\boldsymbol{b} \times \boldsymbol{c}) \neq 0 \tag{7.6}$$

chosen in such a way that they represent the smallest possible translations. We will see later that there is a certain ambiguity in the choice of these vectors although not quite as large as the choices of basic vectors in a vector space.

The general translation is

$$\boldsymbol{T} = n_1 \boldsymbol{a} + n_2 \boldsymbol{b} + n_3 \boldsymbol{c}, \tag{7.7}$$

where $n_1, n_2, n_3 = 0, \pm 1, \pm 2, \ldots$.

The pure translations are not the only possible operations allowed in a

[1] EYRING, WALTER and KIMBALL [1944], page 379.

crystalline structure. There may be certain rotations and reflections possible too. We will indicate these in general by a matrix (R_{ij}). Hence the operations of the space group can be written as

$$\begin{aligned} x_1' &= R_{11}x_1+R_{12}x_2+R_{13}x_3+T_1x_1 \\ x_2' &= R_{21}x_1+R_{22}x_2+R_{23}x_3+T_2x_2 \\ x_3' &= R_{31}x_1+R_{32}x_2+R_{33}x_3+T_3x_3\,. \end{aligned} \tag{7.8}$$

Or in a more convenient notation:

$$\boldsymbol{x}' = (\mathrm{T}|\mathrm{R})\boldsymbol{x} = \mathrm{A}\boldsymbol{x}. \tag{7.8'}$$

(This differs slightly from the convention used by SEITZ [1934] where the rotation is written on the left, the translation on the right in order to preserve the usual order of operations: from right to left.)

A possible representation of A can be obtained as follows: The operation A on any T should result in a general translation. In particular if we operate on a primitive translation vector

$$\mathrm{A}\boldsymbol{a} = n_1^{(a)}\boldsymbol{a}+n_2^{(a)}\boldsymbol{b}+n_3^{(a)}\boldsymbol{c}$$

the integers $n_j^{(a)}$ considered as a three-by-three matrix form a representation of A.

The main characteristics of the elements A are as follows:

1) The Unit element is

$$\mathrm{E} = (0|\delta_{ij}) \quad \text{or} \quad (0|\mathrm{E}) \tag{7.9}$$

2) The product of two or more operations is given by

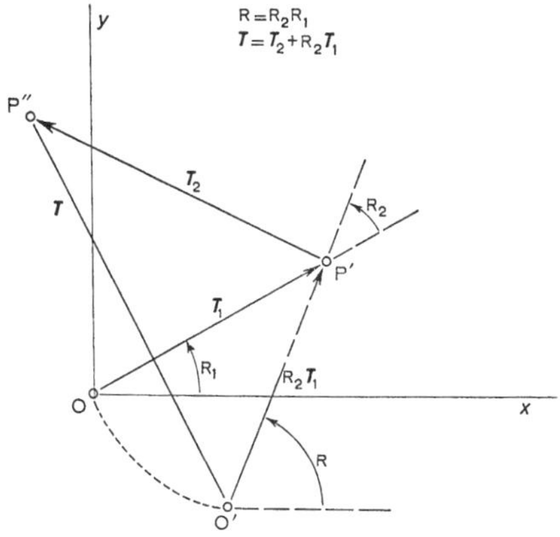

Fig. 7.2. Illustration of the product rule for space group elements.

$$(T_2|R_2)(T_1|R_1) = (T_2+R_2T_1|R_2R_1). \tag{7.10}$$

The product rule can be illustrated with a simple diagram (Fig. 7.2). A certain point taken initially at O and a certain direction in the crystal taken parallel to the x-axis undergo a rotation R_1 and translation T_1. This is followed by a second operation $(T_2|R_2)$. The translation axes *are fixed in the crystal*, hence the second rotation will rotate the vector $\boldsymbol{T}_1$, or to say it differently, the second operation takes place as if the origin had been at O′. Hence the total translation is not $\boldsymbol{T}_2+\boldsymbol{T}_1$ but $\boldsymbol{T}_2+R_2\boldsymbol{T}_1$.

3) The *inverse* element is

$$(T|R)^{-1} = (-R^{-1}T|R^{-1}). \tag{7.11}$$

This rule can be derived from (7.9) and (7.10).

Pure translations are characterized by

$$(n_1\boldsymbol{a}+n_2\boldsymbol{b}+n_3\boldsymbol{c}|\delta) = (n_1\boldsymbol{a}|\delta)(n_2\boldsymbol{b}|\delta)(n_3\boldsymbol{c}|\delta). \tag{7.12}$$

Due to their Abelian character, a group element of this type can be unambiguously written as the product of three group elements, each representing a translation in a certain direction.

4) The associative law holds: This proof will complete the demonstration that the combination of rotations and translations indeed form a group. Let the operator $(T_3|R_3)$ act on the product of two group elements,

$$(T_2|R_2)(T_1|R_1) = (T_2+R_2T_1|R_2R_1).$$

Then the result is,

$$(T|R) = (T_3+R_3(T_2+R_2T_1)|R_3(R_2R_1)). \tag{7.13}$$

Now consider the product of 3 and 2 acting on 1,

$$(T|R) = ((T_3+R_3T_2)+R_3R_2(T_1)|(R_3R_2)R_1). \tag{7.13'}$$

This gives the same result.

The space group will contain elements that are rotations only. These elements form a subgroup: the multiplication of two of these elements

$$(0|R_2)(0|R_1) = (0|R_2R_1)$$

gives an element without translation. This subgroup is not invariant, since

$$\begin{aligned}(T|R)(0|R')(T|R)^{-1} &= (T|R)(0|R')(-R^{-1}T|R^{-1}) \\ &= (T-RR'R^{-1}T|RR'R^{-1})\end{aligned} \tag{7.14}$$

using (7.11) and (7.13). This is clearly *not* an element of the point group (except in the trivial case that $R' = E$). This result is independent of the presence or absence of screw axes.

The triple product (7.14) can also be illustrated with Fig. 7.1. The second operation will transform O into O′ and the third operation will translate P′ backwards to O′, instead of to O with the result that the triple product does not represent a pure rotation.

On the other hand, if pure translations are considered it is immediately seen that they form a subgroup (Abelian), but this subgroup is invariant.

$$(T|R)^{-1}(T'|E)(T|R) = (R^{-1}T'|E) \tag{7.15}$$

using (7.10) and (7.11). The result follows immediately from (7.13) since the total rotation is

$$R_3R_2R_1 = R^{-1}ER = E,$$

without considering the translational part.

The factor group $\mathscr{G}/\mathscr{T}$, (where $\mathscr{G}$ represents the space group and $\mathscr{T}$ the invariant subgroup of pure translations) is isomorphic with the point group.

This holds irrespective of the fact whether the space group contains "screw" elements or not. A screw element is a rotation accompanied by a non-primitive translation: $(\tau(R_i)+T|R_i)$. Here τ is a fraction of T and depends on R_i. Conversely R_i depends on τ and R_i may or may not occur as a pure rotation, i.e. $(0|R_i)$ may exist or not. However, $(\tau|I)$ cannot exist, since τ is not primitive.

Consider the case that $(0|R_i)$ does not exist, that is R_i can only occur in conjunction with a non-primitive translation. In this case the point group to which $\mathscr{G}/\mathscr{T}$ is isomorphic will contain elements that represent rotations around the lattice points T and different elements around $\tau+T$.

The following is an example of a structure containing screw elements.

Consider the so-called diamond or zinc-blende structure, the crystalform of germanium. The conventional way of describing this lattice is to consider it as two face centered lattices displaced over a distance $\frac{1}{4}, \frac{1}{4}, \frac{1}{4}$. The tetrahedral bond arrangement in one lattice can be obtained from the other by an inversion. The "screw" element consists in this case of a translation over $\frac{1}{4}\sqrt{3}$ along the body diagonal (which is not an allowed translation by itself) combined with the improper "rotation": inversion. If we carve a Wigner-Seitz unit cell around these two (non-equivalent) positions by constructing planes halfway perpendicular to the bonds with the nearest neighbors we see easily that this was the only screw element possible. In a zinc-blende crystal, which has the same structure except that one type of site is filled with a Zn and the other with an S, the screw element is excluded.

A tedious analysis shows that there are 230 different space groups. They are a combination of one of the 23 point groups and one of the 14 different translation patterns. A translation pattern is illustrated by a space lattice or a Bravais lattice. These Bravais lattices are the space groups in which there is one atom per primitive unit cell (i.e. the unit cell spanned by the three primitive unit vectors). In order to bring out the *system* (a system is characterized by the number of 2-, 3-, 4- or 6-fold axes) to which these Bravais lattices belong, the so-called *conventional unit cell is used.* This cell has a volume twice or three times the smallest unit cell (compare PHILLIPS [1956]). In quantum mechanics it is convenient to introduce a more symmetrical unit cell, the Wigner-Seitz cell mentioned in § 6.

4. Structure of the Space Group

The synthesis of two groups of order g_1 and g_2 into a group of order $g_1 g_2$ can be accomplished in different ways, depending on the nature of the elements of each group. This problem has its counterpart in physics since it happens in a number of cases that a physical meaning can be assigned to the elements of the first group and a different physical meaning to the elements of the second group. For instance the rotations in space and the permutations of electrons are both connected with a group. The system is invariant under the combined group of spatial rotations and permutations (compare note on page 120).

In many, but by no means all, physical cases the elements of both groups commute which makes them absolutely independent. This type of synthesis, where all elements of $\mathscr{G}_1$ commute with all elements of $\mathscr{G}_2$, is called a *direct product* of two groups. The product group $\mathscr{G}$ contains $\mathscr{G}_1$ and $\mathscr{G}_2$ as invariant subgroups and the factor group with respect to one is isomorphic to the other:

$$\mathscr{G}/\mathscr{G}_1 \sim \mathscr{G}_2; \quad \mathscr{G}/\mathscr{G}_2 \sim \mathscr{G}_1.$$

The converse statement holds also. If (i) $\mathscr{G}_1$ and $\mathscr{G}_2$ are invariant subgroups of $\mathscr{G}$, (ii) the elements of $\mathscr{G}$ consist of the products of the elements of $\mathscr{G}_1$ and the elements of $\mathscr{G}_2$ and (iii) the elements of $\mathscr{G}_1$ and $\mathscr{G}_2$ have only the unit element in common, then the group $\mathscr{G}$ is the direct product of $\mathscr{G}_1$ and $\mathscr{G}_2$. That is a given element of $\mathscr{G}$ can be written as a product:

$$\mathrm{G} = \mathrm{G}_1 \mathrm{G}_2 = \mathrm{G}_2 \mathrm{G}_1,$$

where both G_1 and G_2 are determined by G. (VAN DER WAERDEN [1937] § 47.)

There are less restricted ways to form products of groups: the weak direct product and the semidirect product. Both are generalizations of the direct product mentioned above, but their nature is entirely different.

The weak direct product (MELVIN [1956]) has instead of commutability, the requirement that the product of an element of $\mathcal{G}_1$ with an element of $\mathcal{G}_2$ should give an element of $\mathcal{G}$ such that

$$G_1 G_2 = G = G_2' G_1'$$

where G_2' is another element of $\mathcal{G}_2$ and G_1' another element of $\mathcal{G}_1$, requirements (i) and (iii) are maintained. The weak direct product is the most general product, because it does not require any subgroup to be normal (compare problem).

A third type of direct product is the *semidirect product* (MCINTOSH [1958a].) Let $\mathcal{R}$ and $\mathcal{G}$ be two groups and suppose $\mathcal{R}$ is an operator group for $\mathcal{G}$. That means that an element of $\mathcal{R}$ operating on an element of $\mathcal{G}$ results in an element that belongs again to $\mathcal{G}$. A group can be defined by stating that its elements will consist of a combination of one element of $\mathcal{R}$ and one element of $\mathcal{G}$ such that the product is defined by the rule

$$(G'|R')(G|R) = (G' \cdot R'(G)|R' \cdot R) \tag{7.16}$$

where $R'(G)$ means the operator R' works on G. In the application of this chapter G is the group of translations, an Abelian group, and hence the first product on the right-hand side is usually written as a sum.

In general, that is whether the group $\mathcal{G}$ is Abelian or not we can state that the subgroup $\mathcal{G}$ is invariant and that the subgroup $\mathcal{R}$ is not, except for trivial cases.

The requirements for a group to be a semidirect product are (i) the existence of a non-trivial invariant subgroup. If all cosets of this invariant subgroup are formed it should be possible (ii) to select an element from each coset such that the chosen elements plus the unit element form a group.

Examples of semidirect products are:

(i) The double group (see Chapter 8, § 3). (ii) The space group without screw elements. This group is isomorphic to the semidirect product of the group (T|E) and the group (0|R).

An example of a combination which does not fit any of the categories is the combination of Lorentz transformations without rotations and the rotations without Lorentz transformations. In this case the first set of operations does not even form a subgroup.

5. The Quantum Mechanics of Solid State

In nearly all quantum mechanical problems we want to use wave functions that are adapted as much as possible to the symmetry of the problem. The space groups provide a way to study the transformation properties of the wave functions both with respect to rotations and translations. The structure of the space group provides us not only with some ideas for the best eigenfunctions with which to start in a given perturbation calculation, but with a well-defined set of quantum numbers as well. In such a study two questions arise immediately. First, what kind of space groups are possible, a question considered briefly in § 3; and second, what are the irreducible representations of these groups in unitary space, a question to be considered in §§ 6, etc.

Before we do this we would like to dwell a moment upon the most reasonable choice for the zeroth order wave function in a solid state problem. This question is directly connected with the type of binding [1]: Ionic, Covalent or Metallic. In the case of ionic binding the wave function will be such that the electron from one atom spends most of its time around another atom. In the case of covalent binding the picture is more like an even distribution over the two atomic centers with an increased charge density in between. This density is the result of the exchange integral and responsible for the binding effect of such a wave function. In the case of metallic binding the wave function is spread out all over the crystal. This is called, in chemistry, a non-localized bond, in contrast to the first two which are pair interactions. The famous example of a non-localized bond is the benzene molecule.

The most extensive use of the group theory of translation is made in considerations on metallic binding, since here we are really dealing with wave functions spread out all over the crystal. There are, however, situations in which the application of group theory is of importance to covalent solids; "excitons" or non-localized excitations. As a matter of fact the distinction between these two is vague, since on one hand the metallic binding is associated with bands, but on the other hand covalent pair bond will give rise to a narrow band as well if we take all possible pairs in the crystal. Illustrated below is a simple example where there is no fundamental distinction between the different types of binding or to formulate it differently all degrees of

[1] See for instance SLATER [1953].

covalency, ionicity or metallicity are possible. This is no surprise if it is realized that each of these binding types are related to a certain type of zeroth order wave function.

Consider two electrons 1 and 2 and two centers a and b:

$$\varphi_a(\boldsymbol{r}) = \varphi(\boldsymbol{r}-\boldsymbol{R}_a), \tag{7.17a}$$

$$\varphi_b(\boldsymbol{r}) = \varphi(\boldsymbol{r}-\boldsymbol{R}_b). \tag{7.17b}$$

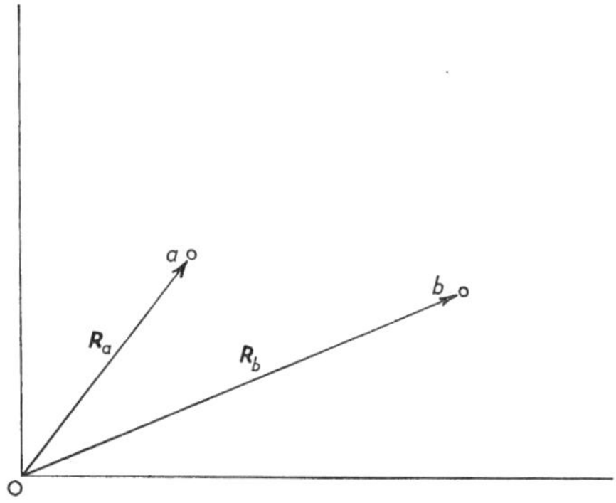

Fig. 7.3. Two center coordinate system used for diatomic molecules.

The two wave functions are the same functions if taken with respect to their centers. The most general zeroth order (i.e. product type) wave function which can be set up is

$$\psi = C_1\varphi_a(1)\varphi_a(2)+C_2\varphi_b(1)\varphi_b(2)+C_3\varphi_a(1)\varphi_b(2)+C_4\varphi_a(2)\varphi_b(1) \tag{7.18}$$

where C_1, C_2, C_3 and C_4 are arbitrary coefficients.

In the ionic case both electrons are at one center, say b. Hence $C_1 \ll C_2$ and, since exchange is disregarded, $C_3 = C_4 = 0$.

In the covalent case take $C_1 = C_2 = 0$ with $C_3 = -C_4$, the only possible combination that is antisymmetric. The binding is the result of a spin-zero state and since the total wave function has to be anti-symmetric, the orbital part has to be anti-symmetric too. This type of wave function gives rise to exchange integrals and as a result of that the charge density at $\frac{1}{2}(\boldsymbol{R}_a+\boldsymbol{R}_b)$ will increase.

In the metallic binding we deal with a product of molecular orbitals. If these are taken as a linear combination of atomic orbitals (L.C.A.O.), which is a simple and convenient choice, we have

$$\varphi_{\mathrm{M}} = (\varphi_a+\varphi_b)/\sqrt{2} \tag{7.19}$$

and

$$\psi = \varphi_M(1)\varphi_M(2) = \tfrac{1}{2}(\varphi_a(1)+\varphi_b(1))(\varphi_a(2)+\varphi_b(2))$$

or

$$C_1 = C_2 = C_3 = C_4 .$$

To summarize this the following list for the two electron two center case is given:

	C_1	C_2	C_3	C_4
Ionic	large	small	0	0
Convalent	0	0	1	1
Metallic	1	1	1	1

It is assumed that the N-electron N-center case shows the same characteristics. This undoubtedly is true for the ionic case, to a lesser extent for the covalent case. The metallic case needs considerable improvement and it is here that the fullest use is made of the space group.

6. Pure Translations

Returning to our main problem, i.e. to what representation of a certain space group does a certain wave function belong, consider an element $\mathrm{S} \equiv (\mathrm{T}|\mathrm{R})$ of the space group $\mathcal{G}$,

$$\mathrm{S}\boldsymbol{r} = \boldsymbol{r}' = \mathrm{R}\boldsymbol{r}+\boldsymbol{T}, \tag{7.20}$$

which induces in the wave function space a transformation that can be written as a linear combination of basis functions

$$\mathrm{S}\psi_i(\boldsymbol{r}) = \psi(\boldsymbol{r}') = \sum_{j=1} S_{ij}\psi_j(\boldsymbol{r}). \tag{7.21}$$

If irreducible representations are sought, it is noticed that the invariant subgroup of pure translations $\boldsymbol{T}_n = n_1\boldsymbol{a}+n_2\boldsymbol{b}+n_3\boldsymbol{c}$ is Abelian. Hence the irreducible representations of this subgroup are one-dimensional. The result of a translation is

$$\mathrm{T}_n\psi_i(\boldsymbol{r}) = \tau\psi_i(\boldsymbol{r}) \tag{7.22}$$

where τ is a number, in general complex. If the three primitive translations are introduced, the vectors are considered as operators,

$$\mathrm{T}_{100}\psi = \mathrm{a}_{\mathrm{op}}\psi(\boldsymbol{r}) = \alpha\psi(\boldsymbol{r}) \quad (7.23a)$$
$$\mathrm{T}_{010}\psi = \mathrm{b}_{\mathrm{op}}\psi(\boldsymbol{r}) = \beta\psi(\boldsymbol{r}) \quad (7.23b)$$
$$\mathrm{T}_{001}\psi = \mathrm{c}_{\mathrm{op}}\psi(\boldsymbol{r}) = \gamma\psi(\boldsymbol{r}) \quad (7.23c)$$

and then we can calculate the three coefficients α, β and γ. If periodic boundary conditions are involked, which is not necessary but convenient, we find

$$(\mathrm{a}_{\mathrm{op}})^{N_1}\psi(\boldsymbol{r}) = \alpha^{N_1}\psi(\boldsymbol{r}) = \psi(\boldsymbol{r}) \rightarrow \alpha^{N_1} = 1 \quad (7.24a)$$
$$(\mathrm{b}_{\mathrm{op}})^{N_2}\psi(\boldsymbol{r}) = \beta^{N_2}\psi(\boldsymbol{r}) = \psi(\boldsymbol{r}) \rightarrow \beta^{N_2} = 1 \quad (7.24b)$$
$$(\mathrm{c}_{\mathrm{op}})^{N_3}\psi(\boldsymbol{r}) = \gamma^{N_3}\psi(\boldsymbol{r}) = \psi(\boldsymbol{r}) \rightarrow \gamma^{N_3} = 1. \quad (7.24c)$$

α^{N_1}, etc. means that the operation of a primitive translation is repeated N_1 times on the wave function. The distance $N_1\boldsymbol{a}$ is the distance over which we consider the lattice periodically repeated in the $\boldsymbol{a}$-direction. Similarly $N_2\boldsymbol{b}$ and $N_3\boldsymbol{c}$ in the $\boldsymbol{b}$- and $\boldsymbol{c}$-direction.

The solutions of these three equations are,

$$\alpha^{N_1} = 1 \rightarrow \alpha = \mathrm{e}^{\mathrm{i}h_1/N_1} \qquad (h_1/2\pi = 0, 1, 2, \ldots, N_1-1) \quad (7.25a)$$
$$\beta^{N_2} = 1 \rightarrow \beta = \mathrm{e}^{\mathrm{i}h_2/N_2} \qquad (h_2/2\pi = 0, 1, 2, \ldots, N_2-1) \quad (7.25b)$$
$$\gamma^{N_3} = 1 \rightarrow \gamma = \mathrm{e}^{\mathrm{i}h_3/N_3} \qquad (h_3/2\pi = 0, 1, 2, \ldots, N_3-1). \quad (7.25c)$$

Now we want to describe the result of an arbitrary translation: say n_1 steps in the $\boldsymbol{a}$ direction, n_2 steps in the $\boldsymbol{b}$ direction and n_3 steps in the $\boldsymbol{c}$ direction, i.e. a translation:

$$\boldsymbol{r} = n_1\boldsymbol{a}+n_2\boldsymbol{b}+n_3\boldsymbol{c}. \quad (7.26)$$

The result is easily obtained from the preceding considerations: it produces an eigenvalue

$$\tau = \exp \mathrm{i}(h_1 n_1/N_1+h_2 n_2/N_2+h_3 n_3/N_3).$$

This is not a very practical way to make this description since (7.26) actually describes $\boldsymbol{r}$ with respect to an oblique set of coordinate vectors $\boldsymbol{a}$, $\boldsymbol{b}$, and $\boldsymbol{c}$. Although n_1, n_2 and n_3 were originally thought of as integers, this restriction can be omitted without any trouble and hence they can be considered as the components of $\boldsymbol{r}$ in this coordinate system which is adapted to the crystal. However, $\boldsymbol{r}$ usually is expressed in rectangular coordinates

$$\boldsymbol{r} = \boldsymbol{e}_1 x+\boldsymbol{e}_2 y+\boldsymbol{e}_3 z. \quad (7.27)$$

Comparing the oblique components (7.26) with the rectangular components (7.27) we find

$$n_1 a_x + n_2 b_x + n_3 c_x = x \tag{7.28a}$$
$$n_1 a_y + n_2 b_y + n_3 c_y = y \tag{7.28b}$$
$$n_1 a_z + n_2 b_z + n_3 c_z = z. \tag{7.28c}$$

The inversion of this formula, i.e. the oblique components expressed in the orthogonal components, gives:

$$n_1 = V^{-1}\{(\boldsymbol{b}\times\boldsymbol{c})_x x + (\boldsymbol{b}\times\boldsymbol{c})_y y + (\boldsymbol{b}\times\boldsymbol{c})_z z\} \tag{7.29a}$$
$$n_2 = V^{-1}\{(\boldsymbol{c}\times\boldsymbol{a})_x x + (\boldsymbol{c}\times\boldsymbol{a})_y y + (\boldsymbol{c}\times\boldsymbol{a})_z z\} \tag{7.29b}$$
$$n_3 = V^{-1}\{(\boldsymbol{a}\times\boldsymbol{b})_x x + (\boldsymbol{a}\times\boldsymbol{b})_y y + (\boldsymbol{a}\times\boldsymbol{b})_z z\} \tag{7.29c}$$

where

$$V = \boldsymbol{a}\cdot(\boldsymbol{b}\times\boldsymbol{c}). \tag{7.30}$$

The coefficients form three vectors [1]

$$\boldsymbol{a}^* = \frac{\boldsymbol{b}\times\boldsymbol{c}}{\boldsymbol{a}\cdot(\boldsymbol{b}\times\boldsymbol{c})}; \quad \boldsymbol{b}^* = \frac{\boldsymbol{c}\times\boldsymbol{a}}{\boldsymbol{a}\cdot(\boldsymbol{b}\times\boldsymbol{c})}; \quad \boldsymbol{c}^* = \frac{\boldsymbol{a}\times\boldsymbol{b}}{\boldsymbol{a}\cdot(\boldsymbol{b}\times\boldsymbol{c})} \tag{7.31abc}$$

which have the following properties as is easily verified by equations (7.31abc)

$$\begin{array}{lll} \boldsymbol{a}^*\cdot\boldsymbol{a} = 1 & \boldsymbol{a}^*\cdot\boldsymbol{b} = 0 & \boldsymbol{a}^*\cdot\boldsymbol{c} = 0 \\ \boldsymbol{b}^*\cdot\boldsymbol{a} = 0 & \boldsymbol{b}^*\cdot\boldsymbol{b} = 1 & \boldsymbol{b}^*\cdot\boldsymbol{c} = 0 \\ \boldsymbol{c}^*\cdot\boldsymbol{a} = 0 & \boldsymbol{c}^*\cdot\boldsymbol{b} = 0 & \boldsymbol{c}^*\cdot\boldsymbol{c} = 1. \end{array} \tag{7.32}$$

Hence the expressions (7.28) can be written as:

$$n_1 = \boldsymbol{a}^*\cdot\boldsymbol{r}; \qquad n_2 = \boldsymbol{b}^*\cdot\boldsymbol{r}; \qquad n_3 = \boldsymbol{c}^*\cdot\boldsymbol{r} \tag{7.28'}$$

and the eigenvalue of an arbitrary translation can now be expressed as,

$$\tau = \exp \mathrm{i}(h_1\boldsymbol{a}^*/N_1 + h_2\boldsymbol{b}^*/N_2 + h_3\boldsymbol{c}^*/N_3)\cdot\boldsymbol{r}. \tag{7.29}$$

The numbers h_1, h_2 and h_3 could be used as labels for the different irreducible representations. Usually it is preferable to use:

$$\boldsymbol{k} = h_1\boldsymbol{a}^*/N_1 + h_2\boldsymbol{b}^*/N_2 + h_3\boldsymbol{c}^*/N_3. \tag{7.30}$$

Obviously this vector $\boldsymbol{k}$ can attain only a limited number of values. This is usually expressed in a geometrical way, i.e. by stating that $\boldsymbol{k}$ can only

[1] The asterisk cannot be confused with complex conjugate, since these vectors are real.

occupy the points of a certain unit cell in an abstract lattice. The lattice is called the inverse lattice, and is generated by multiples of the vectors $\boldsymbol{a}^*$, $\boldsymbol{b}^*$ and $\boldsymbol{c}^*$ in a space called the inverse space since the unit-length is the inverse of the conventional units of length, like cm^{-1} etc. The unit cell is mapped out by the condition for h_1, h_2 and h_3 mentioned in (7.25abc).

Technically speaking this unit cell is a half open point set, i.e. all points at the boundary at one side are included, all points at the opposite boundary plane are excluded. The reason that one should not take both ends is that they differ by a distance $\boldsymbol{a}^*$, $\boldsymbol{b}^*$, or $\boldsymbol{c}^*$ depending on the pair of planes one is considering and hence are equivalent points.

Instead of the condition (7.25 abc) any unit cell in the inverse lattice will do. The most elegant choice is the unit cell which was introduced by Wigner-Seitz in their calculations in the direct lattice. This cell is obtained by considering a lattice point and its neighbours and erecting planes half way, between each atom and its neighbours, and perpendicular to the connecting line. This unit cell is called the (first) Brillouin zone and would have been established directly if we had used the following range of h values:

$$\begin{aligned} h/2\pi &= 0, 1, \ldots, \tfrac{1}{2}(N-1);\quad \tfrac{1}{2}(N+1)-N, \ldots (N-1)-N \\ &= 0, \pm 1, \pm 2, \ldots \pm\tfrac{1}{2}(N-1) \end{aligned} \tag{7.25'}$$

obtained by subtracting N from the second set of values.[1]

The shape of the Brillouin zone is entirely determined by the symmetry of the original lattice and Figs. 7.4 give an illustration of the different lattices.

The general Brillouin zone is the locus determined by the points $\boldsymbol{k}$, such that

$$(\boldsymbol{k}+2\pi\boldsymbol{K})^2 = k^2, \tag{7.31}$$

where $\boldsymbol{K}$ is a vector connecting any pair of points of the reciprocal lattice.

7. Bloch Theorem

The basic point in symmetry considerations of quantum mechanical problems is that although certain operators and their expectation values must be invariants under the operators of the group, the wave functions are not necessarily invariant. Only the absolute value of the wave function (or in case of degeneracy the absolute value of a linear combination) has to be invariant. In the case of a non-degenerate wave function this leaves a certain freedom on the phase of the complex wave function.

[1] The equation (7.25′) holds only for N = odd; similar equations exist for N = even.

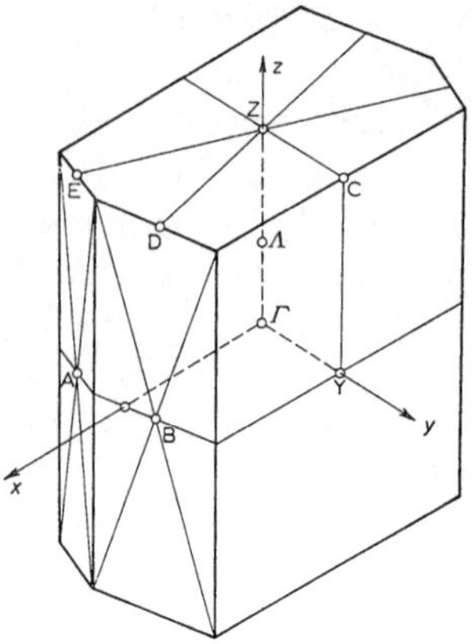

7.4(2)

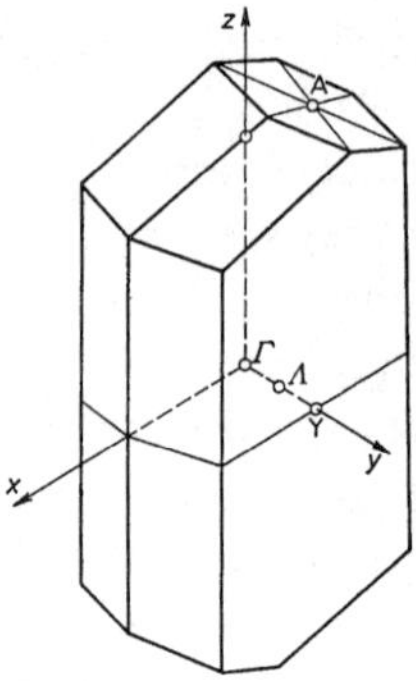

7.4(3)

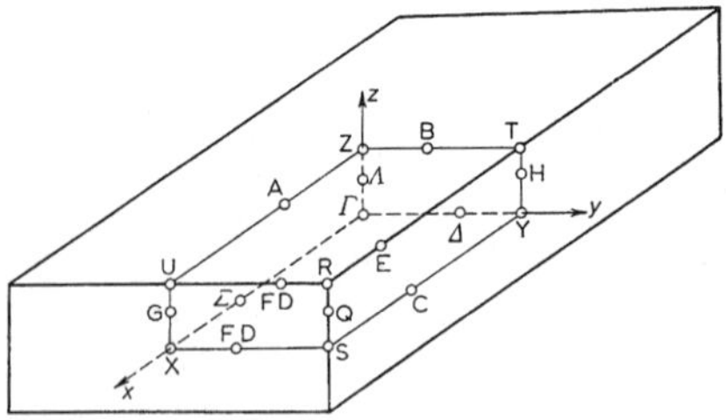

7.4(4)

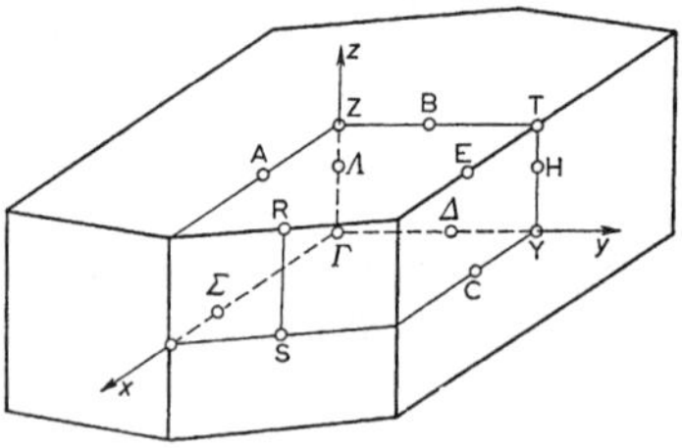

7.4(5)

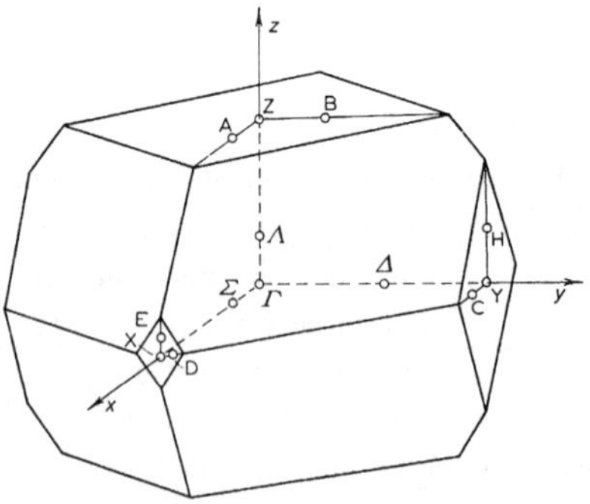

7.4(6a)

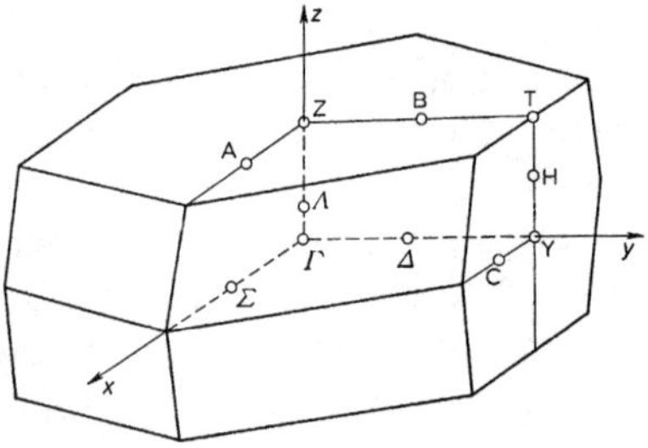

7.4(6b)

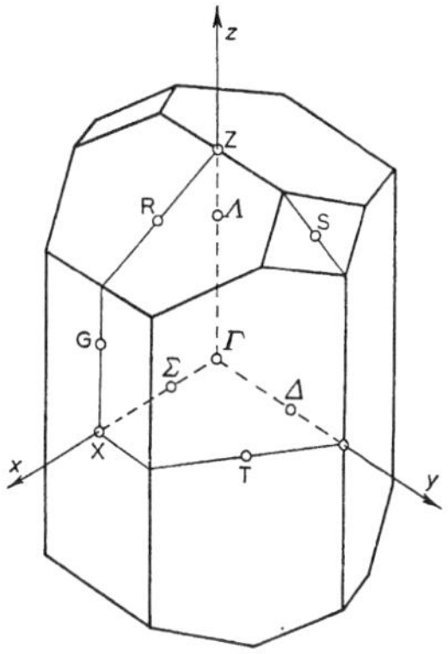

7.4(7)

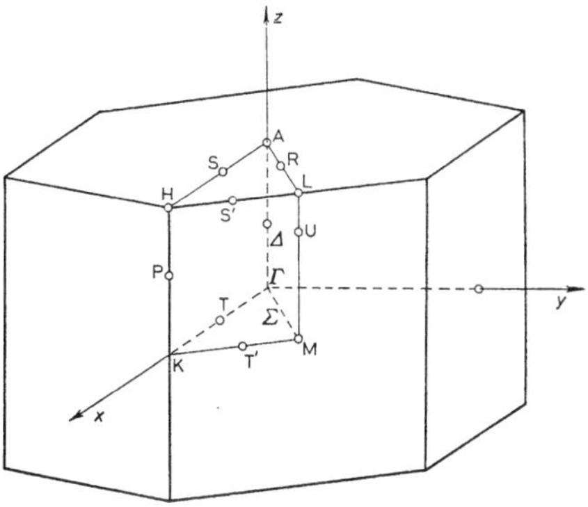

7.4(8)

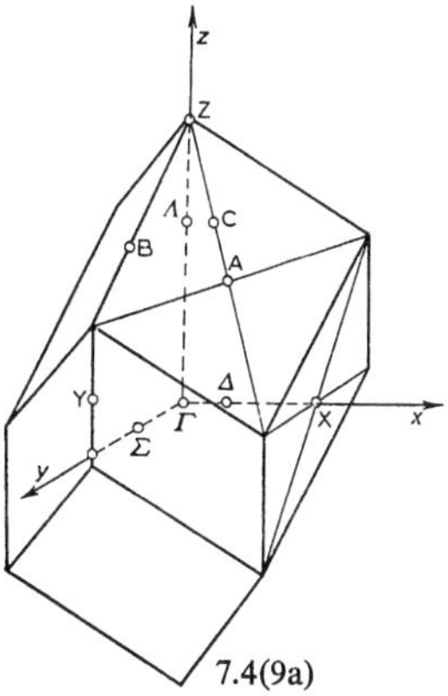

7.4(9a)

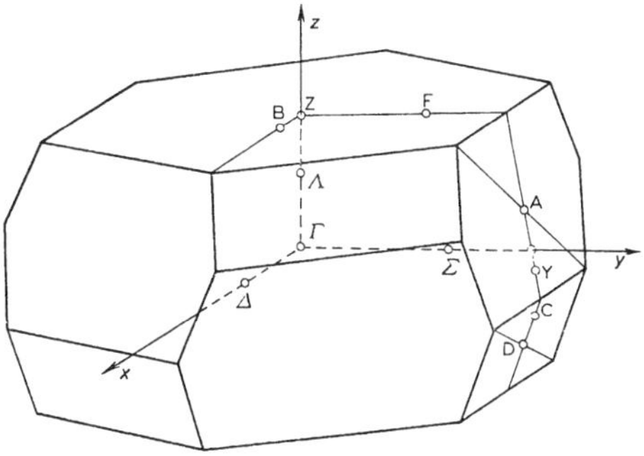

7.4(9b)

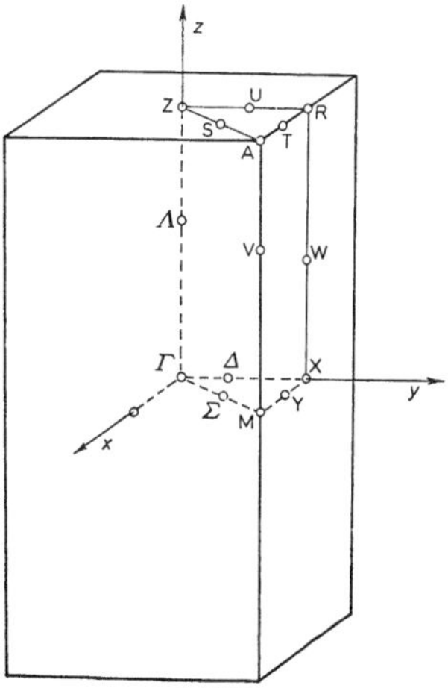

7.4(10)

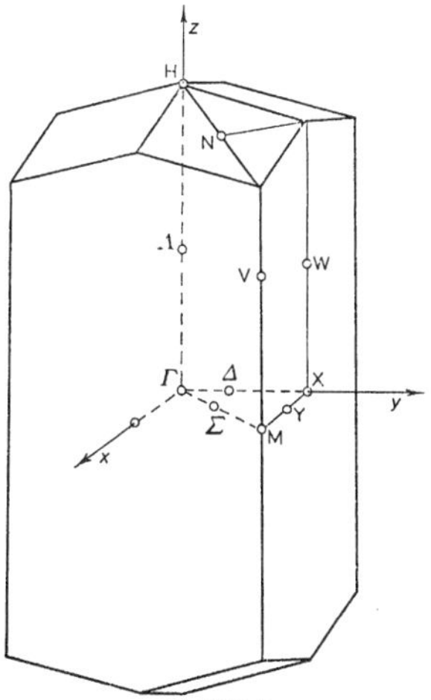

7.4(11a)

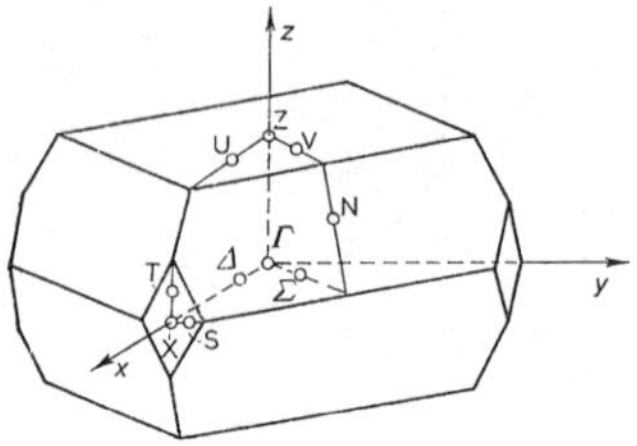

7.4(11b)

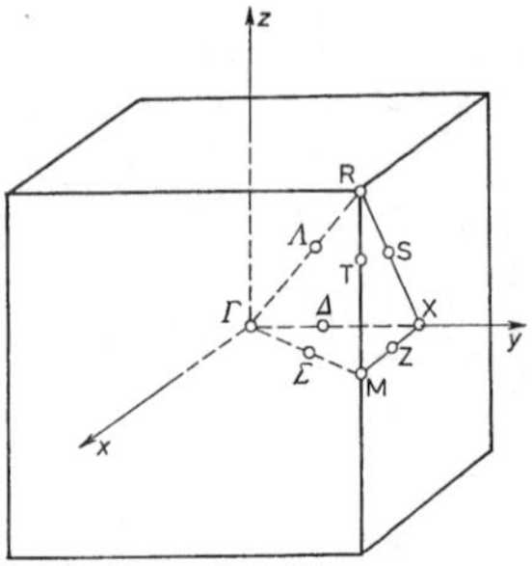

7.4(12)

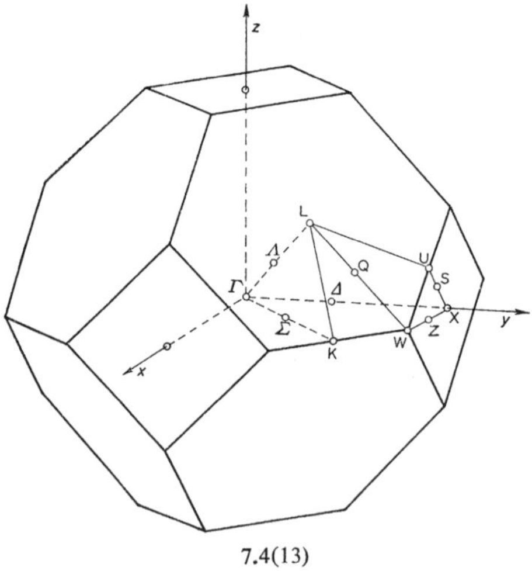

7.4(13)

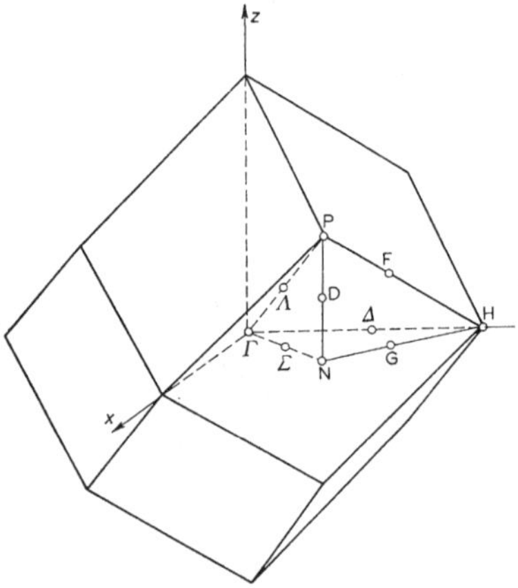

7.4(14)

Fig. 7.4. The 'Wigner-Seitz' or symmetrical unit cell of the fourteen space lattices; (1) Triclinic (not illustrated); (2) Monoclinic, simple; (3) Monoclinic, base centered; (4) Orthorombic, simple; (5) Orthorombic, base centered; (6) Orthorombic, body centered. (a) Height larger than the diagonal of the rectangular base. (b) Height smaller than the diagonal of the rectangular base; (7) Orthorombic, face centered; (8) Hexagonal; (9) Rombohedral, also called trigonal. The three primitive translation vectors are on a cone around the z-axis, making equal angles with one another. (a) Height of the cone larger than $\sqrt{2}$ times the radius of the circle. (b) Height of the cone smaller than $\sqrt{2}$ times the radius of the circle. If equal, the lattice is face centered with z-axis along body diagonal; (10) Tetragonal, simple; (11) Tetragonal, body centered. (a) Height larger than the diagonal of the square, (b) Height smaller than the diagonal of the square. If equal, the lattice is face centered cubic, the diagonal is one of the edges of the cube; (12) Cubic, simple; (13) Cubic, body centered; (14) Cubic, face centered.

If we have one atom per unit cell the corresponding Brillouin zones are the same except for the interchance of 13 and 14. (Compare Koster [1957].)

After these introductory remarks it is obvious that a non-degenerate wave function can be written as

$$\psi(\boldsymbol{r}) = \mathrm{e}^{\mathrm{i}\varphi(\boldsymbol{r})}u(\boldsymbol{r}) \tag{7.32}$$

where $u(\boldsymbol{r})$ has the same translation symmetry as the lattice that is if

$$V(\boldsymbol{r}) = V(\boldsymbol{r}+\boldsymbol{T}) \rightarrow u(\boldsymbol{r}) = u(\boldsymbol{r}+\boldsymbol{T}). \tag{7.33}$$

Since the wave functions in quantum mechanics have to be basis functions for the representation(s) of the symmetry group of the quantum mechanical problem under consideration, the results of the last section can be applied immediately and hence the phase factor $\varphi(r)$ is equal to $\boldsymbol{k} \cdot \boldsymbol{r}$ where $\boldsymbol{k}$ is the label of the representation. The result is that, in full generality, the wave function can be written as

$$\psi(\boldsymbol{r}) = \mathrm{e}^{\mathrm{i}\boldsymbol{k}\cdot\boldsymbol{r}}u_{\boldsymbol{k}}(\boldsymbol{r}) \tag{7.34}$$

where $u_{\boldsymbol{k}}(\boldsymbol{r})$ is periodic with the periodicity of the lattice and $\boldsymbol{u}$ is in general dependent on $\boldsymbol{k}$. For a given value of $\boldsymbol{k}$ there may be more than one periodic function $u_{\boldsymbol{k}}$, hence one has to add another label to distinguish these different "branches".

8. Reduced Wave Vectors

In this section we will show how the symmetry adapted wave functions look for a solid state problem of given symmetry.

If we consider the Bloch Theorem [eq. (7.34)], which indicates the general shape of the wave functions in the periodic lattice we could ask ourselves what the wave functions are for an empty box, that is to say infinite space without any potential energy wells. This question is not quite so trivial as it sounds, since these eigenfunctions will turn out to be the symmetry adapted wave functions for the problem in which the potential energy is not equal to zero. The picture which we are going to describe is essentially a group theoretical idea. Empty space, with infinite boundaries or with periodic boundaries corresponds to a group which consists of an infinite number of translations. Introduce into this empty space a set of potential wells, which are centered around lattice sites, then the corresponding group will consist of a finite number of translations. The Bloch wave functions form the basis for a representation of this group. Hence we are dealing with the problem which we have already mentioned many times: the perturbation requires that the original group of transformations be replaced by a subgroup of transformations. It is helpful to introduce a set of wave functions that already form a irreducible basis for this subgroup.

Hence, the replacement of the free electron wave functions by the following set of Bloch functions

$$e^{i\boldsymbol{k}\cdot\boldsymbol{x}} = e^{i\boldsymbol{k}_{\text{red}}\cdot\boldsymbol{x}} e^{2\pi i\boldsymbol{K}\cdot\boldsymbol{x}} \tag{7.35}$$

is the expression of the above-mentioned procedure. We also see clearly the meaning of the reduced wave vector emerge, that is the wave vector inside the Brillouin zone.

The periodic function

$$u_{\boldsymbol{k}}(\boldsymbol{x}) = e^{2\pi i\boldsymbol{K}\cdot\boldsymbol{x}}; \qquad u(\boldsymbol{x}+\boldsymbol{T}) = u(\boldsymbol{x}) \tag{7.36}$$

is independent of $\boldsymbol{k}$, but bears the label $\boldsymbol{K}$ which numbers the different branches. $\boldsymbol{K}$ is the distance between any two points in the reciprocal lattice. Now if we plot the energy as a function of k we see that for the wave function described above the single parabola is replaced by a large set of parabolas (Fig. 7.5). This description forms the start of our perturbation calculation.

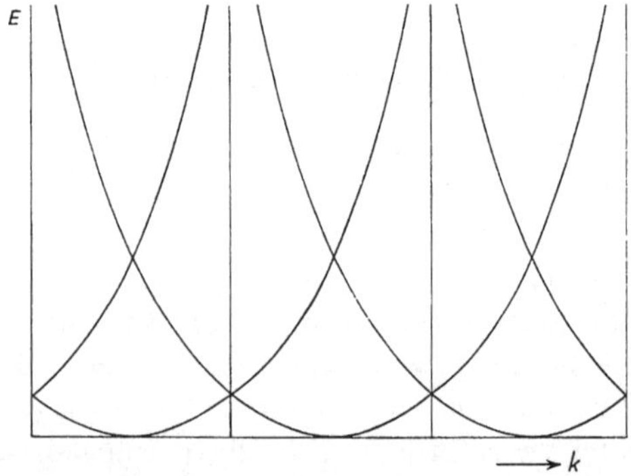

Fig. 7.5. Energy as a function of the unreduced wave vector (or extended Brillouin zone) for a free particle.

The introduction of non-zero periodic potentials will slightly deform the parabolas or in case the perturbation is very strong they may be considerably deformed. In both cases, however, the group theoretical argument which we used stays valid. Suppose for convenience that the perturbation-potential is relatively small. In this case the main shape of the parabola is maintained except in the points near the Brillouin zone. At these points the energy curves intersect and hence we have to refer to the problem of a degenerate or a nearly degenerate level. As we have seen in Chapter 2, § 7.4 this may or may not give rise to the so-called non-crossing rule, depending on whether the off-diagonal elements are zero or non-zero. If we suppose for the moment that the non-diagonal elements in the secular matrices are non-zero, we will have Figure 7.5 changed to Figure 7.6. It is interesting to notice that as a result of

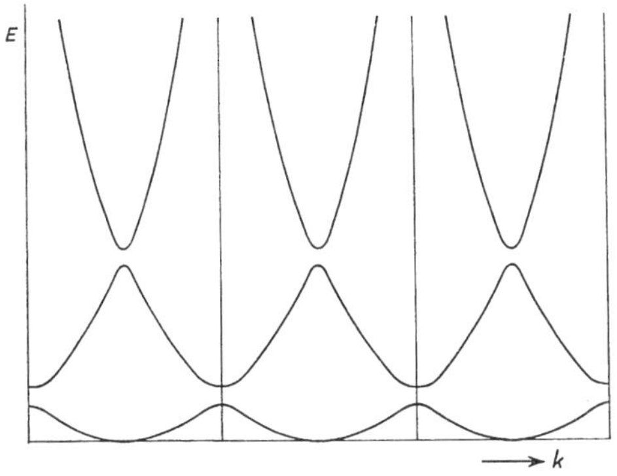

Fig. 7.6. Energy curves in the unreduced wave vector space, for particles in a periodic potential.

symmetry considerations both curves will have a tangent which is horizontal at the "Brillouin-zone" (that is the point $k = \pi/a$ in the one-dimensional case). This leads to two important conclusions. First there will be a gap near the Brillouin zone and second the energy as a function of k can be approximately described by a parabola in this neighborhood. This means in the language of solid state physics that an effective mass can be assigned to the electrons or holes in the neighborhood of the gap. In many cases of interest this simple picture may be complicated by one or more of the following three factors. First, the problem is actually not one-dimensional and hence the energy lines are three-dimensional surfaces in four-dimensional space. Second, the wave function with which we are dealing is not always single, but may be, for instance, three-valued like a p-function in a germanium atom. Third, the conclusion about the parabolic behavior is destroyed if there is spin orbit coupling in the neighborhood of the gap.

The usual group theory in solid state physics can be extended beyond this point. Besides the group of lattice translations which are expressed by the fourteen Bravais lattices there are also point groups in the reciprocal space. These point groups mean that certain sets of $\boldsymbol{k}$ vectors may be the basis for a group of the same or of lower order. This idea was worked out by BOUCKAERT, SMOLUCHOWSKI and WIGNER [1936]. A short review of their ideas will be given in the next section.

9. Little Groups, W.B.S. Method

9.1. LITTLE GROUP THEORY

An irreducible representation of a group $\mathscr{G}$ can be used as a representation of one of its subgroups $\mathscr{H}$ by omitting all matrices that do not belong to $\mathscr{G}$. The representation of $\mathscr{H}$ obtained in this manner is in general reducible. This is expressed as follows:

$$\Gamma_i \to \Xi_i = \sum_k c_{ik}\gamma_k \tag{7.37}$$

where Γ_i represents the ith irreducible representation of $\mathscr{G}$, Ξ_i the representation of $\mathscr{H}$ after omission of the elements of $\mathscr{G}$ that do not belong to $\mathscr{H}$ and γ_k the irreducible representations of $\mathscr{H}$. The coefficients c_{ik} are either zero or integers. This pattern is the major theme in many applications of group theory in physics and the general case will be considered again in the next chapter.

In this chapter the special case where $\mathscr{H}$ is a *normal* subgroup will be considered. This case is, as we shall see below, of principal interest in the application of group theory to solid state problems. From the purely mathematical point of view, the demand that $\mathscr{H}$ be normal is interesting for its own sake since it leads to a number of important considerations.

Let us recall that if the (abstract) elements B belong to $\mathscr{H}$, than ABA^{-1} will also belong to $\mathscr{H}$ (definition normal subgroup). Consider an element A that does *not* belong to $\mathscr{H}$, otherwise the statements would have been trivial. A different way of characterizing the normal subgroup is to say that any "outside" element will induce a permutation of the order of the elements in the subgroup $\mathscr{H}$. The same statement will hold for the representation:

$$\Xi(\text{B}) \to \Xi(\text{ABA}^{-1}) \tag{7.38}$$

which gives rise to another possible representation. We can maintain the notation since we are dealing with the same matrices as before; the difference is that they now correspond to different abstract elements. The important point is that the representation obtained this way (called the conjugate representation) is *not* necessarily equivalent to the original one, since

$$\Xi(\text{ABA}^{-1}) = \Gamma(\text{A})\Gamma(\text{B})\Gamma(\text{A}^{-1}) \equiv \Gamma(\text{A})\Xi(\text{B})\Gamma(\text{A}^{-1}), \tag{7.39}$$

and the matrix $\Gamma(\text{A})$ does not have to belong to the representation Ξ.

If Ξ is decomposed into irreducible representations with respect to $\mathscr{H}$, a set of conjugated representations can again be constructed with help of the different elements A available in $\mathscr{G}$ and "outside" $\mathscr{H}$. These conjugate representations will again be irreducible (they are a permutation of an irreducible representation, hence if they were reducible, the original would also have been), but not necessarily the same; i.e. non-equivalent;

$$\gamma_k(\text{B}) \rightarrow \gamma_k(\text{ABA}^{-1}) = \gamma_{k'}(\text{B}). \tag{7.40}$$

We would like to mention that $\gamma_{k'}$ and γ_k must have the same dimensionality and also, since all irreducible representations originating from Γ_i are each other's conjugate, we find that all representations γ_k coming from one Γ_i necessarily have to have the same dimension. All this stems from the fact that $\gamma_{k'}$ was connected with γ_k by a permutation of the "links" between abstract elements on one hand and the representation matrices on the other hand.

If the element A is considered as a variable which runs through $\mathscr{G}$ a set of irreducible representations conjugate to γ_k is generated, this set is sometimes called an orbit. Conjugation with certain elements A of $\mathscr{G}$ will produce representations which are equivalent to the original one. This will definitely be true if A belongs to $\mathscr{H}$ since in that case the representation remains exactly the same. Also if this condition, i.e.

$$\gamma_{k'}(\text{B}) = \gamma_k(\text{ABA}^{-1}) \equiv \gamma_k(\text{B}) \tag{7.41}$$

is fulfilled for a certain element A, it is also fulfilled for all elements that are in the same coset as A.

All the elements A that fulfil condition (7.41) form a group, the little group of the second kind, $\mathscr{L}^{\text{II}}$. It is easy to show that these elements do indeed form a group. For example the product rule: If A_1 and A_2 are any two elements, then

$$\gamma_k(\text{A}_1\text{BA}_1^{-1}) \equiv \gamma_k(\text{B}) \tag{7.41a}$$

$$\gamma_k(\text{A}_2\text{BA}_2^{-1}) \equiv \gamma_k(\text{B}) \tag{7.41b}$$

and the product element A_1A_2 belongs to $\mathscr{L}^{\text{II}}$ because

$$\gamma_k(\text{A}_1\text{A}_2\text{BA}_2^{-1}\text{A}_1^{-1}) \equiv \gamma_k(\text{A}_1\text{B}'\text{A}_1^{-1}) \equiv \gamma_k(\text{B}''). \tag{7.41c}$$

Similar arguments can be given for the other group postulates.

All elements of $\mathscr{H}$ belong to $\mathscr{L}^{\text{II}}$. All other elements of $\mathscr{L}^{\text{II}}$ can be grouped in cosets, as mentioned before. The elements of $\mathscr{L}^{\text{II}}$ contain $\mathscr{H}$ as a normal subgroup and the factor group $\mathscr{L}^{\text{II}}/\mathscr{H}$ is called the little group of the first kind $\mathscr{L}^{\text{I}}$.

If these ideas are applied to the normal subgroup $\mathscr{T}$ of pure translations in the space group:

$$(\text{T}_0|\text{R})^{-1}(\text{T}|\text{E})(\text{T}_0|\text{R}) = (\text{R}^{-1}\text{T}|\text{E}) \tag{7.42}$$

the right-hand side represents another element of $\mathscr{T}$. If we now take the irreducible representation of the subgroup $\mathscr{T}$ the element $(\text{T}|\text{E})$ will, according to § 6 be represented by the following exponentials

$$(\text{T}|\text{E}) \to e^{i\boldsymbol{k}\cdot\boldsymbol{r}} \in \gamma(\boldsymbol{k}) \tag{7.43}$$

and the conjugate representation of the element T by:

$$(\text{T}' \mid \text{R})\,(\text{T} \mid \text{E})\,(\text{T}' \mid \text{R})^{-1} = (\text{RT} \mid \text{E}) \to e^{i\boldsymbol{k}\cdot\text{R}} \in \gamma_k{}^{(\text{ABA}^{-1})} \tag{7.44}$$

according to (7.40) this is the representation $\gamma_{k'}$. Hence, if we replace R^{-1}T by T we find

$$(\text{RT} \mid \text{E}) \to e^{i\boldsymbol{k}\cdot\text{R}^{-1}\boldsymbol{r}} = e^{i\text{R}\boldsymbol{k}\cdot\boldsymbol{r}} \tag{7.45}$$

All conjugate representations of $\gamma(\boldsymbol{k})$ are the representations $\gamma(\text{R}\boldsymbol{k})$. If we take for instance a two-dimensional square lattice and if we take a representation characterized by a certain $\boldsymbol{k} = (k_x, k_y)$ the conjugate representation will be characterized by the following eight vectors $\boldsymbol{k} = (\pm k_x, \pm k_y)$ and $(\pm k_y, \pm k_x)$ obtained by applying the allowed operations of the point group $\mathscr{R}$ consisting of rotations over $\frac{1}{2}\pi$, π, $\frac{3}{2}\pi$ and reflections with respect to the axis of the lattice. The prong thus formed is called a *star* and is a special case of the "orbit" mentioned before. The $\boldsymbol{k}$-vectors in the star represent the set of all conjugate representations of $\gamma(\boldsymbol{k})$ that are non-equivalent.

From the preceding considerations it is clear that we find the set of vectors that mutually form a star by taking a certain $\boldsymbol{k}$-vector inside the Brillouin zone and have it undergo the point transformations allowed by the lattice symmetry. The $\boldsymbol{k}$-vector is inside the Brillouin zone since the vector refers to the irreducible representations of the pure translations and hence is the reduced vector (§ 6).

The elements of the little group were those matrix elements of the group $\mathscr{G}$ that give rise to conjugate representations that are equivalent to $\gamma(\boldsymbol{k})$. These are given by $\boldsymbol{k}$-vectors that fulfill the condition

$$\text{R}\boldsymbol{k} = \boldsymbol{k} + 2\pi\boldsymbol{K} \tag{7.46}$$

where $\boldsymbol{K}$ is a vector of the reciprocal lattice

$$\boldsymbol{K} = n_1 \boldsymbol{a}^* + n_2 \boldsymbol{b}^* + n_3 \boldsymbol{c}^*; \quad n_1,\ n_2 \text{ and } n_3 = 0, \pm 1, \pm 2 \dots$$

Again using the two-dimensional square lattice as an example, we have indicated in Fig. 7.7 the points (Γ, M, X) and lines (Σ, Δ, Z) of symmetry in

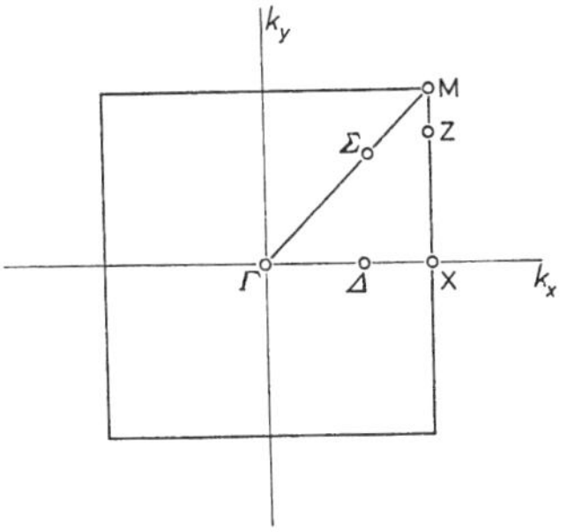

Fig. 7.7. Brillouin zone of a 2-dimensional square lattice.

the Brillouin-zone. The little group belonging to Δ consists of the unit element and the reflection with respect to the k_x-axis.[1] The little group belonging to Γ is the same as the point group of the lattice: D_4. The previous two points were examples of group elements that stay invariant under the condition (7.46) with $\boldsymbol{K} = 0$. The points M, Z, X stay not only invariant under certain rotations or reflections but also under a combination of these with a translation. The irreducible representation of the special points or lines have been worked out for a number of Bravais lattices (see HERMAN [1958] and KOSTER [1957]).

9.2. APPLICATIONS OF THE THEORY OF THE LITTLE GROUP

The first type of question that can be answered with group theory in band theory are of the same nature as the questions answered in crystal field theory. Suppose there are r wave functions all belonging to the same energy level in the unperturbed state, what splitting of these r levels will take place after the interaction is taken into account? The level structure, i.e., the number of separate levels and their degeneracy, on the basis of these symmetry considerations is called the "essential" degeneracy. Whether separation

[1] We mean of course the little group of the first kind. The little group of the second kind consists of $\mathscr{T}$ and A$\mathscr{T}$, where A is the reflection with respect to the x-axis, that is, all pure translations plus all translations combined with reflections around the x-axis.

really takes place remains to be seen from a detailed calculation, and if the separation turns out to be zero, the additional degeneracy is called "accidental" degeneracy. The names are rather unfortunate since there could be other reasons, beyond the symmetry considerations, which made the degeneracy not accidental at all. A famous example is the Fock-argument for the energy levels of hydrogen atoms.[1] The essential or symmetry-induced degeneracies occurring in the special points mentioned above will be illustrated by two examples.

The energy of a plane wave depends only on the magnitude of the vector $\boldsymbol{k}$. Hence all the four points: $(0, \pm k_x) \equiv \mathrm{X}$, $(0, \pm k_y)$ correspond to plane waves that have the same energy. This holds in particular for the two waves corresponding to X. The little group belonging to X, however, has only one-dimensional representations, hence the levels will split under a general perturbation (the potential field of the cores). For symmetry reasons they will have the same value at the points $(0, \pm k_y)$.

A second example is given by considering the four wave functions connected with M. They form a reducible representation of the little group belonging to M. The last group is D_4 and has four one-dimensional and one two-dimensional representations. The representation resulting from the four plane waves decomposes into two one-dimensional representations and one two-dimensional representation (compare Table 7.1), hence under

TABLE 7.1

Character table: $\boldsymbol{D}_4 = \boldsymbol{M}_1 + \boldsymbol{M}_4 + \boldsymbol{M}_5$

	E	C_2	C_3	C_4	C_5
$\boldsymbol{D}_4$	4	0	0	0	2
$\boldsymbol{M}_1$	1	1	1	1	1
$\boldsymbol{M}_4$	1	1	−1	−1	1
$\boldsymbol{M}_5$	2	−2	0	0	0

C_2 : rot. π around z-axis (1 el.)
C_3 : rot. $\frac{1}{2}\pi$ around z-axis (2 el.)
C_4 : reflection with respect to x and y axes (2 el.)
C_5 : reflection with respect to the two diagonals (2 el.)

[1] The wave equation of the non-relativistic hydrogen atom can be written in a form identical to the integral equation for spherical harmonics in four-dimensional space. Invariance under the operations of the last group leads to the conclusion that the energy levels with different L, but the same n, are degenerate. (Compare McINTOSH [1958b].)

influence of the perturbation the four-fold level will split in two single levels and one two-fold level. This is often expressed by stating that two of the four surfaces "stick together" at the point M. This is based on the fact that Γ is a continuous variable of the function $\boldsymbol{k}$ and a slight increase in $\boldsymbol{k}$, say $\boldsymbol{k} \rightarrow \boldsymbol{k}+\Delta\boldsymbol{k}$, will correspond to an arbitrary point in the Brillouin-Zone and hence the four energy surfaces will, in general, be separated if one moves away from M.

Compatibility relations and "accidental" degeneracy. The special points in the Brillouin-Zone fall into two classes. Points like Σ, Z, and Δ form a locus given by the lines ΓM, MX, and ΓX. The little group of Δ should be a subgroup of the little group of Γ as well as of X. Suppose we know the irreducible representation of the degenerate wave functions in X. If we move away from X but stay along Δ the wave functions will span a reducible set and the representation of the wave functions in X will decompose into a number of irreducible representations of Δ. Conversely, this implies that a given representation of Δ can go over into a limited number of representations of X or Γ. These so-called compatibility relations between the irreducible representations have been worked out by BOUCKAERT, SMOLUCHOWSKI and WIGNER [1936] and for the two-dimensional (pedagogical) example we give Table 7.2 taken from HEINE [1960].

TABLE 7.2.

Compatibility relations

Representation	Compatible with
Δ_1	$\Gamma_1\Gamma_3\Gamma_5$; X_1X_4
Δ_2	$\Gamma_2\Gamma_4\Gamma_5$; X_2X_3
Σ_1	$\Gamma_1\Gamma_4\Gamma_5$; $M_1M_4M_5$
Σ_2	$\Gamma_2\Gamma_3\Gamma_5$; $M_2M_3M_5$
Z_1	X_1X_3; $M_1M_3M_5$
Z_2	X_2X_4; $M_2M_4M_5$

Γ_5 reduces into $\Delta_1+\Delta_2$ or $\Sigma_1+\Sigma_2$
M_5 reduces into $\Sigma_1+\Sigma_2$ or Z_1+Z_2.

Group theory is not only helpful in degeneracies attributed to symmetry as has been pointed out by HERRING [1937b]. This type of predictable degeneracy is most easily illustrated by an example.

TABLE 7.3

Character tables of Γ, M, X, Δ, Σ, Z

Γ, M	E	2_z	4_z, 4_z^3	m_y m_x	m_d $m_{d'}$
$\Gamma_1 M_1$	1	1	1	1	1
$\Gamma_2 M_2$	1	1	1	−1	−1
$\Gamma_3 M_3$	1	1	−1	1	−1
$\Gamma_4 M_4$	1	1	−1	−1	1
$\Gamma_5 M_5$	2	−2	0	0	0

$\Delta\ \Sigma\ Z$	E	m_x, m_y, m_d
$\Delta_1\Sigma_1Z_1$	1	1
$\Delta_2\Sigma_2Z_2$	1	−1

X	E	2_z	m_x	m_y
X_1	1	1	1	1
X_2	1	1	−1	−1
X_3	1	−1	1	−1
X_4	1	−1	−1	1

Note: m_x denotes a reflection in a line perpendicular to the x-axis.

Calculations have been made by SLATER [1934] with the Wigner-Seitz method on sodium. This metal has a body centered cubic structure and the corresponding unit cell has the shape of a cube from which all eight corners have been cut off perpendicular to the body diagonals. The faces are hexagonal and the point in the center is called G. The Wigner-Seitz method consists of solving the Schrödinger equation inside the cell under the restriction that it should fulfill the Bloch condition at the surface. Slater was the first to include wave functions with angular momentum (p, d, etc. wave functions) besides the s functions used by Wigner and Seitz.

The wave function will be, in general, a mixture of the s, p, d, etc. wave functions. There is no longer any reason to retain these labels because the problem no longer has spherical symmetry. It turns out, however, in particular in the approximation used, that the wave functions are s-like, p-like, d-like, etc. Calculations show that at the origin, the lowest level is s-like, the next d-like, and at the point G the lowest is p-like, and the next s-like. The lowest level at G lies higher than the lowest level at the origin.

This is already sufficient information to draw the conclusion about non-accidental degeneracy. The little groups belonging to O and G are the same as the point group of the lattice. Hence the s-like level has the unit representation, the p-like a three-dimensional representation, i.e., it stays degenerate, and the d-like splits into a three- and a two-fold degenerate level. (Compare, for example, Chapter 8, § 2.2.) Using now the compatibility relations we can state that from the s-like level originates a Δ_1 representation (along the 111 axis), from the p-like level a Δ_2' and a Δ_5 representation; (This last is two-fold degenerate.), from the d-like level a three-fold representation giving rise to a Δ_5 and Δ_2' and a two-fold level giving Δ_1 and Δ_2'. If curves are drawn between the end points, making sure that every curve has always the same representation, we see that Δ_2' and Δ_1 must cross each other. Considerations

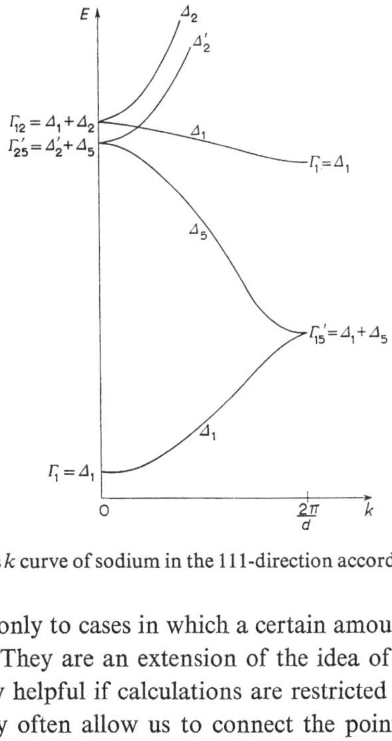

Fig. 7.8. Energy versus k curve of sodium in the 111-direction according to HERRING [1937].

of this type apply only to cases in which a certain amount of information is already available. They are an extension of the idea of compatibility. This idea is particularly helpful if calculations are restricted to certain points in $\boldsymbol{k}$-space, since they often allow us to connect the points in an ambiguous way. From the connections the conclusions about overlapping in a certain direction can be drawn, one of the first questions asked if one studies a particular solid.

The implication of time reversal on the energy band structure has been discussed by HERRING [1937a].

CHAPTER 8

FINITE GROUPS

1. Rotational Crystal Symmetry

In this section we are interested in the so-called point groups. This is a collection of symmetry elements applied about a point that leaves the structure under investigation invariant. The most common elements of symmetry are the n-fold rotation axes about a point, that is the n rotations around a lattice point over angles $2\pi/n$. Other possibilities are for instance mirror reflections with respect to a plane or inversions with respect to a point. A simple example was given in Chapter 3, § 7.4. The example was a triangle having a three-fold axis as well as three mirror lines, altogether six operations. More sophisticated examples will be referred to below. As has already been pointed out in the preceding chapter the translation requirements of the space group mean that only a restricted set of point groups can be dealt with in solid state. This is in contrast to the situation in molecules where the only restriction is of internal consistency between the different symmetry elements of the group. It is only in the last section of this chapter that molecules will be considered again.

The restrictions on point groups resulting from the translational requirements will be illustrated by using the two-dimensional case as a pedagogical example. We have already seen in Chapter 7, § 2 that in solids only one-, two-, three- and six-fold axis are permitted, a proof which can be given in a less formal way by considering the translational vector; having it undergo a certain rotation and demanding that the translation and the rotated translations differ by a certain translational factor. This again leads to the result that the possible angles of rotation are limited to the sources mentioned before (compare KITTEL [1957]).

In case these five possibilities are combined with a mirror line, five more possibilities arise. This is shown in Table 8.1, which indicates the equivalent points in a circle together with the notation used in crystallography for the two-dimensional point group.

TABLE 8.1

Two-dimensional point groups

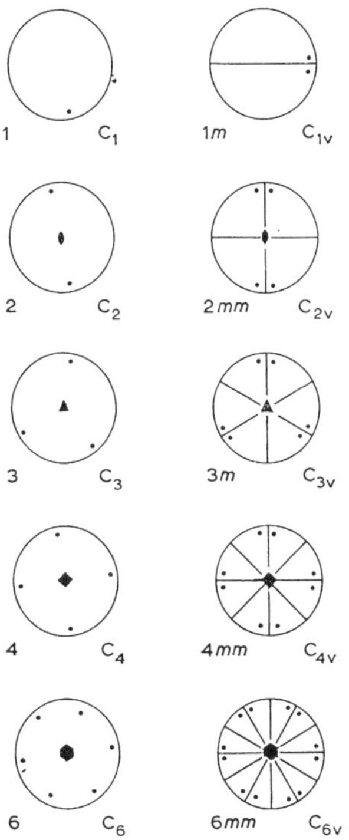

The study of three-dimensional point groups brings besides the rotation axis mentioned before, the possibility of a reflection plane, inversion center, and the so-called rotation-inversion axis. The last are combinations of a rotation and an inversion, such that the rotation by itself is not an allowed transformation, but the rotation is allowed if it is combined with an inversion. All possible combinations of these symmetry elements lead to 32 different three-dimensional point groups. In Table 8.2 the stereograms as well as the figures equivalent to the diagrams are indicated. An open circle refers to a point below the plane of the paper, a dot represents a point above the plane of the paper. The plane of the paper is usually considered to be horizontal so that the z-axis is normal to the paper. In case the plane of the

TABLE 8.2
Three-dimensional point groups

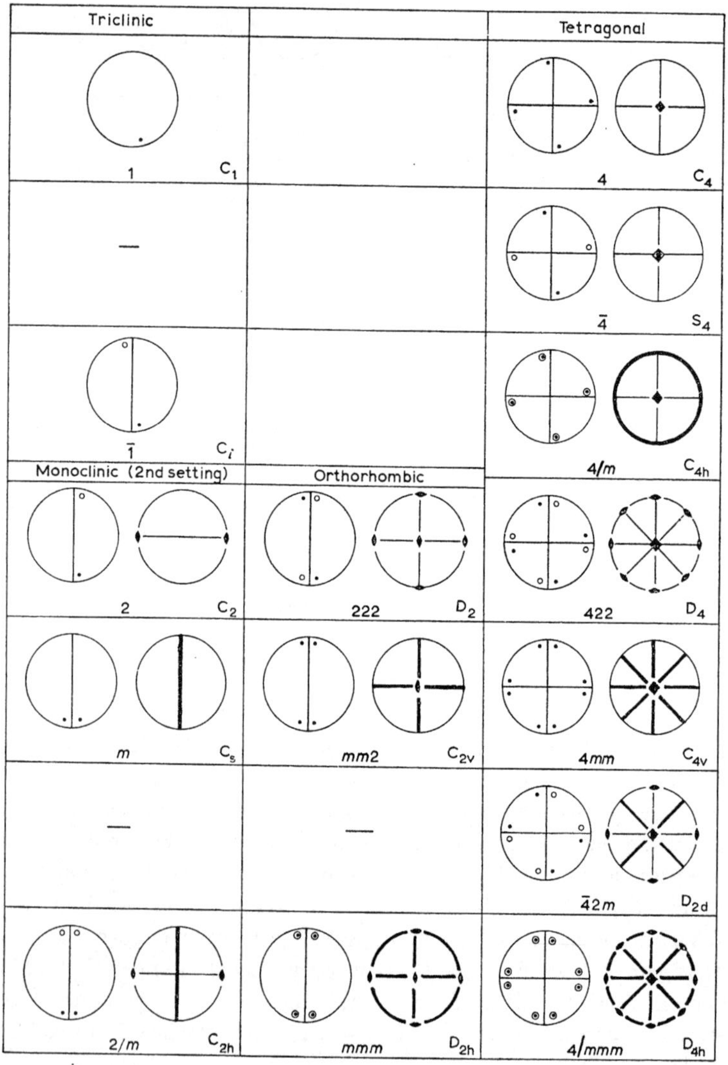
Triclinic
Tetragonal
1 C_1
4 C_4
—
$\bar{4}$ S_4
$\bar{1}$ C_i
4/m C_{4h}
Monoclinic (2nd setting)
Orthorhombic
2 C_2
222 D_2
422 D_4
m C_s
mm2 C_{2v}
4mm C_{4v}
—
—
$\bar{4}2m$ D_{2d}
2/m C_{2h}
mmm D_{2h}
4/mmm D_{4h}

TABLE 8.2
(*Continued*)

Trigonal	Hexagonal	Cubic
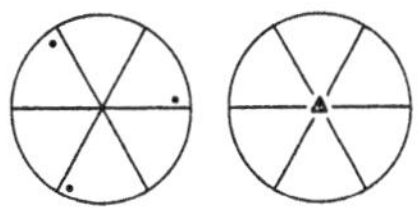 3 C_3	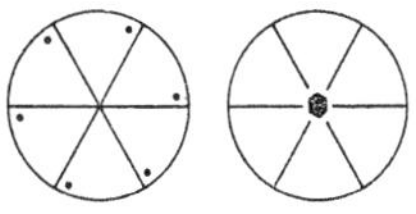 6 C_6	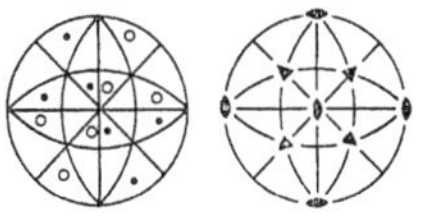 23 T
—	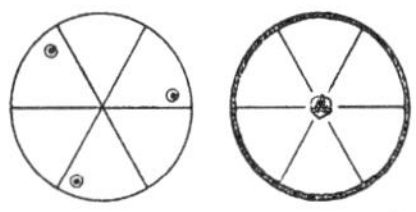 $\bar{6}$ C_{3h}	—
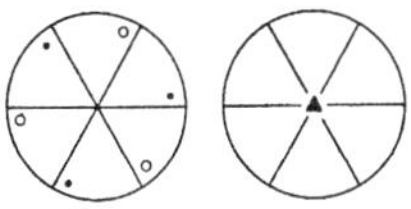 $\bar{3}$ C_{3i}	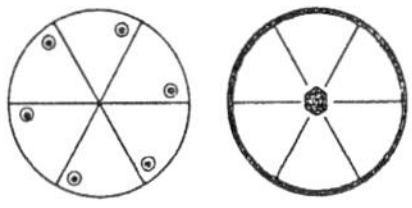 $6/m$ C_{6h}	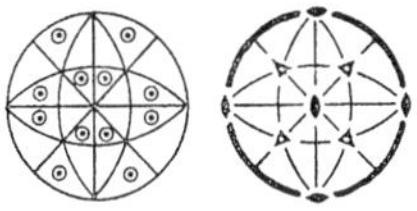 $m3$ T_h
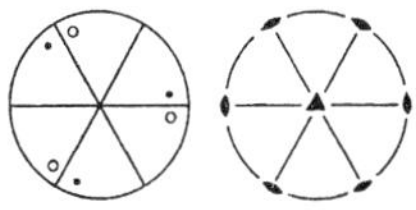 32 D_3	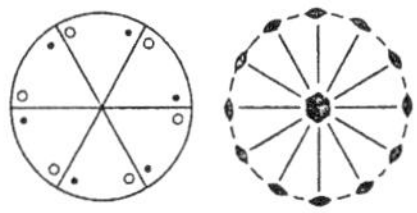 622 D_6	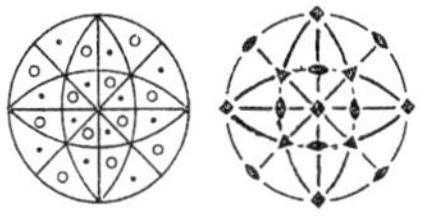 432 O
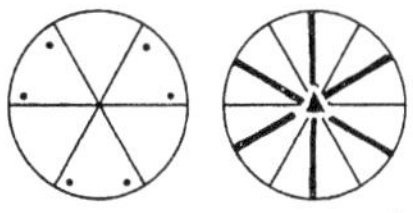 $3m$ C_{3v}	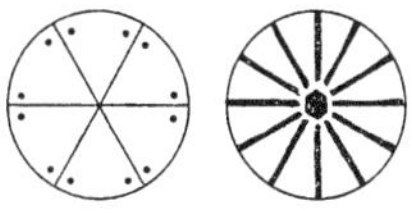 $6mm$ C_{6v}	—
—	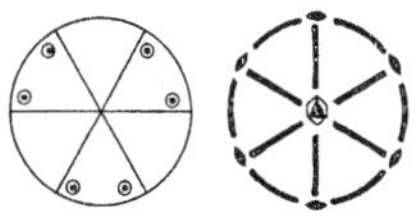 $\bar{6}m2$ D_{3h}	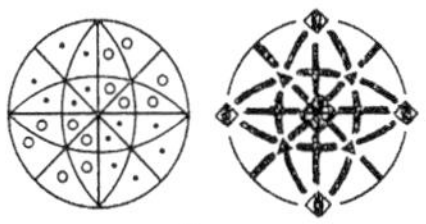 $\bar{4}3m$ T_d
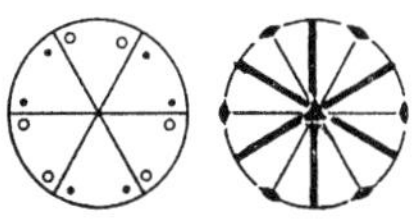 $\bar{3}m$ D_{3d}	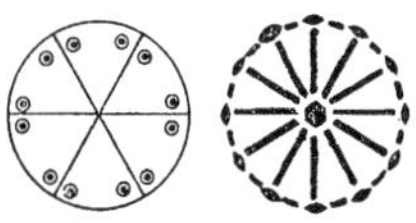 $6/mmm$ D_{6h}	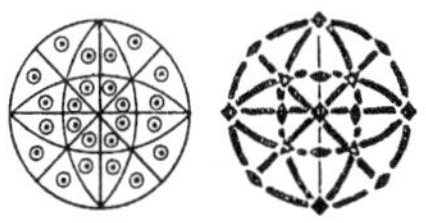 $m3m$ O_h

TABLE 8.2

The thirty-two crystal point groups

International Symbol		
Short	Full	Schoenflies Symbol
1	1	C_1
1	1	$C_i(S_2)$
2	2	C_2
m	m	$C_s(C_{1h})$
$2/m$	$\frac{2}{m}$	C_{2h}
222	222	$D_2(V)$
$mm2$	$mm2$	C_{2v}
mmm	$\frac{2}{m}\frac{2}{m}\frac{2}{m}$	$D_{2h}(V_{2h})$
4	4	C_4
$\bar{4}$	$\bar{4}$	S_4
$4/m$	$4m$	C_{4h}
422	422	D_4
$4mm$	$4mm$	C_{4v}
$\bar{4}2m$	$\bar{4}2m$	$D_{2d}(V_d)$
$4/mmm$	$\frac{4}{m}\frac{2}{m}\frac{2}{m}$	D_{4h}
3	3	C_3
$\bar{3}$	$\bar{3}$	$C_{3i}(S_6)$
32	32	D_3
$3m$	$3m$	C_{3v}
$\bar{3}m$	$\bar{3}\frac{2}{m}$	D_{3d}
6	6	C_6
$\bar{6}$	$\bar{6}$	C_{3h}
$6/m$	$\frac{6}{m}$	C_{6h}
622	622	D_6
$6mm$	$6mm$	C_{6v}
$\bar{6}m2$	$\bar{6}m2$	D_{3h}
$6/mmm$	$\frac{6}{m}\frac{2}{m}\frac{2}{m}$	D_{6h}
23	23	T
$m3$	$\frac{2}{m}\bar{3}$	T_h
432	432	O
$\bar{4}3m$	$\bar{4}3m$	T_d
$m3m$	$\frac{4}{m}\bar{3}\frac{2}{m}$	O_h

paper is a symmetry plane, the circle is drawn with a heavy line. The stereograms are very useful except in the cubic case where it is much easier to picture a cube instead of this rather artificial device.

In case of a purely ionic crystal, that is to say the case where the electrons are completely localized and belong to a certain ion or complex, the wave function of these electrons or complexes will differ from the free ion wave functions in such a way that the environment, in particular the nearest neighbors, will produce a perturbation on the wave function. This result may be hard to calculate particularly if the perturbation is strong, but it is always necessary that the final wave function should have the proposed symmetry. If the symmetry of the crystal is really known the number of unknown parameters in the perturbation calculation may be greatly reduced. Hence if the symmetry symbol of a certain point in the crystal is known from the literature, it may be possible with the help of the theory or representations to set up the symmetry adapted eigenfunctions. This will be demonstrated in the next section.

2. Crystal Field Theory

2.1. ANGULAR WAVE FUNCTIONS UNDER FINITE ROTATIONAL SYMMETRY

Consider a free atom or an ion, which has the symmetry of the full rotation group, and place it inside the lattice of a solid. The allowed symmetry is now restricted to the point group of that particular place in the solid. The physical influence of the environment will be due to 1) electrical fields, 2) magnetic fields, and 3) exchange interaction or covalent binding.

In general it is very difficult to give a complete calculation of these influences. In certain cases, the so-called diluted paramagnetic salts, it can be assumed that the first influence, the electric field, is the most dominant of the three. In this case a reasonable description can be made by assuming electrostatic fields and calculating the influence of this perturbation on the wave functions of the particular atom or ion. Fortunately, however, it is possible to make a number of statements which are based on symmetry considerations only and are completely independent of the actual physical mechanism of interaction.

These qualitative considerations can be given with the help of group theory or more correctly with the theory of representations. They were given the first time by Bethe and simultaneously by Kramers.

Assume that a certain level in the free atom or ion was characterized by a certain value of L or J and assume, in order to take a specific example in mind, that the symmetry of the crystal is cubical. The Hamiltonian operator H_{op} will no longer be invariant under all possible rotations. The symmetry of the crystal has reduced the number of rotations (which was ∞^2) to only those 24 rotational transformations which bring the cube over into itself.

Again apply the theorem of Wigner, that is, if ψ belongs to E then also $S\psi$ belongs to E, except that S now refers to a restricted set of rotations. To formulate it more precisely, the group of transformations S forms a subgroup of the original group, the full rotation group. Again we state that dimension of the reducible representation is equal to the degree of degeneracy of the level, since the number of dimensions of the boxes of the reduced representation is nothing but the number of linearly independent wave functions which belong to that particular eigenvalue. However, the representation characterized by L or J is now in general reducible with respect to the cubical group. That is to say, if there are no common unitary transformations which are able to bring the infinite set of matrices of the full rotation group simultaneously into a form with smaller boxes, there may be transformations which are able to do this for the 24 rotations which correspond to the spacial rotations which bring the cube over into itself. At first we are only interested in the *number* and *kind* of irreducible representations of the finite group which are contained in the original $2L+1$-dimensional irreducible representation which is reducible with respect to the cubical group.

This program can be solved with the help of the characters of the matrices. One of the important conclusions from this method is that only the symmetry determines the pattern of the splitting of the levels. This, however, will hold only as long as the splitting of the level is small compared to the distance to the next higher (or next lower) multiplet. In such a case the wave functions of such a neighboring multiplet will have to be considered before starting the considerations formulated above.

Before considering the general theory we would like to describe an example. In how many levels will an F-state ($L = 3$) split if placed in a cubical field and how large is the residual degeneracy? To answer such a question the characters of the 5 classes of the cubical group in the $(2L+1)$-dimensional representation of the full rotation group have to be calculated. These are easily calculated with the help of the following formula:

$$X_J(\omega) = \mathrm{Tr}\, R = \sum_{M=-J}^{J} \mathrm{e}^{\mathrm{i}m\omega_z} = \mathrm{e}^{-\mathrm{i}J\omega_z} \frac{1-\mathrm{e}^{\mathrm{i}(2J+1)\omega_z}}{1-\mathrm{e}^{\mathrm{i}\omega_z}} = \frac{\sin(2J+1)\omega_z/2}{\sin \omega_z/2}. \quad (8.1)$$

Note that the formula is derived for a rotation around the z-axis (compare (5.28)) but the result holds for a rotation ω around an *arbitrary* axis, since het trace is invariant under a similarity transformation (1.7) or (1.27a).

We have to choose for the unitary matrix S (or U) a transformation that lets the rotation axis coincide with the z-axis.

The five classes of the group of cubic symmetry correspond to the following operations.

E Unit operation.
C_2 Rotation over π around each of the three edges (3 elements).
C_3 Rotation over $\frac{1}{2}\pi$ around each of the three edges (6 elements).
C_4 Rotation over π around the face diagonals (6 elements).
C_5 Rotations over $\frac{2}{3}\pi$ around the body diagonals (8 elements).

Since each class always refers to one and the same angle of rotation, that particular ω in (8.1) has to be substituted. The result is, for general values of L, as follows[1]:

$$X_E = 2L+1; \qquad X_{C_2} = X_{C_4} = (-1)^L$$

$$X_{C_3} = (-1)^{[\frac{1}{2}L]}; \quad X_{C_5} = \begin{cases} 1 & (L = 3n) \\ 0 & (L = 3n+1) \\ -1 & (L = 3n+2). \end{cases} \tag{8.2}$$

In this specific case where $L = 3$ the equations become,

$$X_E = 7; \quad X_2 = X_4 = -1; \quad X_3 = -1; \quad X_5 = 1. \tag{8.3}$$

Now if this set of characters is compared with the 5 irreducible representations of the cubical group, then a linear combination of these five must be found such that each class has the values as given by equation (8.3). This is done simply by solving the five equations in five unknowns. The result is given in the following table:

X	E	C_2	C_3	C_4	C_5
Γ_2	1	1	−1	−1	1
Γ_4	3	−1	1	−1	0
Γ_5	3	−1	−1	1	0
$L = 3$	7	−1	−1	−1	1

(8.4)

Hence the conclusion is that the F level decomposes in two levels which are three-fold degenerate and one level which is not degenerate. From this simple example, which can be worked out in a much more sophisticated way with the help of projection operators, we see that knowledge of the symmetry is sufficient to determine the number of separate levels as well as their degeneracy.

[1] $[\frac{1}{2}L]$ means the largest integer contained in $\frac{1}{2}L$.

A word of warning is necessary. It is, of course, not strictly necessary that the three levels mentioned above are really separated. It could be for instance that certain matrix elements of the perturbation are zero and hence that the perturbation does not give rise to a splitting. If this treatment had been used for a level which was characterized by L, it could be asked now what happens if the spin is taken into account (spin-value unequal to zero) and if a spin-orbit coupling term is assumed in the Hamiltonian. In this case it is necessary to start with the product representation of one of the afore-mentioned Γ's with the $2S+1$-dimensional representation of the spin space and reduce the resulting product space. This again is done with the help of characters of the respective classes. The characters of the product representation are equal to the products of the characters of the composing representations. For example, if it is assumed that the afore-mentioned F-level has a spin $S = 1$, that is a ^{3}F state such as is found in Ni^{++}. Consider the Γ_4 level which is contained in the F state. The characters of the representations D_1 of the spin are found with the help of the formulas indicated above equation (8.2).

Γ_4	3	-1	-1	1	0
D_1	3	-1	1	-1	0
$\Gamma_4 \times D_1$	9	1	-1	-1	0

(8.5)

as a result we find the character: 9, 1, -1, -1 and 0 and have to investigate which combinations of the characters of the cubic group will lead to this set of numbers. It turns out to be $\Gamma_1+\Gamma_3+\Gamma_4+\Gamma_5$. Hence the conclusion is that the three-fold degenerate ground level which after multiplication by the spin variables became nine-fold degenerate, disintegrates in a single level, a two-fold degenerate level and two three-fold degenerate levels. The situa-

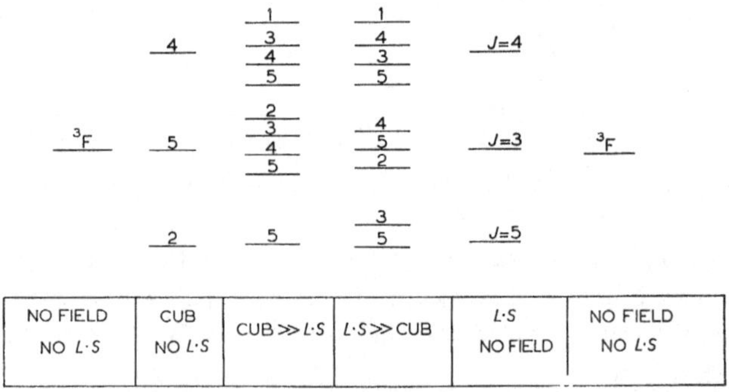

Fig. 8.1. The splitting of a ^{3}F-level under various conditions.

tion is sketched in Figure 8.1. All the unspecified numbers refer to the representations of the cubic group. On the right-hand side is the fictional case in which the spin orbit coupling is dominant. The figure indicates the proper order as found in nickel but the distances are arbitrary. Before going over to general considerations let us mention one more specific case. Suppose there are two levels which are very close together. In this case it is necessary to start out with the product representations of this particular set of levels. Hence take the direct product of, say, $\boldsymbol{L}_1$ and $\boldsymbol{L}_2$ and calculate the characters of this product representation. The next step is again to see which linear combinations of irreducible characters lead to the set of characters found from the product.

2.2. EXPLICIT CALCULATION OF WAVE FUNCTIONS

We want to give an example of how a reduction actually is performed. Suppose an atom with an $L = 2$ (D-state) is imbedded in a cubical field. It is known from the character table that $\boldsymbol{D}_2 = \boldsymbol{\Gamma}_3 + \boldsymbol{\Gamma}_5$, i.e. $L = 2$ will decompose in *the* two-dimensional representation and one of the three-dimensional representations.

Before treating the straightforward method, with the projection operator, we will illustrate the guess-technique usually employed for the lower values of L.

The $L = 2$ level has a basis which can be described by x^2, y^2, z^2, yz, xz, xy excluding $x^2+y^2+z^2$. The three last products transform like x, y, and z, except for inversion. Hence they correspond to one of the two three-dimensional representations. The possibilities with the squares are restricted since only two are independent x^2+y^2 does not work, since it goes over into y^2+z^2, using the rotation around the body diagonal which makes $x \to y$, $y \to z$ and $z \to x$. If this transformation is applied again, the result is z^2+x^2. These three functions together form the combination $2(x^2+y^2+z^2)$ which had to be excluded. If the combination x^2-y^2 is tried, it leads to y^2-z^2 and z^2-x^2. The third is a linear combination of the first two: $z^2-x^2 = -(y^2-z^2)-(x^2-y^2)$. After checking all the other matrix elements it is found that the first two indeed form a (non-orthogonal) set of basis functions.

Slightly less haphazard is the use of spherical harmonics [1] (compare Table 8.3):

[1] In the following calculation the proportionality constant is of no importance. In specific applications it should be noticed that different conventions are used for the sign or phase factor of the spherical harmonics. These differences are listed by EDMONDS [1957] on p. 21.

$$\begin{aligned} Y_2^0 &= 3z^2 - r^2 = 2z^2 - x^2 - y^2 \\ Y_2^{\pm 1} &= -(x \pm iy)z \qquad Y_2^{\pm 2} = (x \pm iy)^2. \end{aligned} \tag{8.6}$$

The three-dimensional representation is obtained by taking $Y_2^2 - Y_2^{-2}$ and $Y_2^1 \pm Y_2^{-1}$. The only choices left are $Y_2^2 + Y_2^{-2}$ and Y_0^2. These correspond to the polynomials $x^2 - y^2$ and $2z^2 - x^2 - y^2$ and these are orthogonal.

The straightforward method is to use the projection operator (Chapter 3, § 12.3). For these it is necessary to calculate the sum of a class.

$$\mathrm{E} = \begin{pmatrix} 1 & 0 & 0 & 0 & 0 \\ 0 & 1 & 0 & 0 & 0 \\ 0 & 0 & 1 & 0 & 0 \\ 0 & 0 & 0 & 1 & 0 \\ 0 & 0 & 0 & 0 & 1 \end{pmatrix} \qquad \mathrm{C}_2 = \begin{pmatrix} 1 & 0 & 0 & 0 & 2 \\ 0 & -1 & 0 & 0 & 0 \\ 0 & 0 & 3 & 0 & 0 \\ 0 & 0 & 0 & -1 & 0 \\ 2 & 0 & 0 & 0 & 1 \end{pmatrix}$$

$$\mathrm{C}_3 = \begin{pmatrix} -1 & 0 & 0 & 0 & 1 \\ 0 & -2 & 0 & 0 & 0 \\ 0 & 0 & 0 & 0 & 0 \\ 0 & 0 & 0 & -2 & 0 \\ 1 & 0 & 0 & 0 & -1 \end{pmatrix} \qquad \mathrm{C}_4 = \begin{pmatrix} 1 & 0 & 0 & 0 & -1 \\ 0 & 2 & 0 & 0 & 0 \\ 0 & 0 & 0 & 0 & 0 \\ 0 & 0 & 0 & 2 & 0 \\ -1 & 0 & 0 & 0 & 1 \end{pmatrix}$$

$$\mathrm{C}_5 = \begin{pmatrix} -2 & 0 & 0 & 0 & -2 \\ 0 & 0 & 0 & 0 & 0 \\ 0 & 0 & -4 & 0 & 0 \\ 0 & 0 & 0 & 0 & 0 \\ -2 & 0 & 0 & 0 & -2 \end{pmatrix}. \tag{8.7}$$

If they are used in

$$\varepsilon^{(\mu)} = \frac{n_\mu}{h} \sum_i X_\mu(i) \mathrm{C}_i, \tag{8.8}$$

which is actually eq. (3.32) except that the character is real, we find for $\mu = 3,5$ (all others give zero matrices)

$$\begin{aligned} &\langle \pm 2 | \varepsilon^{(3)} | \pm 2 \rangle = \langle \mp 2 | \varepsilon^{(3)} | \pm 2 \rangle = \tfrac{1}{2}; \\ &\langle 0 | \varepsilon^{(3)} | 0 \rangle = 1; \qquad \langle \pm 1 | \varepsilon^{(5)} | \pm 1 \rangle = 1; \\ &\langle \pm 2 | \varepsilon^{(5)} | \pm 2 \rangle = -\langle \mp 2 | \varepsilon^{(5)} | \pm 2 \rangle = \tfrac{1}{2}. \end{aligned} \tag{8.9}$$

In this case the state vectors are already orthogonal. We find after normalization that,

$$\Gamma_3 \begin{cases} (|2\rangle+|-2\rangle)/\sqrt{2}; \\ |0\rangle \end{cases} \qquad \Gamma_5 \begin{cases} |1\rangle \\ (|2\rangle-|-2\rangle)/\sqrt{2}. \\ |-1\rangle. \end{cases} \tag{8.10}$$

The symbols, $|m\rangle$, refer to the spherical harmonics; for practical purposes a table for $l \geqq 4$ is included (Table 8.3).

TABLE 8.3

Spherical harmonics

$$Y_{00} = \frac{1}{\sqrt{4\pi}},$$

$$Y_{10} = \sqrt{\frac{3}{4\pi}}\cos\vartheta, \qquad Y_{11} = \sqrt{\frac{3}{8\pi}}\sin\vartheta\, e^{i\varphi},$$

$$Y_{20} = \sqrt{\frac{5}{4\pi}}\left(\tfrac{3}{2}\cos^2\vartheta-\tfrac{1}{2}\right), \qquad Y_{21} = \sqrt{\frac{15}{8\pi}}\sin\vartheta\cos\vartheta\, e^{i\varphi}, \qquad Y_{22} = \tfrac{1}{4}\sqrt{\frac{15}{2\pi}}\sin^2\vartheta\, e^{2i\varphi}$$

$$Y_{30} = \sqrt{\frac{7}{4\pi}}\left(\tfrac{5}{2}\cos^3\vartheta-\tfrac{3}{2}\cos\vartheta\right), \qquad Y_{31} = \tfrac{1}{4}\sqrt{\frac{21}{4\pi}}\sin\vartheta\,(5\cos^2\vartheta-1)e^{i\varphi},$$

$$Y_{32} = \tfrac{1}{4}\sqrt{\frac{105}{2\pi}}\sin^2\vartheta\cos\vartheta\, e^{2i\varphi}, \qquad Y_{33} = \tfrac{1}{4}\sqrt{\frac{35}{4\pi}}\sin^3\vartheta\, e^{3i\varphi},$$

$$Y_{40} = \sqrt{\frac{9}{4\pi}}\left(\frac{35}{8}\cos^4\vartheta-\frac{15}{4}\cos^2\vartheta+\frac{3}{8}\right), \qquad Y_{41} = \tfrac{3}{4}\sqrt{\frac{5}{4\pi}}(7\cos^3\vartheta-3\cos\vartheta)\sin\vartheta\, e^{i\varphi},$$

$$Y_{42} = \tfrac{3}{4}\sqrt{\frac{5}{8\pi}}\sin^2\vartheta(7\cos^2\vartheta-1)\, e^{2i\varphi}, \qquad Y_{43} = \tfrac{3}{4}\sqrt{\frac{35}{4\pi}}\sin^3\vartheta\cos\vartheta\, e^{3i\varphi},$$

$$Y_{44} = \tfrac{3}{8}\sqrt{\frac{35}{8\pi}}\sin^4\vartheta\, e^{4i\varphi}.$$

We want to include, although this is superfluous, a semi-quantitative method. The reason is that on one hand guess work may fail and on the other hand the projection operators although straightforward are rather tedious to compute. Hence it may be worthwhile to look at the secular determinant directly. As an example take $L = 3$ (F-state) and again a cubical field.

A cubical field can be described by

$$\begin{aligned} \mathrm{H}_{\mathrm{cub}} &= ar^{-2}(x^4+y^4+z^4-3r^2) \\ &= \frac{a}{12}\left[2Y_4^0\sqrt{21}+(Y_4^4+Y_4^{-4})\sqrt{30}\right] \end{aligned} \tag{8.11}$$

if the four-fold axes are chosen as the x, y and z-axis.

In general the potential is of the type

$$\mathrm{H}_{\mathrm{el}} = \sum_{\mu,\lambda} a_\lambda^\mu Y_\lambda^\mu. \tag{8.12}$$

This satisfies Laplace's equation, since it is assumed, that the charges that generate the field are outside the atom. The constants, a_λ^μ are the parameters which one hopes to determine experimentally.

The calculation of the matrix elements with respect to a set of angular wave functions of given L-value is found by integration over three spherical harmonics.

$$\langle L, M'|Y_\lambda^\mu|L, M\rangle = (4\pi)^{\frac{3}{2}} \int_{\substack{\text{unit}\\\text{sphere}}} Y_L^{M'} Y_\lambda^\mu Y_L^M \mathrm{d}\omega. \tag{8.13}$$

The integral is only $\neq 0$ if; (i) $M' = M+\mu$ and (ii) the triangular rule is fulfilled: $\lambda \leqq 2L$. This means that the power series (8.12) actually is of no importance beyond a certain term. The coefficients (8.13) are similar to (5.42) (where $\lambda = 1$) and in general to (5.A24) the Wigner or Clebsch-Gordan coefficient. They are tabulated in the literature,[1] but for low L and μ they can be computed quickly by employing the following integral

$$\overline{(ax+by+cz)^{2n}} \equiv (4\pi)^{-1} \int_{\substack{\text{unit}\\\text{sphere}}} (ax+by+cz)^{2n} \mathrm{d}\omega = (2n+1)^{-1}(a^2+b^2+c^2)^n. \tag{8.14}$$

Comparing coefficients of $a^u b^v c^w$ the expressions are found for $\overline{x^n y^m z^l}$. Take for example $L = 3$, then for $\mathrm{H}_{\mathrm{cub}}$ the result is

$$\frac{2}{33}\begin{pmatrix} 3 & 0 & 0 & 0 & \sqrt{15} & 0 & 0 \\ 0 & -7 & 0 & 0 & 0 & 5 & 0 \\ 0 & 0 & 1 & 0 & 0 & 0 & \sqrt{15} \\ 0 & 0 & 0 & 6 & 0 & 0 & 0 \\ \sqrt{15} & 0 & 0 & 0 & 1 & 0 & 0 \\ 0 & 5 & 0 & 0 & 0 & -7 & 0 \\ 0 & 0 & \sqrt{15} & 0 & 0 & 0 & 3 \end{pmatrix} = \langle M|\mathrm{H}_{\mathrm{cub}}|M'\rangle. \tag{8.15}$$

If rows and columns are interchanged this matrix can be reduced to one one-dimensional and three two-dimensional matrices. The roots are $\lambda = 6$ (3 times); $\lambda = -2$ (3 times) and $\lambda = -12$ (once) in accordance with the

[1] See systematic bibliography at the end of this book.

decomposition in § (2.1). This leads immediately to the irreducible set of basis functions. However, this example was facilitated by the easy way in which the matrix can be decomposed.[1] This is generally not the case.

3. Double Groups

In Chapter 5 it was shown that representations of the full rotation group could be obtained with even dimensional (irreducible) bases, corresponding to half-integer values of j. This leads easily to the problem of manufacturing a similar set of representations for a finite group. The irreducible representations of the full rotation group are usually reducible with respect to a finite group and with the techniques of the projection operations the problem is solved in principle.

Before such methods are applied it is necessary to have the character tables of the double groups. In case of the full rotation group it was seen that the characters of the double group are different from representations with the help of spherical harmonics (if they were the same the representations would be equivalent!).

The same is expected to happen for the finite groups. At first inspection it looks strange that additional representations must be added to the character table, since a well-known theorem states that the number of representations is equal to the number of classes and the latter is given by the abstract group table. It follows from this that we are actually not dealing with the same abstract group.

The reason is that the representations derived from the spinors are not in a one-to-one correspondence with the different rotations of the finite group. A careful distinction must be made between a so-called non-faithful representation and the case we are discussing here. In the non-faithful representations a one-to-many correspondence is dealt with in the sense that to one representation matrix there correspond several elements of the abstract group. Here the case is just the opposite. We have two representation matrices, one with the plus and one with the minus sign, which correspond to one abstract group element.

As a result the double group $\mathscr{G}'$ actually consists of $2g$ elements and the number of classes is larger than the number of classes in the original group. Since the abstract group is different, of course there can be a larger number

[1] The factorisation of the secular equation has nothing to do with reducibility. The last term only refers to a set of matrices. An individual (normal) matrix can always be completely diagonalized.

of classes. However, it is not necessary that the number of classes be twice as large. This is an important point which will be discussed below.

The unit element E forms a class by itself; in the double group it will split in two classes E and $R = -E$. These two elements have the same character in those representations which the double group has in common with $\mathscr{G}$, they will have opposite sign in the additional representations. These two elements E and R form an invariant subgroup, hence $\mathscr{G}$ and $R\mathscr{G}$ form a factor group.

We will now show that the opposite is not true. With the opposite we mean that $\mathscr{G}$ is not an invariant subgroup of the abstract double group. This is easy to show since $\mathscr{G}$ is not even a subgroup. The two spinor variables ξ and η will transform under a rotation π around the x-axis according to:

$$a = \begin{pmatrix} 0 & 1 \\ -1 & 0 \end{pmatrix}, \qquad a^{-1} = \begin{pmatrix} 0 & -1 \\ 1 & 0 \end{pmatrix} \tag{8.16}$$

(compare (5.7b) with $\varphi = \frac{1}{2}\pi$; $\psi = -\frac{1}{2}\pi$ and $\vartheta = \omega_x = \pi$).

Hence if the elements of $\mathscr{G}$ are all the elements of $\mathscr{G}'$ with only one of the two signs, the inverse element of A cannot be included, which shows that $\mathscr{G}$ is not a subgroup. This simple illustration helps to stress the fact that the double group is not simply a direct product of the single group and the group consisting of the two elements 1 and -1 as is for instance the case if inversion symmetry is added to a group. Hence the structure of the double group is an interesting subject of study. The structure of the group $2\mathscr{G}$ can be described as follows. The elements E and R form a normal subgroup $\mathscr{N}$ of order 2; the factor group $2\mathscr{G}/\mathscr{N}$ is isomorphic to the ordinary point group. This structure is similar to the structure of the space group $\mathscr{S}$. There too a normal subgroup exists, the group of translations $\mathscr{T}$, and the factor group $\mathscr{S}/\mathscr{T}$ is isomorphic to the group of the rotations. The group of rotations is not a subgroup of $\mathscr{S}$, similar to the fact mentioned above that the elements of $\mathscr{G}$ do not form a subgroup of $2\mathscr{G}$. Both cases are examples of semidirect products (Chapter 7, § 4) of a group and an (operator) group. In both cases we start out with the point group. The operator groups are either the translations with the resulting semidirect product of the space group, or the group consisting of E and R with the resulting semidirect product of the double group.

In order to establish the character table of the double group from the original group $\mathscr{G}$, we first notice that if the classes C_i and RC_i are different their characters should be the same in those representations that $\mathscr{G}'$ has in

TABLE 8.4

Eulerian angles and values of α and β for the elements of the cubical and hexagonal double group

Cubical

No.	Element	$\frac{\psi}{\pi}$,	$\frac{\vartheta}{\pi}$,	$\frac{\varphi}{\pi}$	α	β	Class
1	xyz	0	0	0	1	0	$E(R=-E)$
2, 2′	$x\bar{y}\bar{z}$	0	1	0	0	$\pm i$	
3, 3′	$\bar{x}y\bar{z}$	1	1	0	0	± 1	C_2
4, 4′	$\bar{x}\bar{y}z$	1	0	0	± 1	0	
5	$x\bar{z}y$	0	$\frac{1}{2}$	0	$1/\sqrt{2}$	$i/\sqrt{2}$	
6	$zy\bar{x}$	$\frac{1}{2}$	$\frac{1}{2}$	$-\frac{1}{2}$	$1/\sqrt{2}$	$1/\sqrt{2}$	
7	$\bar{y}xz$	$\frac{1}{2}$	0	0	$(1+i)/\sqrt{2}$	0	C_3
8	$xz\bar{y}$	0	$-\frac{1}{2}$	0	$1/\sqrt{2}$	$-i/\sqrt{2}$	$(C_3' = -C_3)$
9	$\bar{z}yx$	$-\frac{1}{2}$	$\frac{1}{2}$	$\frac{1}{2}$	$1/\sqrt{2}$	$-1/\sqrt{2}$	
10	$y\bar{x}z$	$-\frac{1}{2}$	0	0	$(1-i)/\sqrt{2}$	0	
11, 11′	$\bar{x}\bar{z}\bar{y}$	0	$\frac{1}{2}$	1	$\mp i/\sqrt{2}$	$\pm 1/\sqrt{2}$	
12, 12′	$\bar{z}\bar{y}\bar{x}$	$-\frac{1}{2}$	$\frac{1}{2}$	$-\frac{1}{2}$	$\mp i/\sqrt{2}$	$\pm i/\sqrt{2}$	
13, 13′	$\bar{y}\bar{x}\bar{z}$	$\frac{1}{2}$	1	1	0	$\mp(1-i)/\sqrt{2}$	C_4
14, 14′	$\bar{x}zy$	0	$-\frac{1}{2}$	1	$\pm i/\sqrt{2}$	$\pm 1/\sqrt{2}$	
15, 15′	$z\bar{y}x$	$\frac{1}{2}$	$\frac{1}{2}$	$\frac{1}{2}$	$\pm i/\sqrt{2}$	$\pm i/\sqrt{2}$	
16, 16′	$yx\bar{z}$	$\frac{1}{2}$	1	0	0	$\pm(1+i)/\sqrt{2}$	
17	zxy	$\frac{1}{2}$	$\frac{1}{2}$	0	$\frac{1}{2}(1+i)$	$\frac{1}{2}(1+i)$	
18	$z\bar{x}\bar{y}$	$\frac{1}{2}$	$\frac{1}{2}$	1	$\frac{1}{2}(1-i)$	$\frac{1}{2}(1-i)$	
19	$\bar{z}x\bar{y}$	$-\frac{1}{2}$	$\frac{1}{2}$	1	$\frac{1}{2}(1+i)$	$-\frac{1}{2}(1+i)$	
20	$\bar{z}\hbar y$	$-\frac{1}{2}$	$\frac{1}{2}$	0	$\frac{1}{2}(1-i)$	$-\frac{1}{2}(1-i)$	C_5
21	yzx	0	$-\frac{1}{2}$	$-\frac{1}{2}$	$\frac{1}{2}(1-i)$	$-\frac{1}{2}(1+i)$	$(C_5' = -C_5)$
22	$\bar{y}\bar{z}x$	1	$-\frac{1}{2}$	$-\frac{1}{2}$	$\frac{1}{2}(1+i)$	$-\frac{1}{2}(1-i)$	
23	$y\bar{z}\bar{x}$	1	$-\frac{1}{2}$	$\frac{1}{2}$	$\frac{1}{2}(1-i)$	$-\frac{1}{2}(1+i)$	
24	$\bar{y}z\bar{x}$	0	$-\frac{1}{2}$	$\frac{1}{2}$	$\frac{1}{2}(1+i)$	$\frac{1}{2}(1-i)$	

Hexagonal

Set 1: $\beta = 0$, $\alpha = \exp(i\pi n/6)$			Set 2: $\alpha = 0$, $\beta = \exp(i\pi n/6)$		
Class	Element	n	Class	Element	n
E	1	0		7, 7′	3, 9
R	1′	6	C_5	8, 8′	1, 7
C_2	2, 2′	3, 9		9, 9′	5, 11
C_3	3	2		10, 10′	0, 6
	4	10	C_6	11, 11′	2, 8
C_3'	3′	8		12, 12′	4, 10
	4′	4			
C_4	5	1			
	6	11			
C_4'	5′	7			
	6′	5			

common with $\mathscr{G}$ and should be the opposite sign for the additional representations.

The interesting question when the classes will double and when they will not has been studied by BETHE [1929] and OPECHOWSKI [1940]. Opechowski's major result is: if there is a rotation around an angle π then the elements C_π and RC_π belong to the same class if, and only if, there is also another rotation through π around an axis perpendicular to the axis of the first rotation in the group.

On the basis of this result he shows that the knowledge of the character table of the "single" group is sufficient to construct the character table of the double group. Almost all character tables can be found in the literature (BETHE [1929]: hexagonal and cubic; OPECHOWSKI [1940]: rhomboedric and tetrahedral; ELLIOTT [1954]: for the space groups; JAHN [1938]: for the $\mathscr{C}$ groups).

In Table 8.4 the necessary ingredients to construct explicit representations of the cubic and hexagonal groups (and their subgroups) have been indicated. The two-dimensional representation is obtained by writing the matrix

$$\begin{pmatrix} \alpha^* & \beta^* \\ -\beta & \alpha \end{pmatrix} \quad \text{or} \quad \begin{pmatrix} \alpha & \beta \\ -\beta^* & \alpha^* \end{pmatrix}.$$

Representations of higher dimensions can be obtained by the direct product procedure explained in Chapter 5, § 2.2. The resulting representations are in general reducible and can always be reduced with the help of the projection operator method described in Chapter 3, § 12.3. The most elaborate task is the calculation of the elements containing rotations around an axis perpendicular to the z-axis. These matrices have been tabulated by EDMONDS [1957] and MEIJER [1954] ($j < 5$).

4. Operator Hamiltonians

In order to study the influence of a magnetic field on the spin of an electron belonging to an ion inside a crystal, the perturbation theory will have to be extended. The problem is one of degeneracy but not quite the same as was dealt with before. The difference is that all the levels dealt with are degenerated with the same degeneracy while in the previous considerations only the level under consideration was degenerated, while all others were single. Another difference is that degeneracy is lifted only in the second order.

In Chapter 2 the perturbed energy up to the second order is given by

$$E_{n,\,\text{pert.}} = E_n + \lambda H_{nn} + \lambda^2 \frac{|H_{nm}|^2}{E_n - E_m} \tag{8.17}$$

where H_{nm} represents the matrix elements of the perturbation operator with respect to the unperturbed eigenfunctions and E_n the unperturbed eigenvalues.

We will show below that a similar formula can be given for a set of levels each of which have the same degeneracy r. The difference is that the symbols H_{nm} are now r by r matrices instead of numbers. The energy $E_{n,\,\text{pert}}$ is hence expressed as a polynomial of matrices: the operator Hamiltonian. In § 4.3 we will see that, unless we are interested in numerical values of the constants, perturbation theory is not necessary and the operator Hamiltonian can be constructed on the basis of transformation properties only.

The situation, particularly in transition-element ions, is such that each level is characterized by an orbital quantum number L, a representation label Γ_i and by a label that numbers the different components (irreducible basis functions) of that representation. Since the $2L+1$ wave functions belonging to L usually decompose into several Γ_i which will in general be relatively far apart (about 10 000 cm^{-1}), the set of levels E_i, each belonging to a certain Γ_i, can be considered the unperturbed set in the same way as above. The eigenvalues E_n, E_m, etc. were supposed to be far apart.

If the total spin function is introduced, we will find that the degeneracy factor of every level is multiplied by $2S+1$. The perturbation created by the magnetic field will not be any different from a free ion, i.e. $2S+1$ equidistant levels, unless the spin-orbit coupling is introduced. This interaction will result in the symmetry restriction being "carried over" into spin space. It was already demonstrated how a certain orbital symmetry Γ_i and the representation of the full rotation group in spin space $\mathscr{D}_s$ can form a direct product space in § 2.1.

The spin-orbit interaction has no effect in the first order since all spin and all orbit matrices have trace zero. We shall see in the next section that the second order perturbation can be calculated in a way that bypasses the first order and gives a closed expression for the energy splitting and the g-factor.

4.1. VAN VLECK PERTURBATION THEORY

We want to formulate the perturbation theory with reference to the transition element ions where the crystal field effect can be assumed to be larger than the spin-orbit coupling. The first step in this problem is to diagonalize the crystal electric field in the space spanned by the orbital part of the wave function. It is at this moment that the considerations of group theory from the preceding section enter. The symmetries involved make it possible to calculate the projection operator (compare Chapter 3, § 2.3) that will lead to the proper linear combinations of wave functions. Although this procedure is straightforward and does not involve any diagonalization of matrices it may nevertheless lead to some practical difficulties, particularly if the symmetry is low. In this case the chances that a certain irreducible component occurs more than twice are large and most of the advantage of the method is lost since it is again necessary to diagonalize matrices of order larger than two. Hence it may be profitable to separate the electric field components into a large contribution and a small contribution, because the large part usually represents the main symmetry of the ion site in the crystal.

Suppose the wave function adapted to the main symmetry is obtained in this way:

$$|i, k_i\rangle = |L_g, M_L\rangle \langle L_g M_L|i, k_i\rangle + \sum_{L \neq L_g} |L, M_L\rangle \langle LM_L|i, k_i\rangle \qquad (8.18)$$

where the label i refers to the different energy eigenvalues in the electric crystal field and the label $k_i = 1, \ldots, n_i$ refers to the components of that particular energy level. The degeneracy of the level i is equal to n_i. The subscript g refers to the ground state. The ground states are known from Hund's rules. It could happen that the large electric field terms give rise to a single ground state, in this case the orbital angular momentum is considered to be *quenched*, since the thermal energy is usually much smaller than the distance to the next higher level or group of levels. The result is that only the lower level i is occupied and since the expectation value for the angular momentum of a single level is zero, the orbital motion does not manifest itself anymore, or is "frozen in". The other possibility, namely that the lowest electric field level is not single, occurs less often. It can be described by a pseudo angular momentum (ABRAGAM and PRYCE [1951]).
We will restrict ourselves from now on to the first case of a non-degenerate ground state.

If every electron wave function is multiplied by $2S+1$ spin wave functions, the degeneracy of all levels is multiplied by a factor $2S+1$. It seems at first sight that this problem is related to the degenerate perturbation problem mentioned in Chapter 2, §§ 7.2 and 7.3. The point is, however, that this problem is slightly more complicated since the spin-orbit coupling has the property that it does not lift the degeneracy in first order perturbation. Hence a formalism has to be used that handles the removal of the degeneracy in the second order. This is conveniently done with help of the so-called interaction representation. Transformation matrices are introduced such that the spin orbit coupling is transformed away in first order and a result is obtained which expresses the Hamiltonian directly in the second order contribution of the spin-orbit coupling. In the next section it will be seen that the result thus obtained is actually nothing but a special case of the tensor coupling between the orbital angular momentum operator and the spin angular momentum operator such that the internal product is an invariant. The method, originally invented by Van Vleck (see SCHIFF [1949]) is as follows:

Let the eigenvalue to be solved be expressed as:

$$(\mathrm{H}_{\mathrm{el}}+\mathrm{H}_{\mathrm{so}})|i, M_s\rangle = E|i, M_s\rangle, \tag{8.19}$$

or, if the matrix element of these operators is taken (the electrical field contribution is already diagonalized):

$$\langle j, M_s'|\mathrm{H}_{\mathrm{el}}+\mathrm{H}_{\mathrm{so}}|i, M_s\rangle = E_i^{(0)}\langle j, M_s'|\delta_{ij}|i, M_s\rangle+\langle j, M_s'|\mathrm{H}_{\mathrm{so}}|i, M_s\rangle. \tag{8.20}$$

Consider the following formal transformation of the state vector U into $V = (\exp - \mathrm{S})U$ which leads to the transformed Hamiltonian,

$$\mathrm{H}' = \mathrm{e}^{-\mathrm{S}}\mathrm{H}\mathrm{e}^{\mathrm{S}} = \mathrm{H}_{\mathrm{el}}+\mathrm{H}_{\mathrm{so}}+[\mathrm{H}_{\mathrm{el}}, \mathrm{S}]+[\mathrm{H}_{\mathrm{so}}, \mathrm{S}]+\tfrac{1}{2}[[\mathrm{H}_{\mathrm{el}}, \mathrm{S}], \mathrm{S}]+\ldots. \tag{8.21}$$

In the above development use has already been made of the fact that the spin-orbit coupling is assumed to be small compared to the large electric field terms, an assumption which is reasonably well fulfilled in the case of transition elements. It is not necessary to make a similar restriction for the small electric field terms since only the ground level has been taken into consideration and this was assumed to be single so that the only contribution of the small terms will be a slight shift. Now choose the transformation operator S in such a way that it removes the first order terms

$$\mathrm{H}_{\mathrm{so}}+[\mathrm{H}_{\mathrm{el}}, \mathrm{S}] = 0. \tag{8.22}$$

If this relation is substituted in the previous equation the result is

$$\mathrm{H}' = \mathrm{H}_{\mathrm{el}}+[\mathrm{H}_{\mathrm{so}}, \mathrm{S}]+\tfrac{1}{2}[[\mathrm{H}_{\mathrm{el}}, \mathrm{S}], \mathrm{S}] = \mathrm{H}_{\mathrm{el}}-\tfrac{1}{2}[\mathrm{H}_{\mathrm{so}}, \mathrm{S}]. \tag{8.23}$$

Equation (8.22) can be solved very simply because the matrix elements of the electrical part are diagonal. Hence we have:

$$\langle i|\mathrm{S}|j\rangle = -\langle i|\mathrm{H}_{\mathrm{so}}|j\rangle/(E_i - E_j) \tag{8.24}$$

where E_i and E_j are the diagonal elements of $\mathrm{H}_{\mathrm{electric}}$. If this result is inserted in the transformed matrices H′ the following result is found for the lowest level g:

$$\begin{aligned}\langle g|\mathrm{H}'|g\rangle &= E_g^{(0)} - \tfrac{1}{2}\langle g|[\mathrm{H}_{\mathrm{so}}, \mathrm{S}]|g\rangle \\ &= E_g^{(0)} - \sum_{i \neq g} \langle g|\mathrm{H}_{\mathrm{so}}|i\rangle\langle i|\mathrm{H}_{\mathrm{so}}|g\rangle/(E_i - E_g).\end{aligned} \tag{8.25}$$

After introducing the explicit expression for the spin-orbit coupling

$$\mathrm{H}_{\mathrm{so}} = \zeta \sum_{k=1}^{3} L_k S_k \tag{8.26}$$

the final result for the so-called spin-Hamiltonian is found to be

$$\langle g|\mathrm{H}'(S)|g\rangle = -\zeta^2 \sum_{k,l} A_{kl} S_k S_l + E_g^{(0)}$$

with

$$A_{kl} = \sum_i \langle g|L_k|i\rangle\langle i|L_l|g\rangle/(E_i - E_g). \tag{8.27}$$

The general symmetry considerations imply that this linear combination of spin operators has to have the same transformation properties as the electric field potential. Hence the important question may be asked, how much of the expression could have been constructed from the knowledge of the symmetry alone? This question will be treated in the next subsection.

The expression obtained for H′ as a function of the spin operator may or may not be diagonal with respect to the spin quantum number. After diagonalization, if necessary, the second order correction to the energy levels due to the spin orbit interaction is found. We would like to repeat this result in physical terms. A spin by itself (free spin) does not interact with an electrical field, because a spin is of magnetic nature only. For a bound spin the statement still holds in the first order, but in the next approximation the picture changes. The field interacts with the orbit, that is to say (using classical language) the orbit is not entirely free to orient itself and hence the spin, which is to a certain extent attached to the orbit, also partially loses its orientational freedom. Such an indirect interaction can, in quantum mechanics, only be obtained through at least second order perturbation since, in the language of time dependent perturbation theory, energy has to be "borrowed" temporarily and hence we deal with a second order process.

We want to illustrate the spin-Hamiltonian perturbation treatment with the case of the aforementioned alums. The main electric field is cubic and there is a small electrical field added along the body diagonal of the cube, which is taken as the axis of quantization. It turns out that in this case the spin-Hamiltonian is already diagonal for the following reason. The spin-orbit coupling operator in second order has the selection rules $\Delta M_S = 0, \pm 1, \pm 2$ and $\Delta M_L = 0, \pm 1, \pm 2$. The potential mentioned has $\mu = \pm 3, 0$, which gives a selection rule $\Delta M_L = \pm 3, 0$. Hence the only contribution will come from $\Delta M_L = 0$, this gives $\Delta M_S = 0$ and hence diagonal spin operators. This conclusion does not hold for the third order calculation, or for a potential of lower symmetry.

Since most information is obtained in experiments that are done in a magnetic field, replace the operator H_{so} by

$$\begin{aligned} H_{so}+H_{mf} &= \zeta \boldsymbol{L}\cdot\boldsymbol{S}+\mu(\boldsymbol{L}+g_0\boldsymbol{S})\cdot\boldsymbol{H} \\ &= (\zeta\boldsymbol{S}+\mu\boldsymbol{H})\cdot\boldsymbol{L}+g_0\mu\boldsymbol{S}\cdot\boldsymbol{H}. \end{aligned} \tag{8.28}$$

where H_{mf} represents the contribution to the Hamiltonian due to external magnetic field. This field is, in almost all experiments, of such an order of magnitude that $H_{mf} \ll H_{so}$ and hence can be considered as a small perturbation.

Assuming again that the spin-Hamiltonian is diagonal we see from the previous equation that $\zeta\boldsymbol{S}$ must be replaced by $\zeta\boldsymbol{S}+\mu\boldsymbol{H}$ and the spin-field term must be added. For the magnetic field dependent spin-Hamiltonian the result is

$$\begin{aligned} H_{so}+H_{mf} &= g_0\mu\boldsymbol{S}\cdot\boldsymbol{H}-\sum_{ij} A_{ij}(\zeta S_i+\mu H_i)(\zeta S_j+\mu H_j) \\ &= \sum_{ij}(\mu g_{ij}S_iH_j-\zeta^2 A_{ij}S_iS_j+\mu^2 A_{ij}H_iH_j) \end{aligned} \tag{8.29}$$

The dimensionless coefficient of the first order term in H is the so-called g-factor

$$g_{ij} = g_0\,\delta_{ij}-\zeta A_{ij}. \tag{8.30}$$

This factor can be determined experimentally. If we have a good estimate for ζ it allows us to determine A_{ij}. Since the result is the difference between two terms with the same order of magnitude the accuracy is usually low.

4.2. TENSOR OPERATOR

To illustrate how spin-Hamiltonians can be constructed from symmetry considerations only, take the simple, but often occurring example, of a crystal in which cylindrical symmetry around a certain axis can be assumed. If this axis is taken as the axis of spin-quantization (i.e. the z-axis of the spin matrix is taken along the axis of the cylinder) the spin-Hamiltonian up to second powers in S has to have the form:

$$\mathrm{H} = E_0 + D[3\mathrm{S}_z^2 - S(S+1)] \tag{8.31}$$

where E and D are two constants. The next term would be the polynomial equivalent of Y_3^0, or Y_4^0 in case we require inversion symmetry. How high a power of S is included depends on the number of levels, that is on the total spin value, of the system we are describing. In general it can be stated that:

The number of invariant tensor operators necessary to describe a perturbation on a state of $2S+1$ components is finite, since the total number of independent Hermitian matrices is $(2S+1)(S+1)$.

The generalization of the idea illustrated above is of particular importance for the work in rare earth ions. In this case the spin orbit coupling is of such a strength that L and S are not good quantum numbers and the electric field will have the effect that the levels characterized by an orbital momentum J, which has a degeneration $2J+1$, will split into a certain number of groups of levels. It would be impossible to calculate this by the previous perturbation theory since in this case the $\boldsymbol{L} \cdot \boldsymbol{S}$ coupling is not a small perturbation compared to the electric field, but the reverse is true. However, it is possible, although rather tedious, to take the field as a perturbation but since it acts only on the orbital part of J, it is first necessary to decompose the J.

In many problems, particularly if the components of the multiplet [1] are far apart, it is sufficient to give symmetry considerations and postpone the determination of the constants involved.

Although the principle for such a treatment was already indicated by Wigner and Kramers, the first practical evaluation was made by Stevens who calculated the operator polynomials corresponding to Y_2^0, Y_4^0, Y_6^0 as well as the multiplying factors for the ground state of each of the rare earth ions. The first results are indicated in Table 8.5.

TABLE 8.5

$$\Sigma(3z^2-r^2) \equiv \alpha \overline{r^2}[3J_z^2+J(J+1)]$$

$$\Sigma(35z^4-30r^2z^2+3r^4) \equiv \beta \overline{r^4}[35J_z^4-30J(J+1)J_z^2+25J_z^2+6J(J+1)+3J^2(J+1)^2]$$

$$\Sigma(231z^4-315r^2z^4-105r^4z^2-5r^6) \equiv \gamma \overline{r^6}[231J_z^6-315J(J+1)J_z^4+735J_z^4+105J^2(J+1)^2J_z^2$$
$$-525J(J+1)J_z^2+294J_z^2-5J^3(J+1)^3+40J^2(J+1)^2-60J(J+1)].$$

[1] A *multiplet* is a group of levels, originating from the same L and S value. The components are $J = L+S, \ldots |L-S|$.

We will now give a short account of KORRINGA'S [1954] treatment showing how operator invariants as well as operator covariants can be constructed. An operator covariant is comparable to an irreducible set of basis functions as shown below. A well-known operator covariant is the operator vector (**S**, **L** or **J**), the three components form an irreducible set of basis vectors with respect to the full rotation group.

If an operation G_i corresponding to an element g_i of the point group of the crystal is taken it will induce a transformation in a certain set of basis functions. If the basis functions are changed, the operator G will undergo a similarity transformation

$$\mathrm{G}_i' = \boldsymbol{S}^{-1}\mathrm{G}_i\boldsymbol{S}$$

Instead of using an arbitrary transformation on the basis functions one of the transformations of the group itself could be used. The number of different operators obtained by repeated transformation, cannot exceed the order of the group h.

$$\mathrm{G}_i'(j) = \mathrm{g}_j^{-1}\mathrm{G}_i\mathrm{g}_j \qquad (j = 1 \ldots h). \tag{8.33}$$

Since these operators are linear and since there are only a finite number of different operators available then each operator generated in this way from a given G_i can be written as a finite sum of operators.

$$\mathrm{G}_i'(j) = \sum_{l=1}^{m} a(j)_{il}\mathrm{G}_l \qquad m \leqq h. \tag{8.34}$$

A second condition on m is that it cannot exceed the number $\frac{1}{2}n(n+1)$, where n is the dimension of the space, since this is the total number of independent Hermitian matrices.

Now if j runs through all the elements of the group a representation with matrices (a_{il}) is created. The operators play the role of basis functions. If certain linear combinations are chosen the representations may be irreducible and an "irreducible set of basis operators" or a "covariant set of basis operators" is found.

Besides the terms in the full Hamiltonian that describe the interaction of the electric field and the electrons (1) we have terms that describe the coupling between two sets of degrees of freedom (2). Following Korringa we notice that these two sets of terms have counterparts in the spin-Hamiltonian:

(1) Operators that are invariant under the group $\mathscr{G}$ that corresponds to the

(main) symmetry of the crystal. These operators usually operate in a space that forms an irreducible representation of the full rotation group.

This is not strictly necessary, one could imagine the use of representations irreducible with respect to an intermediate group; i.e. a group that is a subgroup of the full rotation group. This choice is sometimes important in order to secure an unambiguous set of quantum numbers (compare HERZFELD and MEIJER [1960], Chapter 4, § 8).

(2) Operators in a product space of two (or three) of the following spaces: the space of the orbital electronic wave function $\mathfrak{R}_e$, the space of the electronic spin functions $\mathfrak{R}_s$, the space of the nuclear spin $\mathfrak{R}_n$. The bases used in these spaces are irreducible with respect to one of the three groups mentioned under (1). Take for instance the electron spin-nuclear spin coupling in a system embedded in a field of cylindrical symmetry. The three electron spin operators $S_{+1} = S_x + iS_y$, $S_0 = S_z$, $S_{-1} = S_x - iS_y$ each form a three-dimensional, and hence irreducible, representation of the crystal symmetry-group. (However, the first and the last are representations of the second kind.) The three nuclear spin operators I_{+1}, I_0 and I_{-1} form a three-dimensional representation irreducible with respect to the full rotation group. The reduction of the last set of basis operators, in the case of the subgroup of cylindrical symmetry, is obvious and the most general coupling, invariant under cylindrical operations, will be

$$H = A'S_{+1}I_{-1} + B'S_0 I_0 + C'S_{-1}I_{+1} \tag{8.35}$$

where A', B', and C' are constants, or actually functions of the radial coordinate.

However, this is not real (Hermitian), hence ignoring the imaginary part we find

$$H = AS_z I_z + B(S_x I_x + S_y I_y) \tag{8.36}$$

for the coupling. This result is rather obvious in this simple example, but for more sophisticated symmetries group theory is needed in order to construct the linear combinations of operators that transform like an irreducible basis for a certain representation.

Now to each irreducible representation of a certain dimensionality n belongs another, such that the direct product of the bases functions generates the unit representation, i.e. a sort of generalized internal product is taken as in vector analysis and an invariant is obtained. This adjoint set of basis functions forms a space of the same dimensionality and it is also irreducible.

In vector language it is the contravariant set of components. The same holds in operator space and by taking such an "internal product" that part of the operator Hamiltonian is obtained that describes, for example, the coupling between the nuclear and electron spin.

5. Kramers' Theorem and Time Reversal

A level which is characterized by J has a degeneracy $2J+1$. In case J is a half-integer this degeneracy is even. Kramers has discovered and proved that an electric field will split such a level at most in a number of two-fold degenerate levels. In the literature these levels are referred to as Kramers-pairs or Kramers-doublets and the expression Kramers-degeneracy is also found. Only a magnetic field, which has a lower symmetry than an electric field, as mentioned in Chapter 3, § 1 is able to lift the last part of this degeneracy. The difference between an electric and a magnetic field can be illustrated classically as follows. If the time is reversed in a magnetic field the electrons will flow in the opposite direction. The force exerted by the electrical field is in both cases the same since the force is independent of the velocity. In a magnetic field, however, the situation is different, the Lorentz force will change its sign if the velocity has the opposite direction. In quantum mechanics the reversal of time is connected with complex conjugation, as a simple look at the time dependence of the Hamiltonian shows. The wave functions ψ and ψ^* are both eigenfunctions of the H-operator belonging to the same energy value, provided the time dependent part is disregarded for the moment. This is also true if the H-operator contains the electric field strength $\boldsymbol{E}$. It is not true, however, if a magnetic field term is found in the Hamiltonian since these terms are pure imaginary with the result that ψ and ψ^* do not have to belong to the same eigenvalue. First the proof of this theorem will be given in the same way as the original proof by Kramers and later we shall return to the considerations about time reversal.

5.1. KRAMERS' THEOREM

Kramers indicated two proofs of his theorem. The first one (1930) is the most general and is based on the properties of the coefficients of $\boldsymbol{E}$ and $\boldsymbol{H}$ in the Hamiltonian just mentioned. The second proof (1933) is less abstract and has the advantage that it is closely connected with actual calculations. Suppose there is a single level characterized by J. The state vector is a linear combination of $2J+1$ wave functions φ_M:

$$\psi = \sum_M a_M \varphi_M \tag{8.37}$$

or in the Dirac-notation:

$$\langle | = \langle | M \rangle \langle M |. \tag{8.37a}$$

Previous considerations have shown us that the matrix elements of the electrical potential are proportional to a Clebsch-Gordan coefficient:

$$\langle J, M | V_\lambda^\mu | JM' \rangle \propto (JM', \lambda\mu | J\lambda JM)$$
$$M = M' + \mu. \tag{8.38}$$

The Clebsch-Gordan coefficients have the following symmetry relation

$$(JM', \lambda\mu | J\lambda JM) = (-1)^{2J+M+M'} (J \quad -M, \lambda\mu | J\lambda J \quad -M'). \tag{8.39}$$

The exponent of (-1) is equal to:

$$2J + M + M' = 2(J + M') + M - M'. \tag{8.40}$$

The first term of this expression is always even since if J is half-integer, M is also half-integer. Combining (8.38) and (8.39) gives the following symmetry relation for the matrix elements:

$$\langle -M | V | -M' \rangle = (-1)^{M'-M} \langle M' | V | M \rangle. \tag{8.41}$$

The matrix elements of the magnetic field,

$$\langle M | W | M \rangle = M g \mu_B H_z$$
$$\langle M | W | M \pm 1 \rangle = \tfrac{1}{2} g \sqrt{(J \pm M)(J \mp M + 1)} \mu_B (H_x \mp i H_y),$$

obey a similar relation, that is:

$$\langle -M | W(\boldsymbol{H}) | -M' \rangle = -(-1)^{M'-M} \langle M' | W(\boldsymbol{H}) | M \rangle \tag{8.42}$$

as one can see readily from the equations above. The magnetic field behaves under the transformation from M to $-M'$ and M' to $-M$ in the opposite way as the electric field. The secular equation for the coefficients $\langle M | \rangle$ are:

$$E \langle M' | \rangle = \sum_M [\langle M' | V | M \rangle + \langle M' | W(\boldsymbol{H}) M \rangle] \langle M | \rangle. \tag{8.43}$$

As a result of the properties (8.41) and (8.42) a second secular equation can be formulated as follows:

$$E(-1)^{J+M}\langle|-M\rangle = \sum_{M}[\langle M'|V|M\rangle + \langle M'|W(-\boldsymbol{H})|M\rangle](-1)^{J+M}\langle|M\rangle. \quad (8.44)$$

The whole equation was multiplied by $(-1)^J$ in order to avoid fractional exponentials.

The set of eigenfunctions of (8.43) is given by

$$|a\rangle = |M\rangle\langle M|a\rangle \quad (8.45a)$$

and the eigenfunctions of (8.44) are given by

$$|b\rangle = |M\rangle\langle M|b\rangle. \quad (8.45b)$$

If $\boldsymbol{H}$ is equal to zero, the following conclusion can be drawn:

$$\langle M|b\rangle = C(-1)^{J+M}\langle a|-M\rangle. \quad (8.46)$$

If (8.45a) and (8.45b) are eigenfunctions belonging to the same eigenvalue then $E_a = E_b$, because solutions of a homogeneous set of equations are proportional to each other. If (8.45a) and (8.45b) would be identical, except for a proportionality constant, we would have:

$$\langle M|b\rangle = C(-1)^{J+M}\langle -M|a\rangle^* = c\langle M|a\rangle. \quad (8.47)$$

Complex conjugation and substitution of M for $-M$ gives

$$C^*(-1)^{J-M}\langle M|a\rangle = c^*\langle -M|a\rangle^* \quad (8.48)$$

which cannot be fulfilled for a half-integer J, because the equation

$$c^*c = (-1)^{2J}C^*C \quad (8.49)$$

cannot be satisfied by a half-integer.

The final conclusion is that for $\boldsymbol{H}$ is equal to zero the eigenvalues E_a and E_b are equal as long as J is a half-integer. The usefulness of the above considerations is not, however, restricted only to $\boldsymbol{H} = 0$ but some conclusions can be made in case there is also a magnetic field. We want to mention that the reversal of the sign of the actual quantum number is, of course, related to the rotation of the electrons in the opposite direction. Hence as will be seen later, there is a close relation between the properties of the matrix element indicated above and time reversal. As a matter of fact it is possible to give this proof in such a way that only the time reversal property is used, without going into explicit calculations of the matrix elements.

5.2. TIME REVERSAL

In classical mechanics a system will traverse a certain trajectory in space and if the time, which serves as a measure of arc length, is reversed, the orbit will in general be the same but the direction in which it is traversed will be the opposite. This statement ceases to be true in case there are velocity dependent forces as for instance the Lorentz-force or the Corioli's force. If the orbit curves to the right for forward speeds, it will of course curve to the left if the motion is reversed. Inspection of Newton's equation immediately tells the whole story. The acceleration term is time (reversal) invariant, the force term is always invariant in case it contains only position coordinates. In case it contains velocity components it depends on whether the function is even or odd in these components. For instance, air friction, which is proportional to the speed or absolute value of the velocity, gives rise to an even function, while the Lorentz force does not.

In case we work with an Hamiltonian formalism similar statements can be made. The magnetic field is now represented by a vector potential. The result is the same, the orbit is invariant if one reverses the time, provided the magnetic field is inverted also.

Considering the time dependent Schrödinger equation without a magnetic field, we see that complex conjugation and time reversal leaves the operators on the right and on the left side invariant. Time reversal is now postulated equivalent to complex conjugation of the Schrödinger equation. In case a magnetic field is present invariance can only be expected, as illustrated above, in case time reversal is accompanied by a field reversal. This is a reasonable procedure since if the magnetic field producing device is incorporated in the system, the generating currents would also be reversed under time reversal and hence the field is inverted. The case in which the spins are the field producing agent (a permanent magnet) will be discussed next. So far the Hamiltonian was considered to be without spin.

The Hamiltonian for a system with spin will be of the form:

$$\mathrm{H} = \frac{1}{2m}\left(\boldsymbol{p}+\frac{e}{c}\boldsymbol{A}\right)^2 + e\varphi + \mu \boldsymbol{H}\cdot\boldsymbol{S}. \tag{8.50}$$

The operation of $\boldsymbol{A} \to -\boldsymbol{A}$ and $\boldsymbol{p} \to \boldsymbol{p}^*$ will leave the first term invariant, but field reversal in the second term has to be accompanied by a reversal in sign of all three spin components. The result of these considerations is that the equations of quantum mechanics will be invariant under time reversal plus complex conjugation plus reversal of the sign of the spin components. Hence it is stated that time reversal is equivalent to the application of the last two. The operator which reverses the spin components is

$$\pm \mathrm{iS}_y \begin{pmatrix} \xi \\ \eta \end{pmatrix} = \begin{pmatrix} \mp\eta^* \\ \pm\xi^* \end{pmatrix}. \tag{8.51}$$

Kramers calls this the spin conjugated spinor. This idea can easily be extended to arbitrary spin values.

In dealing with a n-spin system product spaces can be easily constructed from ξ and η and the corresponding operator will be a product of iS_y's, each acting on one of the spins.

Having established the n-particle time reversal operator:

$$\mathrm{K} = \mathrm{i}^n \mathrm{S}_y^{(1)} \mathrm{S}_y^{(2)} \dots \mathrm{S}_y^{(n)} \cdot \mathsf{C} \tag{8.52}$$

where C is the complex-conjugation operator.

It is of course necessary to demand that "double-time-reversal" be equivalent to the unit transformation. If this operator is applied to a non-degenerate wave function it is found that

$$\psi = \mathrm{K}^2\psi = (-1)^n\psi \tag{8.53}$$

hence that the wave function would always vanish for odd values of n. For even values of n time reversal corresponds to multiplication with a phase factor

$$\mathrm{K}^2\psi = \psi \to \mathrm{K}\psi = c\psi; \quad |c^2| = 1. \tag{8.54}$$

If there is at least two-fold degeneracy in the case of odd n-values one may have

$$\mathrm{K}\psi_1 = c\psi_2; \quad \mathrm{K}\psi_2 = -c\psi_1 \tag{8.55}$$

with $|c^2| = 1$ and the wave function is not equal to zero.

Returning to the case that $\boldsymbol{H} \neq 0$ the following properties can be shown: Applying the operator K to the representation of the magnetic moment of the system

$$\mathscr{M} = \mu \sum_{i=1}^{n} \boldsymbol{r}_i \times \boldsymbol{p}_i + g_0 \boldsymbol{S}_i \tag{8.56}$$

gives, if the wave function is non degenerate:

$$(\psi, \mathscr{M}\psi) = (\mathrm{K}\psi, (\mathrm{K}\mathscr{M}\mathrm{K}^{-1})\mathrm{K}\psi) = -(\psi, \mathscr{M}\psi). \tag{8.57}$$

Hence the expectation value of $\mathscr{M}$ is zero. The equations (8.54) and $\mathrm{K}\mathscr{M}\mathrm{K}^{-1} = -\mathscr{M}$ were used since the orbital part of (8.56) is purely imaginary and the spin reverses sign under this operation. For a Kramers doublet we find in the same way

$$(\psi_1, \mathscr{M}\psi_1) = -(\psi_2, \mathscr{M}\psi_2) \tag{8.58}$$

hence the trace of the two by two matrix is zero and the "center of gravity" of the two corresponding levels will not shift from the zero field value as long as we consider the terms proportional to $\boldsymbol{H}$:

$$E_\pm = E_0 \pm \mathscr{M} \cdot \boldsymbol{H}. \tag{8.59}$$

For an even number of particles it was seen that the K-operator has the property (8.54) and if $\psi' = c^{\frac{1}{2}}\psi$ is introduced we have

$$\mathrm{K}\psi' = \psi' \tag{8.54a}$$

which implies that the function is real. This idea can be generalized in case the function is degenerate.

FROBENIUS and SCHUR [1906] have shown that there are in principle three different kinds of representations possible for a finite group. The first kind of representation consists of matrices that are real, or that are equivalent to real matrices. The second kind consists of matrices that are complex, but complex in such a way that the complex conjugate matrices will transform in a way equivalent to the original. The third kind of representation consists of complex matrices, but the complex conjugate transformations belong to an irreducible representation which is different from the original. It is clear that the characters of the representations of the first and the second kind are real, while the characters of the representation of the third kind are imaginary. In the full rotation group representations of integer J value have the same property as the representations of the first kind, while the representations of the second kind correspond to the representations which have a half-integer J value. Since in the finite groups the distinction between the integer and half-integer values of J is lost, in this case the distinction of the different "kinds" of representations has to be used.

The generalization of equation (8.54a) now is that all systems with an even number of electrons have representations of the first kind. The generalization of (8.55) is that all odd electron systems have representations of the second kind. The representations of the third kind are excluded since the bilinear form

$$E = \sum_{n=1}^{r} (\psi_n^*, \mathrm{H}\psi_n) \tag{8.60}$$

has to be an invariant and this excludes the possibility of such a representation.

6. Jahn-Teller Effect

6.1. INTRODUCTION, EXAMPLES

In [1937] Jahn and Teller discovered the following interesting and useful theorem: Unless a molecule is linear, stability and (orbital) degeneracy are not possible simultaneously. The importance of this theorem is of course that certain configurations for molecules can be disqualified "a priori" and applications of this theorem are found in many places in the literature (see bibliography).

In order to get an insight into the idea behind this theorem consider a nuclear configuration that will have a certain symmetry which is maintained if all distances are multiplied by a certain factor. This type of configuration is called a *similar configuration.* In the case of vibrations, these similarity vibrations are usually called breathing vibrations. Consider a configuration with all the configurations similar to it and take the binding energy as a function of the scale factor (or configuration coordinate). This function will have a minimum, otherwise there is no possibility of obtaining a molecule this way. Take the configuration coordinate corresponding to this minimum as a point of departure and study all other displacements of the nuclei to see whether they are stable or unstable.

Following Jahn and Teller, let us first give an example. A linear triatomic molecule can undergo vibrations in which the center atom is displaced perpendicular to the cylindrical axis of symmetry. Since the displacement d, and $-d$ are identical, we will have $E(d) = E(-d)$. To understand this remember that the wave function is characterized by $\lambda = 0, \pm 1, \pm 2, \ldots$ (respectively a σ, π, δ etc. state) where λ measures the angular momentum around the z-axis, i.e. the cylindrical axis (compare Chapter 4, § 4). The states π and δ, etc. are two-fold degenerate, corresponding to right and left circular orbits. A left circular motion stays a left circular motion whether the molecule is bent, straight or bent the opposite way. The first goes over into the last if the molecule is rotated 180° around the z-axis, hence the wave function rotated over $-180°$ will undo this and since the energy, being an expectation value, will not depend on the phase factor we should have $E(d) = E(-d)$. This means that, assuming a continuous energy curve, a power series around the origin $d = 0$ will have only even terms, or that for small displacements this curve is a parabola. For $\lambda \neq 0$ there are actually two parabolas, since the degeneracy was due to the cylindrical symmetry which is destroyed for $d \neq 0$. (Fig. 8.2.)

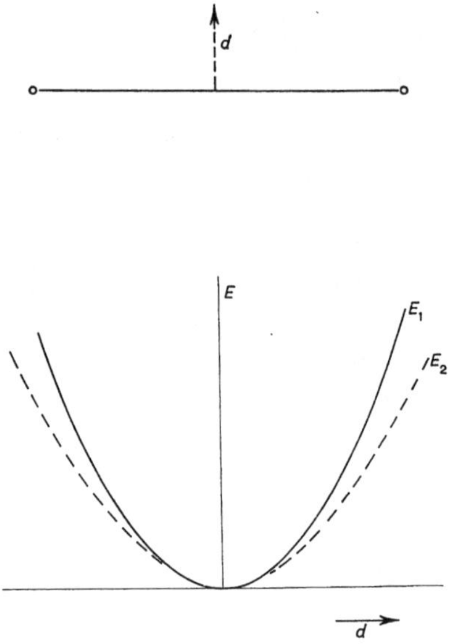

Fig. 8.2. Energy as a function of the displacement coordinate in a linear molecule.

In this case the wave function is of the type

$$\psi(r, z, \theta) = \varphi(r, z) \exp i\lambda\theta \tag{8.61}$$

where θ has any value. If we consider a different example, i.e. of a square molecule, we will have again an angular dependence $\exp i\lambda\theta$ but $\theta = 0$, $\pm\frac{1}{2}\pi$, π and the quantum number λ is equal to 0, ± 1. We are only interested in the degenerate state $\lambda \neq 0$ and describe the two possibilities by their nodelines, (position 1 and 2 in Fig. 8.3) instead of by their quantum number λ. From the picture we learn again that the wave function is degenerate since a transformation *belonging to the group of this molecule* will bring ψ_1 into ψ_2. If $d \neq 0$ the degeneracy will be lifted, if we compare $E(+d)$ with $E(-d)$ we find the following scheme:

$$E_1(+d) = E_2(-d) \tag{8.62a}$$

$$E_2(+d) = E_1(-d) \tag{8.62b}$$

because "Position$+d$" and nodeline 1, go over into "Position$-d$" and nodeline 2 if Fig. 8.3 is rotated 90°. The resulting energy diagram (indicated in Fig. 8.3) excludes a minimum, except for accidental degeneracies. That part of the degeneracy which comes from the symmetry of the molecule will be removed in such a way that one of the levels is lowered if the $d \neq 0$. Hence we deal with an unstable configuration. We want to stress that this example, and also the general theorem, shows that there is at least one type of displacement which is unstable, and that this does not exclude other displacements that are stable. As a matter of fact a stable displacement has already been postulated, the similarity displacement (or breathing vibration). No need to say that *one* unstable displacement is sufficient to make the molecule go away from the initial configuration. This displacement is usually referred to as a Jahn-Teller distortion.

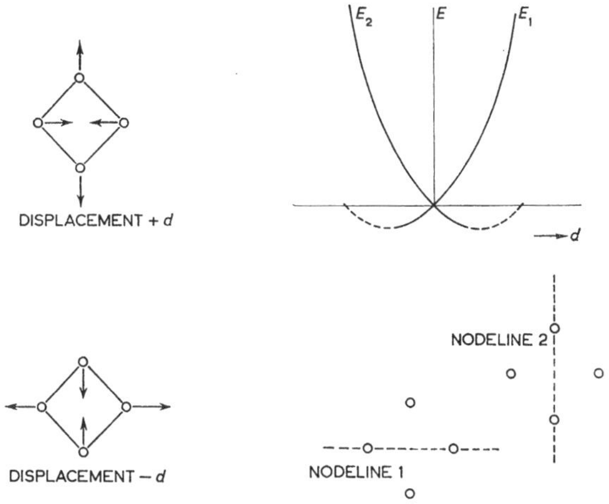

Fig. 8.3. Energy as a function of the displacement coordinate in a square molecule. The nodelines refer to the nodes of the electronic wave function.

6.2. NORMAL COORDINATES

In order to look into the general case it is necessary to make a slight digression into "normal coordinates" of vibrations in molecules. This subject actually can also be considered as an application of group theory, and hence is of interest on its own, but since this is a purely classical subject it actually falls outside the scope of this book.[1] The potential energy of a molecule under nuclear displacements is

[1] A similar group theoretical problem exists in lattice vibrations. The normal coordinates are labelled by two types of quantum numbers, one type referring to the branches (similar to the bands in wave mechanics of solids) and the other type referring to the wave vector. (See PHILLIPS [1956].)

$$V = \sum_{i,j} b_{ij} q_i q_j \tag{8.63}$$

where q are the $3N$ coordinates of the nuclei measured from the equilibrium positions (or any linear combination thereof) and b_{ij} represent either the spring constants in case the molecule is characterized by a set of masses coupled by weightless springs or the coefficients of the second term in the multivariable power series around the equilibrium position in case a more physical picture is preferred. The generalized kinetic energy is

$$T = \tfrac{1}{2} \sum_{ij} a_{ij} \dot{q}_i \dot{q}_j, \tag{8.64}$$

where the a_{ij} are dependent on the masses. The equations of motion obtained, for instance, from the Lagrangian[1] (the Hamiltonian is just as good, but associated with momenta instead of the $\dot{q}$) are

$$\sum_j a_{ij} \ddot{q}_j + \sum_j b_{ij} q_j = 0. \tag{8.65}$$

We take a linear combination of q's:

$$Q = \sum_i h_i q_i \tag{8.66}$$

(which, as we will see below, is again a special case of (1.3a)) in order to solve the problem by diagonalization. Multiply (8.65) by c_i and sum over i. This gives by comparing coefficients

$$\sum_i c_i a_{ij} = h_j \quad \text{and} \quad \sum_i c_i b_{ij} = -\lambda h_j. \tag{8.67}$$

With the c_i determined this way the original set of equations becomes:

$$\ddot{Q} + \lambda Q = 0, \tag{8.68}$$

which has the solution

$$Q(t) = Q(0) \exp i\omega t, \quad (\omega^2 = \lambda).$$

The coefficients c_i are determined by

$$\sum_i (\lambda a_{ij} - b_{ij}) c_i = 0. \tag{8.69}$$

These equations are basically the same as (1.14) since by multiplication with (a_{kl}^{-1}) the first term will be diagonal. Solving the secular problem will give a set of eigenvalues λ_k and to each a set of coefficients $c_i^{(k)}$ which in turn deter-

[1] If we start with Newton's law $F = ma$, and there is actually no compelling reason to use the Lagrangian formalism, the kinetic energy will contain only diagonal terms.

mine the coefficients $h_j^{(k)}$. The coordinates

$$Q_j^{(k)} = \sum_j h_j^{(k)} q_j \tag{8.70}$$

are the normal coordinates of the problem. Each of these coordinates vibrates independent of the other, with its own frequency and amplitude. This amplitude is determined by the initial condition only. This result is applicable to a molecule but also to an infinite solid and leads in the last case to the well-known lattice waves, characterized by k, with their corresponding frequencies $\omega(k)$.

The degeneracies in λ are connected with the symmetry of the molecule. The same statements hold for solids where different k-values may correspond to the same ω. The study of these degeneracies can again be undertaken by group theory since they are the result of the fact that those transformations R in laboratory space which are allowed by the symmetry of the molecule (lattice) will induce a transformation in the normal coordinates. The result of a transformation will be

a) if λ is non-degenerate:

$$\mathrm{R}Q_k = \pm Q_k \text{ [1]} \tag{8.71}$$

b) if λ is r-fold degenerate:

$$\mathrm{R}Q_k = \sum_{l=1}^{r} a_{kl} Q_l \tag{8.72}$$

and the coefficients (a_{kl}) form an r-dimensional representation of the symmetry group. This representation will be irreducible, excluding accidental degeneracy.

Take for example a triangular molecule in a two-dimensional plane. The number of actual degrees of freedom is 3 (i.e. 6 minus 2 translations minus one rotation). One is the "breathing" vibration and the other two form a degenerate pair, called $v_{2a}(0)$ and $v_{2b}(0)$ (HERZBERG, [1945], p. 84). The elements of the group (rotations of $\frac{2}{3}\pi$ around the center) transform one into the other, or into a linear combination of these two. This (real) two-dimensional representation has non-symmetrical matrices. It can of course be symmetrized, but then the matrices will no longer be real.

[1] Notice the difference with quantum mechanics where all coefficients are complex and all matrices unitary. The only phase factor in this problem is ± 1.

6.3. GENERAL DESCRIPTION

Normal coordinates are used as a new system of axes in which the displacements of the molecule are described. In order to do this they have to be normalized, and a positive direction has to be indicated for each of them. If the magnitude of the normal displacement along the r-th coordinate is called η_r, the deformed molecule has coordinates,

$$\boldsymbol{Q} = \boldsymbol{Q}^0 \eta_0 + \sum_{r \neq 0} \boldsymbol{Q}_r \eta_r, \tag{8.73}$$

where $\boldsymbol{Q}^0$ refers to the normal coordinate representing the original shape of the molecule (the corresponding vibration is the breathing vibration) and η_0 is the scale factor and has the value corresponding to a minimum of the energy. The Hamiltonian is a function of these η, and close to equilibrium (i.e. the equilibrium of the $\boldsymbol{Q}^0$ coordinate) it may be expanded as follows:

$$\mathrm{H} = \mathrm{H}_0 + V = \mathrm{H}_0 + \sum_r V_r(q)\eta_r + \sum_{r,s} V_{rs}(q)\eta_r\eta_s \tag{8.74}$$

where $V = V(\boldsymbol{Q})$ and hence $V(q)$. If we want to determine the energy by perturbation calculation of a p-fold degenerate level the secular matrix has to be solved; the elements are given by:

$$V_{nm} = \sum_r \eta_r \int \psi_n^* V_r(q) \psi_m \mathrm{d}\tau \qquad (n, m = 1 \ldots p) \tag{8.75}$$

and the first order correction on the energy is found after diagonalization of this matrix

$$E = E^0 + (\mathrm{S}^{-1} V \mathrm{S})_n = E^0 + V_n'. \tag{8.76}$$

The energy is a scalar, i.e. has to transform like the unit representation. If the matrix elements (8.75) are studied they transform according to:

$$\boldsymbol{\Gamma}_{\psi^*} \times \boldsymbol{\Gamma}_r \times \boldsymbol{\Gamma}_\psi = \sum_i a_i \boldsymbol{\Gamma}_i \tag{8.77}$$

a direct product of the p-dimensional representation of the wave function $\boldsymbol{\Gamma}_\psi$, the representation of the complex conjugate wave function $\boldsymbol{\Gamma}_{\psi^*}$(usually the same basis, but not in systems with an odd number of spins) and the r-dimensional representation described by (8.72). The right-hand side of (8.77) are the irreducible representations contained in this product. If $a_1 = 0$ there is no "scalar" representation possible and hence $V_{nm} = 0$. This happens to be the case in linear molecules. In the D_4-case $a_1 \neq 0$ and since, apart from accidental degeneracy $V_{nm} \neq 0$, there will be a linear term in the energy. The main part of the Jahn-Teller proof consists in showing that for all non-linear molecules, the direct product (8.77) contains the unit representation.

Again take the example of a two-dimensional square molecule. There are $8-3 = 5$ vibrations possible. That is the breathing or totally symmetrical one, plus 2 non-degenerate (called B_{1g} and B_{2g}) and one 2-fold degenerate vibration called E_u. (Compare HERZBERG [1945], p. 92.) The wave function can belong to one of the five representations of this group. (Compare BETHE [1929] or HERZFELD-MEIJER [1961]) but only one is degenerate $\Gamma_4 \equiv E$. Multiplying $[\Gamma_4]^2$ by B_{2g} we find that it contains $\Gamma_1 \equiv A_g$. (The same holds for B_{1g}). Hence these two modes lead to decomposition of the molecule, or at least to a non-square distortion.

How much the molecule deforms depends on the actual shape of the potential curves in Fig. 8.3. It may, for instance, be that the curves will go through a minimum and increase again as suggested by the dotted part. The distance from this minimum to the origin is the final amount of distortion. This distance may range from infinite, in which case the molecule will dissociate along this mode, to very small. In the last case the effect may not be detectable. This means that the application of the theorem has to be accompanied by an order of magnitude calculation. The theorem also holds for spin-degeneracy, a statement that brings up some interesting details, but Teller (See JAHN [1938]) estimated that the order of magnitude of the distance from the origins is so small that the implications are not important.

PROBLEMS

1\. 1. Show that the eigenvalues of a unitary matrix have absolute value one.

1\. 2. Show the invariance of the trace by transforming the definition (1.17) by a similarity transformation (§ 3.5).

1\. 3. Is the product of two Hermitian matrices a Hermitian matrix? The same question for a unitary matrix.

1\. 4. Show that every unitary matrix U in a unitary vector space can be written in the form

$$\mathsf{U} = e^{i\mathsf{S}}$$

where S is an Hermitian matrix. The exponential of a matrix is defined by its power series expansion.

1\. 5. Show that, if $\boldsymbol{x}_n$ and $\boldsymbol{x}'_n$ are orthogonal if they are two eigenvectors in the unitary space of a Hermitian matrix that belong to different eigenvalues.

1\. 6. Is the *product* of two Hermitian matrices a Hermitian matrix? Answer the same question with respect to the commutator and the double commutator. Apply the results to equation (8.21).

1\. 7. Show that an arbitrary matrix A can be written as

$$\mathsf{A} = \mathsf{HU}$$

where H is an hermitian matrix and U a unitary matrix. (This so called "Polar factorisation" is described in Murnaghan "Theory of Group Representations", page 27. The proof consists of two steps. Step one is the proof that $\mathsf{A^*A}$, which is obviously Hermitian, is positive definite Hermitian. Hence we can write

$$\mathsf{A^*A} = \mathsf{P}^2$$

where P is also positive definite. The second step is to show that $\mathsf{U} = \mathsf{AP}^{-1}$ is unitary.)

1\. 8. Show that the necessary and sufficient condition that A is normal is: $\mathsf{UP} = \mathsf{PU}$.

1. 9. The matrix

$$\mathsf{U} = \frac{1-\mathrm{i}\mathsf{H}}{1+\mathrm{i}\mathsf{H}}$$

is unitary if and only if H is hermitian (CAYLEY).

1. 10. Using the matrix S defined in problem 1.4 show that

$$\mathsf{H} = \mathrm{tg}\,\tfrac{1}{2}\mathsf{S}.$$

1. 11. Show that

$$\mathsf{VU} = \frac{1-\mathrm{i}\mathsf{H}'}{1+\mathrm{i}\mathsf{H}'}\,\mathsf{V}$$

where $\mathrm{H}' = \mathsf{V}^{-1}\mathrm{HV}$.

1. 12. Show that

$$\frac{1}{1+\mathsf{A}-\mathsf{B}} = \frac{1}{1+\mathsf{A}} + \frac{1}{1+\mathsf{A}}\,\mathsf{B}\,\frac{1}{1+\mathsf{A}} + \frac{1}{1+\mathsf{A}}\,\mathsf{B}\,\frac{1}{1+\mathsf{A}}\,\mathsf{B}\,\frac{1}{1+\mathsf{A}} + \ldots$$

and

$$\frac{1}{1+\mathsf{A}-\mathsf{B}} = \frac{1}{1+\mathsf{A}}\left(1+\mathsf{B}\,\frac{1}{1+\mathsf{A}-\mathsf{B}}\right).$$

1. 13. Show that the eigen values of the Hermitian matrix

$$\begin{pmatrix} c_{11}-\lambda & c_{12} & \ldots \\ c_{21} & c_{22}-\lambda & \ldots \\ \ldots & \ldots & \ldots \end{pmatrix} \qquad c_{ij} = a_{ij}+ib_{ij}$$

are the same as the eigen values of the matrix

$$\begin{pmatrix} a_{11}-\lambda & 0 & a_{12} & -b_{12} & \ldots \\ 0 & a_{11}-\lambda & b_{12} & a_{12} & \ldots \\ a_{21} & -b_{21} & a_{22}-\lambda & 0 & \ldots \\ b_{21} & a_{21} & 0 & a_{22}-\lambda & \ldots \\ \ldots & \ldots & \ldots & \ldots & \ldots \end{pmatrix}$$

if we disregard certain multiple roots. This property is useful for machine calculations of complex matrices.

2. 1. Suppose that a particle is described by the wave function:

$$\psi = f(r) \quad (2Y_1^1 + 2Y_0^1 + iY_0^0),$$

calculate: $L^2\psi$, $L_z\psi$, the probability that a measurement of L_z will give zero, and the expectation value of L^2.

2. 2. The operator A is defined by the equation:

$$A\psi(x) = \psi(x+a)$$

where a is a real constant. Verify that the function $u_k(x) \exp(ikx)$ is an eigenfunction of A if $u_k(x)$ is a periodic function with the period a, i.e. $u_k(x+a) = u_k(x)$. Is A an Hermitian operator? Express the commutator of x and A in terms of A. Consider the operator $(A-1)/a$ as a goes to zero, and give it a physical interpretation.

---

3. 1. Determine from the group table of the symmetry group $\mathscr{S}_3$ the minimum number of elements necessary to generate the complete group.

3. 2. Make a group table similar to the example in Ch. III, § 2 for the tetrahedral group. A geometrical representation of this group is found by taking the four points $(1, -1, -1)$; $(-1, 1, -1)$; $(1, 1, 1)$ and $(-1, -1, 1)$ in three dimensional space and performing all spacial rotations that bring these points into coincidence. Is this group the same abstract group as $\mathscr{S}_4$?

3. 3. The ordinary complex numbers are a special case of hypercomplex numbers. The basis of this algebra is related to the Abelian group of order four. The hypercomplex numbers are

$$\xi = a(1) + b(i) + c(-1) + d(-i).$$

Construct the regular representation.

3. 4. Show that the regular-representation matrix A given as example in § 10.3 satisfies the multiplication rules:

$$A^{-1}E = B, \quad A^{-1}A = E, \quad A^{-1}B = A, \quad A^{-1}C = F,$$

$$A^{-1}D = C \text{ and } A^{-1}F = D.$$

3. 5. Obtain from the regular representation of $\mathscr{S}_3$, as constructed in problem 4, the two dimensional irreducible representation by

means of the projection operators. The character table is given in § 12.

3. 6. Show that the orthogonality relations (3.16), (3.17) and (3.18) are fulfilled for the example given in § 7.4.

3. 7. It is stated in § 11.2 that the matrix elements a_{ik}, a'_{ik} $\cdots$ are linearly independent, i.e. the form

$$\sum_{ik} \lambda_{ik}\, a_{ik} + \sum_{jl} \lambda'_{jl}\, a'_{jl} + \ldots + \sum_{\mu\nu} \lambda^{(p)}_{\mu\nu}\, a^{(p)}_{\mu\nu} = 0$$

is only equal to zero if all λ, λ' . . . are zero. Show that this is correct. The matrix λ is independent of the operation a. (Hint: multiply by $a^{*(p)}_{\mu\nu}$ and sum over a.)

3. 8. Calculate the characters of a regular representation and show with the help of (3.23) that the regular representation contains every irreducible representation as many times as the dimensionality of that irreducible representation.

3. 9. The 24 operations that bring a cube to final positions indistinguishable from the initial positions form a group. This group is isomorphic with the permutation group of four objects (the body diagonals, e.g. we number the corners such that two opposite corners carry the same label). The symmetries of a tetrahedron imbedded in the cube (compare Fig. 7.1) correspond to the elements of a subgroup of the cubical group. Establish the corresponding permutations. Does this set of permutations form a symmetric group of lower order?

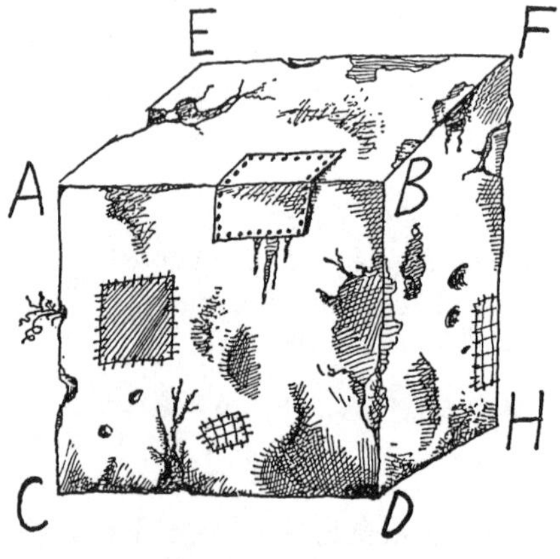

Fig. 3P.1

3. 10. A cube in the conventional sense need not to have cubical symmetry (compare Fig. 3P.1), conversely there are objects that do not look like a cube, and have the cubical symmetry. To find examples, check through the literature: Wigner-Seitz Unit cell for b.c.c. and f.c.c. lattices (compare KITTEL [1957], p. 286). Brillouin zones and Fermi surfaces for cubic lattices. (Compare HARRISON [1960].)

3. 11. Let any two numbers a and b be equivalent in $m-n = 4K$ where K is some integer. Under this equivalence any number is equivalent to either $K = 1, 2, 3,$ or 4. Notation

$$m = n \pmod 4.$$

(This is pronounced m is equal to n modulo 4.) Show that the numbers 1, 2, 3, 4 form an additive group by constructing the group table of the group. How many classes does this group have?

3. 12. (Compare Ch. III § 7.5.) If the bilinear form is indicated by $F = (\boldsymbol{x} \cdot \boldsymbol{x})$ and the transformed version $(\mathrm{A}\boldsymbol{x} \cdot \mathrm{A}\boldsymbol{x})$ by $F^{(a)}$, show that

$$I = \sum_a F^{(a)} \quad (a \text{ are all elements of the set})$$

is invariant under the operations of the set. If a linear transformation [1] is introduced that brings I into the diagonal form, the new set of coordinates form the basis for a unitary representation of the set of matrices. Show this and apply this procedure for "unitarisation" to the example of the two dimensional representation in § 7.4.

3. 13a. Perform the reduction of the regular representation of the group $\mathscr{S}_3$ with the help of the Young tableaux. The three tableaux will be labelled H (for horizontal), L (for L-shaped) and V (for vertical). Show that PQ for the first and last is equal to the group itself and that PQ for the L-tableau consists of four elements.
Show that $\rho = 6$ (for H and V) and $\rho = 3$ for L.

3. 13b. Use the representation resulting from the basis e_1, e_2, e_3 as discussed in § 7.4 to calculate the projection operators corresponding to the following tableaux:

1	2	3

1	2
3	

1	3
2	

Show that projection operators, when acting on the basis functions

[1] This transformation is not necessarily unitary.

give rise to the following three linear combinations: $e_1+e_2+e_3$; e_1-e_3; e_1-e_2. (Note that this new basis will result in exactly the same representation as the example worked out in § 7.4).

3\. 14. Show that each irreducible representation of the factor-group $\mathscr{G}/\mathscr{H}$ is also an irreducible representation of $\mathscr{G}$.

3\. 15. Show that the group table for the regular n–gone with two sides (the so called dieder groups) are given by

$$\mathrm{A}^n = \mathrm{E};\ \mathrm{B}^2 = \mathrm{E};\ \mathrm{BA} = \mathrm{A}^{-1}\mathrm{B}$$

and show that the double dieder group has the group table

$$\mathrm{A}^n = \mathrm{R};\ \mathrm{B}^2 = \mathrm{R};\ \mathrm{R}^2 = \mathrm{E};\ \mathrm{BA} = \mathrm{A}^{-1}\mathrm{B}.$$

4\. 1. Wave function for equilateral triangular molecules are constructed by taking linear combinations of "atomic orbitals". An atomic orbital is a wave function described in polar coordinates, where the origin of the coordinate system is the nucleus of that particular atom. If spherical harmonics are used for the angular part, rotations of the molecule will mean that the spherical harmonic orbital attached to a certain corner will go over possibly into another spherical harmonic around another center (i.e. another orbital) or a linear combination thereof.
Construct the irreducible linear combination of atomic orbitals when the orbitals are s-functions. Same if they are p-functions.

5\. 1. Show that the quantities X, Y, X', Y', as introduced in § 2.2, have indeed the same transformation rules as ξ, η, ξ', η', i.e. they satisfy (5.4) as well.

5\. 2. Perform the calculation to obtain (5.7a) *Hint*: Introduce the following quantity,

$$(\xi - i\eta)^*(\xi - i\eta) = C$$

and show that C is a real constant. Next show that

$$\left(\frac{\alpha+\alpha^*}{2}\right)^2 + \left(\frac{\beta+\beta^*}{2}\right)^2 = 1.$$

Introduce a formal angle which makes that this equation is automatically satisfied and prove that this angle corresponds to ω_y.

5. 3. Show that the coupling of 2 electrons gives rise to a ^{1}S, a ^{3}P and a ^{1}D-state, and determine the bases for these three representations. There are 15 possible antisymmetrized products of two wave functions (two by two Slater determinants). For instance:

$$(1^+ \, 0^-) \equiv \begin{vmatrix} 1^+_{(1)} 0^-_{(1)} \\ 1^+_{(2)} 0^-_{(2)} \end{vmatrix}$$

where the numbers refer to the m_l-values of the p-electrons, the $\pm$ signs to their m_s-values and the subscripts are the electron labels. The "standard order" will be one of decreasing (or actually non-increasing) m_s-numbers. Construct the 15 basis functions for the representations mentioned above.
Hint: the $M_L = 2$; $M_S = 0$ wave function of the representation ^{1}D is unambiguously determined: $(1^+ \, 1^-)$. If we now operate with the operator μ_q of equation (5.48) where J and j are equal to $L = 2$, we can create the $m_l = 1$; $M_s = 0$ linear combination of wave functions of the D representation. The remaining part of this linear combination has to belong to the ^{3}P representation. Further operation with μ_p and μ_q acting either on M_L or M_S will create all further basis functions. (This is the so-called Gray and Wills method, compare CONDON and SHORTLEY [1935].)

5. 4. The Wigner coefficients can be considered as a matrix which connects Q_M with $q_{M-m_z} q_{m_z}$. Calculate this matrix for

$$D_{\frac{1}{2}} \times D_{\frac{1}{2}} = D_1 + D_0.$$

---

6. 1. Calculate the reduced matrix elements of a spin operator between two spin $\frac{1}{2}$ states.

6. 2. The electric quadruple is an example of T^λ_μ with $\lambda = 2$, $\mu = 0$. Determine the coefficient in front of the reduced matrix element for this operator.

---

7. 1. (Compare Ch. 7, § 4.) Show that the tetrahedral group or the permutation group of four elements is the weak direct product of

the following two subgroups. One is the subgroup of permutations of three elements keeping one fixed. The other is the subgroup of cyclical permutations of all four elements. Show also that this is not a semi-direct product. (McIntosh.)

7. 2. Determine the inverse lattice of the b.c.c.; f.c.c. and hexagonal close packed lattices.

7. 3. (Compare Ch. VII, § 9.1) Show that all group postulates hold for the little group.

---

8. 1. The cubical group corresponds to the permutation group of four objects. These four objects are the body diagonals. The classes can be found by writing down all possible partitions of four numbers. Construct the classes with the method of partition indicated in Chapter 3 and calculate the character Table 8.4.

8. 2. Show that a $J = 2\frac{1}{2}$ level will split into a quartet and a doublet in a cubical field.

SYMBOLS

$\rightarrow$	with .... is associated
$\sim$	transforms as
$(.)$	functional scalar product
$\langle\ \rangle$	(time) average, also: expectation value average over probability density function
$\times$	direct product of matrices or sets of matrices
$+$	addition of representation
δx	small, so-called "virtual", variation of x.
$\boldsymbol{x}$	vector in n-dimensional space
$\boldsymbol{e}_i$	base vector of a vector space
$\mathfrak{R}_n$	vector space spanned by n basis vectors
δ_{ik}	Kronecker symbol (equals 1 if $i = k$, and is zero otherwise)
$\mathsf{A} \equiv (a_{ik})$	Matrix consisting of elements
A^{-1}	Inverse of A that is $\mathsf{A}^{-1}\mathsf{A} = \mathbb{I}$
$\mathbb{I}$	Unit matrix $\mathbb{I} \equiv (\delta_{ik})$
S, U	Unitary matrix used for similarity transformation
(g_{ik})	Metric tensor
$\boldsymbol{x}^*, \mathsf{A}^*$	Complex conjugate vector, matrix
$\tilde{\boldsymbol{x}}, \tilde{\mathsf{A}}$	transposed vector, matrix
$\mathsf{A}^\dagger$	Hermitian conjugate matrix (i.e., complex conjugate and transposed)
A, α	Operator
$\bar{\mathrm{A}}$	Adjoint operator
D	Domain of integration
β_k, γ_k	Fourier component or generalized Fourier component
$\delta(x_1 - x_2)$	Dirac delta function
∇^2	"Del-squared": $\dfrac{\partial^2}{\partial x^2} + \dfrac{\partial^2}{\partial y^2} + \dfrac{\partial^2}{\partial z^2}$
c	velocity of light
$\langle m\vert$	Bra-vector
$\vert m\rangle$	Ket-vector
ν	frequency

$\omega = 2\pi\nu$	angular frequency
$\boldsymbol{k}$	wave (number) vector; $k = 2\pi/\lambda$
$h, \hbar$	h of Planck, $\hbar = h/2\pi$
$\boldsymbol{p}$	momentum vector
ψ, φ	wave function
$d\tau$	volume element of configuration space
E	energy eigen value, (occasionally also used for electric field strength)
H	Hamilton operator
V	potential energy
$\boldsymbol{L}, \boldsymbol{l}$	angular momentum vector or vector-operator
λW	perturbing energy
λ	perturbation parameter
u_i	set of unperturbed eigenfunctions
η	auxiliary parameter in nearly degenerate systems, expressing the splitting in terms of the strength of the interaction
n, l, m	radial, angular and azimuthal quantum numbers
P, S, A	(abstract) group element
E	same: unit element
$\mathscr{G}$	group i.e., set of (abstract) elements that form a group
$\mathscr{S}_n$	Permutation group of n-elements ($\mathscr{S}$ stands for symmetry)
$\mathscr{H}$	Subgroup or invariant subgroup
$\mathscr{A}_n$	Alternating group
$\mathscr{A}_3$	Subgroup of $\mathscr{S}_3$ containing the three-fold rotations
$\boldsymbol{E}, \boldsymbol{A}, \boldsymbol{B}$	Matrices respresenting the group elements E, A, B
$\boldsymbol{\mathscr{G}}$	the set of matrices representing the elements of $\mathscr{G}$
$X(\text{A})$	character of the matrix $\boldsymbol{A}$
$X^{(i)}(\text{C})$	character of class C belonging to the irreducible representation Γ_i
$\tilde{X}_i$	Component of the vector in class-space or "reduced character" $\tilde{X}_i = \sqrt{h_i/g}\ X(\text{C}_i)$
h_i	number of elements in class i
ξ or ξ	hypercomplex number
Γ_i	i-th (irreducible) representation
g	order of the group; i.e., the number of elements
C_i	class i of the group
$\varepsilon^{(\mu)}$	idempotent element projecting the μ-th irreducible representation

$\varepsilon^{\{a\}}$	same (permutation group); irreducible representation corresponding to the partition $\{a\}$
$\boldsymbol{\varepsilon}$	matrix representation ε above
n_μ	dimensionality of the (irreducible) representation μ is equal to character of unit element $X^{(\mu)}(E)$.
ζ	essential idempotent element corresponding to a shape
Y_l^m	spherical harmonic
$\mathscr{D}_3$	three-dimensional space group
$\boldsymbol{D}_l$	$(2l+1)$-dimensional representation of the space group above
$\omega_x, \omega_y, \omega_z$	rotation around the x, y or z axis
$\mathscr{U}_2$	unitary unimodular group in two dimensions
σ	element of this group
ξ, η	basis functions of $\mathscr{U}_2$, spin variables
q_m^j	monomials in ξ and η of the degree j
$\boldsymbol{M}_x, \boldsymbol{M}_y, \boldsymbol{M}_z$	matrices representing infinitesimal rotations
$\boldsymbol{S}_x, \boldsymbol{S}_y, \boldsymbol{S}_z$	Pauli matrices
$\boldsymbol{L}_x, \boldsymbol{L}_y, \boldsymbol{L}_z$	orbital angular momentum
$l_x l_y l_z$	same for individual electrons (usually in dimensionless form)
$\mathscr{M}_z$	Magnetic moment around the z-axis
ω_L	Larmor frequency
g	Landé g-factor
$\boldsymbol{J}$	total angular momentum (in units $\hbar$)
$m \equiv j_z$	z-component of j
$\mathscr{D}_j$	$2j+1$-dimensional representation of the rotation group
κ	parity operator
$C_{m_1 m_2}^J$	Clebsch-Gordan coefficient (equivalent to Wigner $3j$-symbol)
T_μ^λ	tensor operator
$\langle j, m \| \mathrm{T} \| j', m' \rangle$	matrix element of a tensor operator
$\langle j \|\| \mathrm{T} \|\| j' \rangle$	reduced matrix element of a tensor operator
$R_{j'j''}$ or $W(j_1 j_2 j j_3; j' j'')$	Racah coefficient, equivalent to:
$\begin{Bmatrix} j_1 j_2 j' \\ j_3 j j'' \end{Bmatrix}$	Wigner $6j$-symbol
$\left.\begin{matrix} t_1, t_2, t_3 \\ \boldsymbol{a}, \boldsymbol{b}, \boldsymbol{c} \end{matrix}\right\}$	three primitive translations
T	general (allowed) translation
(T|R)	element of the space group consisting of a translation T and a rotation R
$\boldsymbol{a}^*, \boldsymbol{b}^*, \boldsymbol{c}^*$	reciprocal lattice vectors

$\boldsymbol{k}$	label of the irreducible representation of the translation group
$u(\boldsymbol{r})$	periodic part of the wave function
$\boldsymbol{K}$	reciprocal lattice vector
Ξ_i	reducible representation of an invariant subgroup resulting from a irreducible representation Γ_i of the main group
γ_i	irreducible representation of the subgroup
K	time reversal operator
$\mathscr{M}$	total magnetic moment
q_i	nuclear coordinates
Q_i	normal nuclear coordinates
$\boldsymbol{D}$	dielectric displacement
ε	dielectric constant
μ	electric moment (in e.s.u.)

References cited

Abragam, A. and M.H.L. Pryce, 1951, Proc. Roy. Soc. A **230** 169
Bethe, H. A., Ann. Phys. [5] **3** (1929) 133 or [Consultants Bureau (English Translation) New York]
Birkhoff, G., 1950, *Hydrodynamics* (Princeton University Press, 1950; Dover Publications, New York, 1955)
Born, M. and P. Jordan, 1930, *Elementare Quanten Mechanik* (Springer, Berlin)
Bouckaert, L. P., R. Smoluchowski and E. P. Wigner, Phys. Rev. **50** (1936) 58
Brinkman, H. C., 1956, *Applications of Spinor Invariants in Atomic Physics* (North Holland Publishing Co., Amsterdam)
Burckhardt, J. J., 1947, *Die Bewegungsgruppen der Kristallographie* (Birkhäuser, Basel)
Cartan, E., 1938, *Leçons sur la théorie des Spineurs* (Hermann et Cie., Paris)
Cartan, E., 1894, *Concerning the structure of finite and continuous transformation groups* (Thesis Nony, 1894)
Cartan, E., 1913, Bull. Soc. Math. de France **4** 53
Cartan, E., 1914, Journal de Mathématiques **10** 149
Cayley, A., 1854, Phil. Mag., vii (4), 40—57
Condon, E. V. and G. H. Shortly, 1935, *The Theory of Atomic Spectra* (Cambridge University Press)
Dirac, P. A. M., 1958, *The Principles of Quantum Mechanics*, 4th ed. (Clarendon Press, Oxford)
Eckart, C., 1930, Revs. Mod. Phys. **2** 305
Edmonds, A. R., 1957, *Angular Momentum in Quantum Mechanics* (Princeton University Press, New Jersey)
Elliott, R. J., 1954, Phys. Rev. **96** 280
Eyring, H., J. Walter and G. Kimball, 1944, *Quantum Chemistry* (Chapman & Hall, Ltd., London; Wiley & Sons, Inc., New York)
Fano, U. and G. Racah, 1959, *Irreducible Tensorial Sets* (Academic Press, New York)
Frobenius, G. and I. Schur, 1906, Sitzber. Akad. Wiss. Berlin, Phys. Math. Kl., 186
Heine, V., 1960, *Group theory in Quantum Mechanics* (Pergamon Press, London)
Heitler, W., 1957, *The Quantum Theory of Radiation*, 3rd ed. (Clarendon Press, Oxford)
Harrison, W. A., 1960, Phys. Rev. **118** 1190
Herman, F., 1958, Revs. Mod. Phys. **30** 102
Herring, C., 1937a, Phys. Rev. **52** 361
Herring, C., 1937b, Phys. Rev. **52** 365
Herzberg, G., 1945, *Molecular Spectra and Molecular Structure. Infrared and Raman Spectra of Polyatomic Molecules*, D. Van Nostrand, Inc., New York)
Herzfeld, C. M. and P. H. E. Meijer, 1960, Solid State Physics **12** 1
Hilbert, D. and R. Courant, 1930, *Methoden der Mathematischen Physik*, 2nd ed. (Berlin, Springer)
Jahn, H. A. and E. Teller, 1937, Proc. Roy. Soc. (London) A **161** 220
Jahn, H. A., 1938, Proc. Roy. Soc. (London) **164** 117
Kittel, C., 1957, *Introduction to Solid State Physics*, 2nd ed. (Wiley, New York)
Korringa, J., 1954, Techn. Rep. (Ohio State University) and Solid State Physics (to be published)

Koster, G. F., 1957, Solid State Physics **5** 174
Koster, G. F., 1958, Phys. Rev. **109** 227
Kramers, H. A., 1937, *Quantum Mechanics* (North-Holland Publishing Co., Amsterdam)
Lee, T. D., 1960, *Physics Today* **13** (October 1960)
Lighthill, M. J., 1958, *Introduction to Fourier analysis and generalized Functions* (Cambridge University Press)
Mayer, J. E. and M. G. Mayer, 1940, *Statistical Mechanics* (John Wiley, New York)
McIntosh, H. V., 1958a, *Symmetry adapted Functions belonging to the crystallographic lattice groups*, R.I.A.S., report 58–3
McIntosh, H. V., 1958b, *On accidental degeneracy in classical and quantum Mechanics*, R.I.A.S., report 58–4.
Meijer, P. H. E., 1954, Phys. Rev. **95** 1443
Melvin, M. A., 1956, Rev. Mod. Phys. **28** 18
Melvin, M. A., 1960, Rev. Mod. Phys. **32** 477
Molenaar, P. G., 1930, *Eindige Substitutie Groepen* (P. Noordhoff, N.V., Groningen)
Møller, C., 1952, *The Theory of Relativity* (The University Press, Oxford)
Morse, P. M. and H. Feshbach, 1953, *Methods of Theoretical physics* (McGraw-Hill Book Co., Inc., New York)
Neumann, J. v., 1932, *Mathematische Grundlagen der Quanten Mechanik* (Springer, Berlin, 1932; Dover Publications, New York, 1943); also: *Mathematical Foundations of Quantum Mechanics*, Transl. from German by R. T. Beyer, 1955 (New Jersey, Princeton University Press)
Opechowski, W., 1940, Physica **7** 552
Phillips, F. C., 1956, *An Introduction to Crystallography* (Longmans Green, London)
Phillips, J. C., 1956, Phys. Rev. **104** 1263 (See errata Phys. Rev. **105** (1957) 1933)
Racah, G., 1942, Phys. Rev. **62** 438; and **63** (1943) 367
Rosenthal, J. E. and G. M. Murphy, 1936, Rev. Mod. Phys. **8** 317
Schiff, L., 1949, *Quantum Mechanics* (McGraw-Hill Book Co., Inc., New York)
Schiff, L., 1954, Am. J. Phys. **22** 621
Seitz, F., 1934, Z. für Kristallogr. **88** 433
Slater, J. C., 1953, *Electronic Structure of Solids, Solid State and Mol. Theory Reports* No 4. (M.I.T., Cambridge Mass.)
Slater, J. C., 1934, Phys. Rev. **45** 794
Speiser, A., 1937, *Theory of Groups of Finite Order*, (Dover Publications, New York) 4th ed. (Birkhäuser, Basel, 1956)
Sucksmith, W., 1930, Proc. Phys. Soc. **42** 385
Van der Waerden, B. L., 1949, *Modern Algebra* (Frederick Ungar Publishing Co., New York)
Von der Lage, F. C. and H. A. Bethe, 1947, Phys. Rev. **71** 612
Webster, A. G., 1955, *Partial Differential Equations of Mathematical Physics*, 2nd ed. (Dover Publications, New York)
Weyl, H., 1950, *The Theory of Groups and Quantum Mechanics*, Translated from 2nd revised German, 1931 ed. (Dover Publications, New York)
Weyl, H., 1922, *Space-Time-Matter*, Reprint of the 1922 ed. (Dover Publications, New York)
Weyl, H., 1925, Mat. Zeitschr. **23** 275
Wigner, E. P., 1927, Zeitschr. f. Phys. **43** 624
Wigner, E. P., 1931, *Gruppentheorie und ihre Anwendung auf die Quantenmechanik der Atomspektren* (Vieweg & Sohn, Braunschweig)
Wigner, E. P., 1959, *Group Theory and its Application to the Quantum Mechanics of Atomic Spectra* (Academic Press, New York)

SYSTEMATIC BIBLIOGRAPHY

1. MATRICES AND LINEAR VECTOR SPACES

ALBERT, A. A., 1956, *Fundamental Concepts of Higher Algebra* (University of Chicago Press), (Chapter I Groups, Chapter III Vector Spaces and Matrices)

BODEWIG, E., 1956, *Matrix Calculus* (North Holland Publishing Co., Amsterdam)

HALMOS, P. R., 1958, *Finite Dimensional Vector Spaces* (Van Nostrand, Princeton)

MARGENAU, H. and G. M. MURPHY, 1943, *The Mathematics of Physics and Chemistry* (Van Nostrand, New York), (Chapter 10 Matrices and Chapter 15 Group Theory)

VAN DER WAERDEN, B. L., 1949, *Modern Algebra* (Frederick Ungar, New York)

1.1. *Fourier Series*

LIGHTHILL, M. J., 1958, *Introduction to Fourier Analysis* (Cambridge University Press)

WIENER, N., 1933, *The Fourier Integral and certain of its Applications* (Dover Publications, New York)

2. QUANTUM MECHANICS

BOHM, D., 1951, *Quantum Theory* (Prentice Hall, New York)

BORN, M. and P. JORDAN 1930, *Elementare Quanten Mechanik* (Springer, Berlin)

DIRAC, P. A. M., 1958, *Quantum Mechanics* (Oxford University Press)

KRAMERS, H. A., 1937, *Quantum Mechanics* (North-Holland Publishing Co., Amsterdam)

NEUMANN, J. v., 1932, 1943, *Mathematische Grundlagen der Quanten Mechanik* (Springer, Berlin, 1932; Dover Publications, New York, 1943); Also: *Mathematical Foundations of Quantum Mechanics*, Transl. from German by R. T. Beyer, 1955 (Princeton University Press, New Jersey, 1955)

SCHIFF, L., 1949, *Quantum Mechanics* (McGraw-Hill Book, Co., Inc., New York)

3. GROUPS

BOERNER, H., 1955, *Darstellungen von Gruppen mit Berücksichtigung der Bedürfnisse der modernen Physik* (Springer, Berlin); *Representation of Groups with special consideration for the needs of Modern Physics* (translated from German by P. G. Murphy *et al.*) (North-Holland Publishing Co., Amsterdam, 1963)

EYRING, H., J. WALTER and G. E. KIMBALL, 1944, *Quantum Chemistry* (Wiley & Sons, New York) (Chapter 10)

HALL, M., 1959, *The Theory of Groups* (McMillan, New York)

LEDERMAN, W., 1953, *Introduction to the Theory of Finite Groups*, 2nd ed. (Methuen, London)

LITTLEWOOD, D. E., 1950, *The Theory of Group Characters* (Clarendon Press, Oxford)

McINTOSH, H. V., 1957, *Group Theory* (RIAS, Inc. Baltimore, Technical report, No 57)

MURNAGHAN, F. D., 1938, *The Theory of Group Representations* (Johns Hopkins Press, Baltimore)

WEYL, H., 1947, *The Classical Groups* (Princeton University Press)

SPEISER, A., 1956, *Theory of Groups of Finite Order* 4th ed. (Birkhäuser, Basel)

3.1. *Tableaux*

LANDAU, L. D. and E. M. LIFSHITZ, 1958, *Quantum Mechanics* (Pergamon Press, London) Ch. 9

MOLENAAR, P. G., 1930, *Eindige Substitutie Groepen* (Noordhoff, Groningen)

RUTHERFORD, D. E., 1948, *Substitutional Analysis* (Edinburgh University Press, Edinburgh)

YAMANOUCHI, T., 1937, Proc. Phys. Math. Soc. Japan **19**, 436 (On the construction of unitary irreducible representations of the symmetric group)

4. APPLICATIONS OF GROUP THEORY TO PHYSICS IN GENERAL

ALTMANN, S. L., 1961, Group Theory in Volume 2 of Quantum Theory, (D. A. Bates, editor) (Academic Press, New York)

ALTMANN, S. L., in preparation, *Group Theory in Pure and Applied Physics* (Oxford University Press)

BHAGAVANTUM, S. and T. VENKATARAYUDU, 1951, *Theory of Groups and its Application to Physical Problems* (Waltair, India, Andhra University)

ECKART, C., 1930, Revs. Mod. Phys. **2** 305

HEINE, V., 1960, *Group Theory in Quantum Mechanics* (Pergamon Press, London)

HAMMERMESH, M., 1962, *Group Theory and its applications to physical problems* (Addison Wesley, Reading, Mass.)

HIGMAN, B., 1955, *Applied Group-theoretic and Matrix Methods* (Clarendon Press, Oxford)

KOSTER, G. F., 1956, *Notes on Group Theory* (Solid-state and Molecular Theory Technical Rep., No. 8, Cambridge, Mass.)

LOMONT, L. S., 1959, *Applications of Finite Groups* (Academic Press, New York)

LYABARSKII, G. YA., 1960, *The Application of Group Theory in Physics* (Pergamon Press, London)

MELVIN, M. A. and C. M. HERZFELD, *Modern Algebra and Modern Physics* (in preparation)

MELVIN, M. A., 1956, Rev. Mod. Phys. **28** 18

VAN DER WAERDEN, B. L., 1931, *Die Gruppentheoretische Methode in der Quantenmechanik* (Springer, Berlin)

VENKATARYVUDU, T., 1953, *Application of Group Theory to Physical Problems* (New York University)

WEYL, H., 1950, *The Theory of Groups and Quantum Mechanics* (Translated from 2nd revised German 1931 ed.) (Dover Publications, New York)

WIGNER, E. P., 1931, *Gruppentheorie und ihre Anwendung auf die Quantenmechanik der Atomspektren* (Vieweg & Sohn, Braunschweig)

WIGNER, E. P., 1959, *Group Theory and its Application to the Quantum Mechanics of Atomic Spectra* (Academic Press, New York)

WIGNER, E. P., 1927, Zeitschr. f. Phys. **43** 624

4.1. *Application to Wave Guide Junctions*

AULD, B. A., 1952, *Applications of Group Theory in the Study of Symmetrical Wave Guide Junctions* (Stanford University, Microwave Laboratory Report no. 157)

PANNENBORG, A. E., 1952, Philips Research Reports **7** 131, 169 and 270

5. ROTATIONS IN SPACE

COHEN, A., 1931, *An Introduction to the Lie Theory of One-Parameter Groups* (D. C. Heath Co., New York)

CONDON, E. V. and G. H. SHORTLY, 1935, *The Theory of Atomic Spectra* (Cambridge University Press)
EDMONDS, A. R., 1957, *Angular Momentum in Quantum Mechanics* (Princeton University Press, New Jersey)
FEENBERG, E. and G. E. PAKE, 1953, *Notes on the quantum theory of Angular Momentum* (Addison-Wesley, Cambridge, Mass,)
ROSE, M. E., 1957, *Elementary Theory of Angular Momentum* (Wiley & Sons, New York)

5.1. *Spinors*

BRINKMAN, H. C., 1956, *Applications of Spinor Invariants in Atomic Physics* (North-Holland Publishing Co., Amsterdam)
CARTAN, E., 1938, *Leçons sur la theorie des Spineurs* (Hermann & Cie, Paris)

6. CLEBSCH-GORDAN AND RACAH COEFFICIENTS

EDMONDS, A. R., 1957, *Angular Momentum in Quantum Mechanics* (Princeton, University Press, New Jersey)
FANO, U. and G. RACAH, 1959, *Irreducible Tensorial sets* (Academic Press, New York)
ROSE, M. E., 1957, *Elementary Theory of Angular Momentum* (Wiley & Sons, New York)

6.1. *Tables*

BIEDENHARN, L. C., 1952, *Tables of Racah coefficients* (Oak Ridge National Laboratory, Physics Div., ORNL-1501, suppl. 1, Febr. 1952)
BIEDENHARN, L. C., J. M. BLATT and M. E. ROSE, 1952, Revs. Mod. Phys. **24** 249 (These tables give $W(abcd; ef)$ for $e \leqq 2$ and $abcdf$ arbitrary)
CONDON, E. V. and G. H. SHORTLY, 1935, *The Theory of Atomic Spectra* (Cambridge University Press)
EDMONDS, A. R., 1957, *Angular Momentum in Quantum Mechanics* (Princeton University Press, New Jersey)
ROTENBERG, M., R. BIVINS, N. METROPOLIS and J. K. WOOTEN, 1960, The 3-*j* and 6-*j* symbols (Wiley & Sons, New York)

7.1. SPACE GROUPS

BARNES, R. B., R. R. BRATTAIN and F. SEITZ, 1935, Phys. Rev. **48** 582
BELL, D., 1954, Rev. Mod. Phys. **26** 311
BOUCKAERT, L. P., R. SMOLUCHOWSKI and E. P. WIGNER, 1936, Phys. Rev. **50** 58
BRILLOUIN, L., 1953, *Wave Propagation in Periodic Structures* (Dover Publications, New York)
BRAVAIS, A., 1850, *On the systems formed by points regularly distributed on a plane or in a space* (Eng. transl. 1949 by the Crystallography Society of America, Houston, Texas)
BURCKHARDT, J. J., 1947, *Die Bewegungsgruppen der Kristallographie* (Birkhäuser, Basel)
DAVYDOV, A. S., 1951, *Theory of Molecular Excitons* (Translated by M. Kasha and M. Oppenheimer, 1962) (McGraw-Hill, New York)
ELLIOTT, R. J., 1954, Phys. Rev. **96** 280
JOHNSTON, D. F., 1960, Rep. Progress Physics **23** 66
JONES, H., 1960, *The Theory of Brillouin Zones and Electronic States in Crystals* (North-Holland Publishing Co., Amsterdam)
HERMAN, F., 1958, Rev. Mod. Phys. **30** 102
HERRING, C., 1937, Phys. Rev. **52** 365

HERRING, C., 1942, Franklin Institute **233** 525. Example of representation of space group for which no point has the symmetry of the pointgroup
KITTEL, C., 1957, *Introduction to Solid-state Physics*, 2nd ed. (Wiley & Sons, New York)
KOSTER, G. F., 1957, Solid State Physics **5** 174
MARIOT, L., 1962, *Group Theory and Solid State Physics* (Prentice Hall, Englewood Cliffs, New Jersey) Translated by A. Nussbaum
OVERHAUSER, A. W., 1955, Phys. Rev. **101** 1702
PHILLIPS, F. C., 1946, *An Introduction to Crystallography* (Longmans, London)
SEITZ, F., 1934, Z. f. Kristallographie, **88** 433
SEITZ, F., 1934, Z. f. Kristallographie **90** 289
SEITZ, F., 1935, Z. f. Kristallographie **91** 336
SEITS, F., 1936, Z. f. Kristallographie **94** 100
SEITZ, F., 1936, Annals of Math. **37** 17 (On the reduction of space groups)
VON DER LAGE, F. C. and H. A. BETHE, 1947, Phys. Rev. **71** 612
ZAMORZAEV, A. M., 1958, Soviet Physics Crystallography **2** 10
ZACHARIASEN, W. H., 1945, *Theory of X-ray Diffraction in Crystals*, (Wiley, New York) (section II.6)

7.2. *Tables*

International Tables for X-Ray Crystallography 1952 (Eng. Kynoch Press, Birmingham)

8.1. *Finite Groups, Crystal Field Theory*

BETHE, H. A., 1929, Ann. Phys. [5] **3** 133 Translated by The Consultants Bureau, New York
BLEANY, B. and K. W. H. STEVENS, 1953 Rep. Progr. in Phys. **16** 108, continued in: BOWENS K. D. and J. OWEN, 1955, Rep. Progr. Phys. **18** 304; and ORTON J., 1959, Rep. Progr. Phys. **22** 204
FICK, E. and G. JOOS, 1957, *Handbuch der Physik* **28** 205
GRIFFITH, J. S., 1961, *Theory of transition Metal ions* (Cambridge University Press)
HERZFELD, C. M. and P. H. E. MEIJER, 1961 Solid-state Physics **12** 1
KITTEL, C. and J. M. LUTTINGER, 1948, Phys. Rev. **73** 162
LANDAU, L. D. and E. M. LIFSHITZ, 1958, *Quantum Mechanics* (Pergamon Press, London) Ch. 12
MCCLURE, 1959, Solid State Physics **9** 400
LOW, W., 1960, *Paramagnetic Resonance in Solids* (Academic Press, New York)
MOFFITT, W. and C. J. BALLHAUSEN, 1956, Ann. Rev. Phys. Chem. **7** 107
OPECHOWSKI, W., 1940, Physica **7** 552
ORGEL, L. E., 1961, *An Introduction to Transition Metal Chemistry* (Methuen-Wiley & Sons)
PRATHER, J. L., 1961, *Atomic Energy Levels in Crystals* (N.B.S. Monograph No. 19, U.S. Department of Commerce, National Bureau of Standards, Washington, D.C.)
STEVENS, K. W. H., 1952, Proc. Phys. Soc. **65** 209
TAVGER, B. A. and V. M. ZAITSEV, 1956, J.E.T.P. **3** 430

8.2. *Finite Groups, Molecules*

EYRING, H., J. WALTER and G. KIMBALL, 1944, *Quantum Chemistry* (Wiley & Sons, New York)
HEITLER, W., 1928, Z. f. Physik **47** 835
HEITLER, W. and G. RUMER, 1931, Z. f. Physik **68** 12
MULLIKEN, R. S., 1933, Phys. Rev. **43** 279

ROOTHAAN, C. C. J., 1951, Rev. Mod. Phys. **23** 69
ROSENTHAL, J. E. and G. M. MURPHY, 1936, Rev. Mod. Phys. **8** 317
TISZA, L., 1933, Z. f. Physik **82** 48
WIGNER, E., 1930, *On the characteristic elastic vibrations of symmetrical systems*, Göttinger Nachrichten 133

8.3. *Time Reversal*

KLEIN, M. J., 1952, Amer. J. Phys. **20** 65
HERRING, C., 1937, Phys. Rev. **52** 361
WIGNER, E., 1932, Math. Phys. Kl. 546 *Über die Operation der Zeitumkehr in der Quantenmechanik*, Nachr. Akad. der Wiss. Göttingen

8.4. *Jahn-Teller Effect*

BLEANEY, B., K. D. BOWERS and M. H. L. PRYCE, 1955, Proc. Roy. Soc. (London) A **228** 166
CLINTON, W. L. and B. RICE, 1959, J. Chem. Phys. **30** 542
DUNITZ, J. D. and L. E. ORGEL, 1957, J. Chem. Phys. of Solids **3** 20
JAHN, H. A., 1938, Proc. Roy. Soc. (London) **164** 117
JAHN, H. A. and E. TELLER, 1937, Proc. Roy. Soc. (London) A **161** 220
MOFFITT, W. and W. THOMSON, 1957, Phys. Rev. **108** 1251
ÖPIK, U. and M. H. L. PRYCE, 1956, Proc. Roy. Soc. (London) **238** 425; LONGUET-HIGGINS, H. C., U. ÖPIK and M. H. L. PRYCE, 1958, Proc. Roy. Soc. (London) **244** 1
VAN VLECK, J. H., 1939, J. Chem. Phys. **7** 72
WOJTOWICZ, P. J., 1959, Phys. Rev. **116** 32

8.5. *Molecular Vibrations*

HERZBERG, G., 1945, *Molecular Spectra and Molecular Structure. Infrared and Raman Spectra of Polyatomic Molecules* (D. Van Nostrand, Inc., New York)
MATHIEU, J. P., 1945, *Spectres de vibration et symmetrie des molecules et des cristeaux* (Hermann, Paris)
MATOSSI, F., 1961, *Gruppentheorie der Eigenschwingungen von Punktsysteme* (Springer, Berlin)
MULLIKEN, R. S., 1955, J. Chem. Phys. **23** 1997
WILSON, E. B., J. C. DECIUS and P. C. CROSS, 1955, *Molecular Vibrations* (McGraw-Hill, New York)

INDEX

A CATALOG OF SELECTED

DOVER BOOKS

IN SCIENCE AND MATHEMATICS

Astronomy

BURNHAM'S CELESTIAL HANDBOOK, Robert Burnham, Jr. Thorough guide to the stars beyond our solar system. Exhaustive treatment. Alphabetical by constellation: Andromeda to Cetus in Vol. 1; Chamaeleon to Orion in Vol. 2; and Pavo to Vulpecula in Vol. 3. Hundreds of illustrations. Index in Vol. 3. 2,000pp. 6⅛ x 9¼.
Vol. I: 23567-X
Vol. II: 23568-8
Vol. III: 23673-0

EXPLORING THE MOON THROUGH BINOCULARS AND SMALL TELESCOPES, Ernest H. Cherrington, Jr. Informative, profusely illustrated guide to locating and identifying craters, rills, seas, mountains, other lunar features. Newly revised and updated with special section of new photos. Over 100 photos and diagrams. 240pp. 8¼ x 11. 24491-1

THE EXTRATERRESTRIAL LIFE DEBATE, 1750–1900, Michael J. Crowe. First detailed, scholarly study in English of the many ideas that developed from 1750 to 1900 regarding the existence of intelligent extraterrestrial life. Examines ideas of Kant, Herschel, Voltaire, Percival Lowell, many other scientists and thinkers. 16 illustrations. 704pp. 5⅜ x 8½. 40675-X

THEORIES OF THE WORLD FROM ANTIQUITY TO THE COPERNICAN REVOLUTION, Michael J. Crowe. Newly revised edition of an accessible, enlightening book recreates the change from an earth-centered to a sun-centered conception of the solar system. 242pp. 5⅜ x 8½. 41444-2

A HISTORY OF ASTRONOMY, A. Pannekoek. Well-balanced, carefully reasoned study covers such topics as Ptolemaic theory, work of Copernicus, Kepler, Newton, Eddington's work on stars, much more. Illustrated. References. 521pp. 5⅜ x 8½.
65994-1

A COMPLETE MANUAL OF AMATEUR ASTRONOMY: Tools and Techniques for Astronomical Observations, P. Clay Sherrod with Thomas L. Koed. Concise, highly readable book discusses: selecting, setting up and maintaining a telescope; amateur studies of the sun; lunar topography and occultations; observations of Mars, Jupiter, Saturn, the minor planets and the stars; an introduction to photoelectric photometry; more. 1981 ed. 124 figures. 26 halftones. 37 tables. 335pp. 6½ x 9¼.
42820-6

AMATEUR ASTRONOMER'S HANDBOOK, J. B. Sidgwick. Timeless, comprehensive coverage of telescopes, mirrors, lenses, mountings, telescope drives, micrometers, spectroscopes, more. 189 illustrations. 576pp. 5⅝ x 8¼. (Available in U.S. only.)
24034-7

STARS AND RELATIVITY, Ya. B. Zel'dovich and I. D. Novikov. Vol. 1 of *Relativistic Astrophysics* by famed Russian scientists. General relativity, properties of matter under astrophysical conditions, stars, and stellar systems. Deep physical insights, clear presentation. 1971 edition. References. 544pp. 5⅝ x 8¼. 69424-0

Chemistry

THE SCEPTICAL CHYMIST: The Classic 1661 Text, Robert Boyle. Boyle defines the term "element," asserting that all natural phenomena can be explained by the motion and organization of primary particles. 1911 ed. viii+232pp. 5⅜ x 8½. 42825-7

RADIOACTIVE SUBSTANCES, Marie Curie. Here is the celebrated scientist's doctoral thesis, the prelude to her receipt of the 1903 Nobel Prize. Curie discusses establishing atomic character of radioactivity found in compounds of uranium and thorium; extraction from pitchblende of polonium and radium; isolation of pure radium chloride; determination of atomic weight of radium; plus electric, photographic, luminous, heat, color effects of radioactivity. ii+94pp. 5⅜ x 8½. 42550-9

CHEMICAL MAGIC, Leonard A. Ford. Second Edition, Revised by E. Winston Grundmeier. Over 100 unusual stunts demonstrating cold fire, dust explosions, much more. Text explains scientific principles and stresses safety precautions. 128pp. 5⅜ x 8½. 67628-5

THE DEVELOPMENT OF MODERN CHEMISTRY, Aaron J. Ihde. Authoritative history of chemistry from ancient Greek theory to 20th-century innovation. Covers major chemists and their discoveries. 209 illustrations. 14 tables. Bibliographies. Indices. Appendices. 851pp. 5⅜ x 8½. 64235-6

CATALYSIS IN CHEMISTRY AND ENZYMOLOGY, William P. Jencks. Exceptionally clear coverage of mechanisms for catalysis, forces in aqueous solution, carbonyl- and acyl-group reactions, practical kinetics, more. 864pp. 5⅜ x 8½. 65460-5

ELEMENTS OF CHEMISTRY, Antoine Lavoisier. Monumental classic by founder of modern chemistry in remarkable reprint of rare 1790 Kerr translation. A must for every student of chemistry or the history of science. 539pp. 5⅜ x 8½. 64624-6

THE HISTORICAL BACKGROUND OF CHEMISTRY, Henry M. Leicester. Evolution of ideas, not individual biography. Concentrates on formulation of a coherent set of chemical laws. 260pp. 5⅜ x 8½. 61053-5

A SHORT HISTORY OF CHEMISTRY, J. R. Partington. Classic exposition explores origins of chemistry, alchemy, early medical chemistry, nature of atmosphere, theory of valency, laws and structure of atomic theory, much more. 428pp. 5⅜ x 8½. (Available in U.S. only.) 65977-1

GENERAL CHEMISTRY, Linus Pauling. Revised 3rd edition of classic first-year text by Nobel laureate. Atomic and molecular structure, quantum mechanics, statistical mechanics, thermodynamics correlated with descriptive chemistry. Problems. 992pp. 5⅜ x 8½. 65622-5

FROM ALCHEMY TO CHEMISTRY, John Read. Broad, humanistic treatment focuses on great figures of chemistry and ideas that revolutionized the science. 50 illustrations. 240pp. 5⅜ x 8½. 28690-8

Engineering

DE RE METALLICA, Georgius Agricola. The famous Hoover translation of greatest treatise on technological chemistry, engineering, geology, mining of early modern times (1556). All 289 original woodcuts. 638pp. 6¾ x 11. 60006-8

FUNDAMENTALS OF ASTRODYNAMICS, Roger Bate et al. Modern approach developed by U.S. Air Force Academy. Designed as a first course. Problems, exercises. Numerous illustrations. 455pp. 5⅜ x 8½. 60061-0

DYNAMICS OF FLUIDS IN POROUS MEDIA, Jacob Bear. For advanced students of ground water hydrology, soil mechanics and physics, drainage and irrigation engineering, and more. 335 illustrations. Exercises, with answers. 784pp. 6⅛ x 9¼. 65675-6

THEORY OF VISCOELASTICITY (Second Edition), Richard M. Christensen. Complete, consistent description of the linear theory of the viscoelastic behavior of materials. Problem-solving techniques discussed. 1982 edition. 29 figures. xiv+364pp. 6⅛ x 9¼. 42880-X

MECHANICS, J. P. Den Hartog. A classic introductory text or refresher. Hundreds of applications and design problems illuminate fundamentals of trusses, loaded beams and cables, etc. 334 answered problems. 462pp. 5⅜ x 8½. 60754-2

MECHANICAL VIBRATIONS, J. P. Den Hartog. Classic textbook offers lucid explanations and illustrative models, applying theories of vibrations to a variety of practical industrial engineering problems. Numerous figures. 233 problems, solutions. Appendix. Index. Preface. 436pp. 5⅜ x 8½. 64785-4

STRENGTH OF MATERIALS, J. P. Den Hartog. Full, clear treatment of basic material (tension, torsion, bending, etc.) plus advanced material on engineering methods, applications. 350 answered problems. 323pp. 5⅜ x 8½. 60755-0

A HISTORY OF MECHANICS, René Dugas. Monumental study of mechanical principles from antiquity to quantum mechanics. Contributions of ancient Greeks, Galileo, Leonardo, Kepler, Lagrange, many others. 671pp. 5⅜ x 8½. 65632-2

STABILITY THEORY AND ITS APPLICATIONS TO STRUCTURAL MECHANICS, Clive L. Dym. Self-contained text focuses on Koiter postbuckling analyses, with mathematical notions of stability of motion. Basing minimum energy principles for static stability upon dynamic concepts of stability of motion, it develops asymptotic buckling and postbuckling analyses from potential energy considerations, with applications to columns, plates, and arches. 1974 ed. 208pp. 5⅜ x 8½. 42541-X

METAL FATIGUE, N. E. Frost, K. J. Marsh, and L. P. Pook. Definitive, clearly written, and well-illustrated volume addresses all aspects of the subject, from the historical development of understanding metal fatigue to vital concepts of the cyclic stress that causes a crack to grow. Includes 7 appendixes. 544pp. 5⅜ x 8½. 40927-9

Mathematics

FUNCTIONAL ANALYSIS (Second Corrected Edition), George Bachman and Lawrence Narici. Excellent treatment of subject geared toward students with background in linear algebra, advanced calculus, physics, and engineering. Text covers introduction to inner-product spaces, normed, metric spaces, and topological spaces; complete orthonormal sets, the Hahn-Banach Theorem and its consequences, and many other related subjects. 1966 ed. 544pp. 6⅛ x 9¼. 40251-7

ASYMPTOTIC EXPANSIONS OF INTEGRALS, Norman Bleistein & Richard A. Handelsman. Best introduction to important field with applications in a variety of scientific disciplines. New preface. Problems. Diagrams. Tables. Bibliography. Index. 448pp. 5⅜ x 8½. 65082-0

VECTOR AND TENSOR ANALYSIS WITH APPLICATIONS, A. I. Borisenko and I. E. Tarapov. Concise introduction. Worked-out problems, solutions, exercises. 257pp. 5⅝ x 8¼. 63833-2

THE ABSOLUTE DIFFERENTIAL CALCULUS (CALCULUS OF TENSORS), Tullio Levi-Civita. Great 20th-century mathematician's classic work on material necessary for mathematical grasp of theory of relativity. 452pp. 5⅝ x 8¼. 63401-9

AN INTRODUCTION TO ORDINARY DIFFERENTIAL EQUATIONS, Earl A. Coddington. A thorough and systematic first course in elementary differential equations for undergraduates in mathematics and science, with many exercises and problems (with answers). Index. 304pp. 5⅜ x 8½. 65942-9

FOURIER SERIES AND ORTHOGONAL FUNCTIONS, Harry F. Davis. An incisive text combining theory and practical example to introduce Fourier series, orthogonal functions and applications of the Fourier method to boundary-value problems. 570 exercises. Answers and notes. 416pp. 5⅜ x 8½. 65973-9

COMPUTABILITY AND UNSOLVABILITY, Martin Davis. Classic graduate-level introduction to theory of computability, usually referred to as theory of recurrent functions. New preface and appendix. 288pp. 5⅜ x 8½. 61471-9

ASYMPTOTIC METHODS IN ANALYSIS, N. G. de Bruijn. An inexpensive, comprehensive guide to asymptotic methods–the pioneering work that teaches by explaining worked examples in detail. Index. 224pp. 5⅜ x 8½ 64221-6

APPLIED COMPLEX VARIABLES, John W. Dettman. Step-by-step coverage of fundamentals of analytic function theory–plus lucid exposition of five important applications: Potential Theory; Ordinary Differential Equations; Fourier Transforms; Laplace Transforms; Asymptotic Expansions. 66 figures. Exercises at chapter ends. 512pp. 5⅜ x 8½. 64670-X

INTRODUCTION TO LINEAR ALGEBRA AND DIFFERENTIAL EQUATIONS, John W. Dettman. Excellent text covers complex numbers, determinants, orthonormal bases, Laplace transforms, much more. Exercises with solutions. Undergraduate level. 416pp. 5⅜ x 8½. 65191-6

TENSOR CALCULUS, J.L. Synge and A. Schild. Widely used introductory text covers spaces and tensors, basic operations in Riemannian space, non-Riemannian spaces, etc. 324pp. 5⅝ x 8¼. 63612-7

ORDINARY DIFFERENTIAL EQUATIONS, Morris Tenenbaum and Harry Pollard. Exhaustive survey of ordinary differential equations for undergraduates in mathematics, engineering, science. Thorough analysis of theorems. Diagrams. Bibliography. Index. 818pp. 5⅜ x 8½. 64940-7

INTEGRAL EQUATIONS, F. G. Tricomi. Authoritative, well-written treatment of extremely useful mathematical tool with wide applications. Volterra Equations, Fredholm Equations, much more. Advanced undergraduate to graduate level. Exercises. Bibliography. 238pp. 5⅜ x 8½. 64828-1

FOURIER SERIES, Georgi P. Tolstov. Translated by Richard A. Silverman. A valuable addition to the literature on the subject, moving clearly from subject to subject and theorem to theorem. 107 problems, answers. 336pp. 5⅜ x 8½. 63317-9

INTRODUCTION TO MATHEMATICAL THINKING, Friedrich Waismann. Examinations of arithmetic, geometry, and theory of integers; rational and natural numbers; complete induction; limit and point of accumulation; remarkable curves; complex and hypercomplex numbers, more. 1959 ed. 27 figures. xii+260pp. 5⅜ x 8½. 42804-4

POPULAR LECTURES ON MATHEMATICAL LOGIC, Hao Wang. Noted logician's lucid treatment of historical developments, set theory, model theory, recursion theory and constructivism, proof theory, more. 3 appendixes. Bibliography. 1981 ed. ix+283pp. 5⅜ x 8½. 67632-3

CALCULUS OF VARIATIONS, Robert Weinstock. Basic introduction covering isoperimetric problems, theory of elasticity, quantum mechanics, electrostatics, etc. Exercises throughout. 326pp. 5⅜ x 8½. 63069-2

THE CONTINUUM: A Critical Examination of the Foundation of Analysis, Hermann Weyl. Classic of 20th-century foundational research deals with the conceptual problem posed by the continuum. 156pp. 5⅜ x 8½. 67982-9

CHALLENGING MATHEMATICAL PROBLEMS WITH ELEMENTARY SOLUTIONS, A. M. Yaglom and I. M. Yaglom. Over 170 challenging problems on probability theory, combinatorial analysis, points and lines, topology, convex polygons, many other topics. Solutions. Total of 445pp. 5⅜ x 8½. Two-vol. set.
Vol. I: 65536-9 Vol. II: 65537-7

INTRODUCTION TO PARTIAL DIFFERENTIAL EQUATIONS WITH APPLICATIONS, E. C. Zachmanoglou and Dale W. Thoe. Essentials of partial differential equations applied to common problems in engineering and the physical sciences. Problems and answers. 416pp. 5⅜ x 8½. 65251-3

THE THEORY OF GROUPS, Hans J. Zassenhaus. Well-written graduate-level text acquaints reader with group-theoretic methods and demonstrates their usefulness in mathematics. Axioms, the calculus of complexes, homomorphic mapping, *p*-group theory, more. 276pp. 5⅜ x 8½. 40922-8

Math–Decision Theory, Statistics, Probability

ELEMENTARY DECISION THEORY, Herman Chernoff and Lincoln E. Moses. Clear introduction to statistics and statistical theory covers data processing, probability and random variables, testing hypotheses, much more. Exercises. 364pp. 5⅜ x 8½. 65218-1

STATISTICS MANUAL, Edwin L. Crow et al. Comprehensive, practical collection of classical and modern methods prepared by U.S. Naval Ordnance Test Station. Stress on use. Basics of statistics assumed. 288pp. 5⅜ x 8½. 60599-X

SOME THEORY OF SAMPLING, William Edwards Deming. Analysis of the problems, theory, and design of sampling techniques for social scientists, industrial managers, and others who find statistics important at work. 61 tables. 90 figures. xvii +602pp. 5⅜ x 8½. 64684-X

LINEAR PROGRAMMING AND ECONOMIC ANALYSIS, Robert Dorfman, Paul A. Samuelson and Robert M. Solow. First comprehensive treatment of linear programming in standard economic analysis. Game theory, modern welfare economics, Leontief input-output, more. 525pp. 5⅜ x 8½. 65491-5

PROBABILITY: An Introduction, Samuel Goldberg. Excellent basic text covers set theory, probability theory for finite sample spaces, binomial theorem, much more. 360 problems. Bibliographies. 322pp. 5⅜ x 8½. 65252-1

GAMES AND DECISIONS: Introduction and Critical Survey, R. Duncan Luce and Howard Raiffa. Superb nontechnical introduction to game theory, primarily applied to social sciences. Utility theory, zero-sum games, n-person games, decision-making, much more. Bibliography. 509pp. 5⅜ x 8½. 65943-7

INTRODUCTION TO THE THEORY OF GAMES, J. C. C. McKinsey. This comprehensive overview of the mathematical theory of games illustrates applications to situations involving conflicts of interest, including economic, social, political, and military contexts. Appropriate for advanced undergraduate and graduate courses; advanced calculus a prerequisite. 1952 ed. x+372pp. 5⅜ x 8½. 42811-7

FIFTY CHALLENGING PROBLEMS IN PROBABILITY WITH SOLUTIONS, Frederick Mosteller. Remarkable puzzlers, graded in difficulty, illustrate elementary and advanced aspects of probability. Detailed solutions. 88pp. 5⅜ x 8½. 65355-2

PROBABILITY THEORY: A Concise Course, Y. A. Rozanov. Highly readable, self-contained introduction covers combination of events, dependent events, Bernoulli trials, etc. 148pp. 5⅜ x 8¼. 63544-9

STATISTICAL METHOD FROM THE VIEWPOINT OF QUALITY CONTROL, Walter A. Shewhart. Important text explains regulation of variables, uses of statistical control to achieve quality control in industry, agriculture, other areas. 192pp. 5⅜ x 8½. 65232-7

Math–Geometry and Topology

ELEMENTARY CONCEPTS OF TOPOLOGY, Paul Alexandroff. Elegant, intuitive approach to topology from set-theoretic topology to Betti groups; how concepts of topology are useful in math and physics. 25 figures. 57pp. 5⅜ x 8½. 60747-X

COMBINATORIAL TOPOLOGY, P. S. Alexandrov. Clearly written, well-organized, three-part text begins by dealing with certain classic problems without using the formal techniques of homology theory and advances to the central concept, the Betti groups. Numerous detailed examples. 654pp. 5⅜ x 8½. 40179-0

EXPERIMENTS IN TOPOLOGY, Stephen Barr. Classic, lively explanation of one of the byways of mathematics. Klein bottles, Moebius strips, projective planes, map coloring, problem of the Koenigsberg bridges, much more, described with clarity and wit. 43 figures. 210pp. 5⅜ x 8½. 25933-1

CONFORMAL MAPPING ON RIEMANN SURFACES, Harvey Cohn. Lucid, insightful book presents ideal coverage of subject. 334 exercises make book perfect for self-study. 55 figures. 352pp. 5⅝ x 8¼. 64025-6

THE GEOMETRY OF RENÉ DESCARTES, René Descartes. The great work founded analytical geometry. Original French text, Descartes's own diagrams, together with definitive Smith-Latham translation. 244pp. 5⅜ x 8½. 60068-8

PRACTICAL CONIC SECTIONS: The Geometric Properties of Ellipses, Parabolas and Hyperbolas, J. W. Downs. This text shows how to create ellipses, parabolas, and hyperbolas. It also presents historical background on their ancient origins and describes the reflective properties and roles of curves in design applications. 1993 ed. 98 figures. xii+100pp. 6½ x 9¼. 42876-1

THE THIRTEEN BOOKS OF EUCLID'S ELEMENTS, translated with introduction and commentary by Thomas L. Heath. Definitive edition. Textual and linguistic notes, mathematical analysis. 2,500 years of critical commentary. Unabridged. 1,414pp. 5⅜ x 8½. Three-vol. set. Vol. I: 60088-2 Vol. II: 60089-0 Vol. III: 60090-4

GEOMETRY OF COMPLEX NUMBERS, Hans Schwerdtfeger. Illuminating, widely praised book on analytic geometry of circles, the Moebius transformation, and two-dimensional non-Euclidean geometries. 200pp. 5⅝ x 8¼. 63830-8

DIFFERENTIAL GEOMETRY, Heinrich W. Guggenheimer. Local differential geometry as an application of advanced calculus and linear algebra. Curvature, transformation groups, surfaces, more. Exercises. 62 figures. 378pp. 5⅜ x 8½. 63433-7

CURVATURE AND HOMOLOGY: Enlarged Edition, Samuel I. Goldberg. Revised edition examines topology of differentiable manifolds; curvature, homology of Riemannian manifolds; compact Lie groups; complex manifolds; curvature, homology of Kaehler manifolds. New Preface. Four new appendixes. 416pp. 5⅜ x 8½. 40207-X

History of Math

THE WORKS OF ARCHIMEDES, Archimedes (T. L. Heath, ed.). Topics include the famous problems of the ratio of the areas of a cylinder and an inscribed sphere; the measurement of a circle; the properties of conoids, spheroids, and spirals; and the quadrature of the parabola. Informative introduction. clxxxvi+326pp; supplement, 52pp. 5⅜ x 8½. 42084-1

A SHORT ACCOUNT OF THE HISTORY OF MATHEMATICS, W. W. Rouse Ball. One of clearest, most authoritative surveys from the Egyptians and Phoenicians through 19th-century figures such as Grassman, Galois, Riemann. Fourth edition. 522pp. 5⅜ x 8½. 20630-0

THE HISTORY OF THE CALCULUS AND ITS CONCEPTUAL DEVELOPMENT, Carl B. Boyer. Origins in antiquity, medieval contributions, work of Newton, Leibniz, rigorous formulation. Treatment is verbal. 346pp. 5⅜ x 8½. 60509-4

THE HISTORICAL ROOTS OF ELEMENTARY MATHEMATICS, Lucas N. H. Bunt, Phillip S. Jones, and Jack D. Bedient. Fundamental underpinnings of modern arithmetic, algebra, geometry, and number systems derived from ancient civilizations. 320pp. 5⅜ x 8½. 25563-8

A HISTORY OF MATHEMATICAL NOTATIONS, Florian Cajori. This classic study notes the first appearance of a mathematical symbol and its origin, the competition it encountered, its spread among writers in different countries, its rise to popularity, its eventual decline or ultimate survival. Original 1929 two-volume edition presented here in one volume. xxviii+820pp. 5⅜ x 8½. 67766-4

GAMES, GODS & GAMBLING: A History of Probability and Statistical Ideas, F. N. David. Episodes from the lives of Galileo, Fermat, Pascal, and others illustrate this fascinating account of the roots of mathematics. Features thought-provoking references to classics, archaeology, biography, poetry. 1962 edition. 304pp. 5⅜ x 8½. (Available in U.S. only.) 40023-9

OF MEN AND NUMBERS: The Story of the Great Mathematicians, Jane Muir. Fascinating accounts of the lives and accomplishments of history's greatest mathematical minds–Pythagoras, Descartes, Euler, Pascal, Cantor, many more. Anecdotal, illuminating. 30 diagrams. Bibliography. 256pp. 5⅜ x 8½. 28973-7

HISTORY OF MATHEMATICS, David E. Smith. Nontechnical survey from ancient Greece and Orient to late 19th century; evolution of arithmetic, geometry, trigonometry, calculating devices, algebra, the calculus. 362 illustrations. 1,355pp. 5⅜ x 8½. Two-vol. set. Vol. I: 20429-4 Vol. II: 20430-8

A CONCISE HISTORY OF MATHEMATICS, Dirk J. Struik. The best brief history of mathematics. Stresses origins and covers every major figure from ancient Near East to 19th century. 41 illustrations. 195pp. 5⅜ x 8½. 60255-9

Physics

OPTICAL RESONANCE AND TWO-LEVEL ATOMS, L. Allen and J. H. Eberly. Clear, comprehensive introduction to basic principles behind all quantum optical resonance phenomena. 53 illustrations. Preface. Index. 256pp. 5⅜ x 8½. 65533-4

QUANTUM THEORY, David Bohm. This advanced undergraduate-level text presents the quantum theory in terms of qualitative and imaginative concepts, followed by specific applications worked out in mathematical detail. Preface. Index. 655pp. 5⅜ x 8½. 65969-0

ATOMIC PHYSICS: 8th edition, Max Born. Nobel laureate's lucid treatment of kinetic theory of gases, elementary particles, nuclear atom, wave-corpuscles, atomic structure and spectral lines, much more. Over 40 appendices, bibliography. 495pp. 5⅜ x 8½. 65984-4

A SOPHISTICATE'S PRIMER OF RELATIVITY, P. W. Bridgman. Geared toward readers already acquainted with special relativity, this book transcends the view of theory as a working tool to answer natural questions: What is a frame of reference? What is a "law of nature"? What is the role of the "observer"? Extensive treatment, written in terms accessible to those without a scientific background. 1983 ed. xlviii+172pp. 5⅜ x 8½. 42549-5

AN INTRODUCTION TO HAMILTONIAN OPTICS, H. A. Buchdahl. Detailed account of the Hamiltonian treatment of aberration theory in geometrical optics. Many classes of optical systems defined in terms of the symmetries they possess. Problems with detailed solutions. 1970 edition. xv+360pp. 5⅜ x 8½. 67597-1

PRIMER OF QUANTUM MECHANICS, Marvin Chester. Introductory text examines the classical quantum bead on a track: its state and representations; operator eigenvalues; harmonic oscillator and bound bead in a symmetric force field; and bead in a spherical shell. Other topics include spin, matrices, and the structure of quantum mechanics; the simplest atom; indistinguishable particles; and stationary-state perturbation theory. 1992 ed. xiv+314pp. 6⅛ x 9¼. 42878-8

LECTURES ON QUANTUM MECHANICS, Paul A. M. Dirac. Four concise, brilliant lectures on mathematical methods in quantum mechanics from Nobel Prize–winning quantum pioneer build on idea of visualizing quantum theory through the use of classical mechanics. 96pp. 5⅜ x 8½. 41713-1

THIRTY YEARS THAT SHOOK PHYSICS: The Story of Quantum Theory, George Gamow. Lucid, accessible introduction to influential theory of energy and matter. Careful explanations of Dirac's anti-particles, Bohr's model of the atom, much more. 12 plates. Numerous drawings. 240pp. 5⅜ x 8½. 24895-X

ELECTRONIC STRUCTURE AND THE PROPERTIES OF SOLIDS: The Physics of the Chemical Bond, Walter A. Harrison. Innovative text offers basic understanding of the electronic structure of covalent and ionic solids, simple metals, transition metals and their compounds. Problems. 1980 edition. 582pp. 6⅛ x 9¼. 66021-4

QUANTUM MECHANICS: Principles and Formalism, Roy McWeeny. Graduate student–oriented volume develops subject as fundamental discipline, opening with review of origins of Schrödinger's equations and vector spaces. Focusing on main principles of quantum mechanics and their immediate consequences, it concludes with final generalizations covering alternative "languages" or representations. 1972 ed. 15 figures. xi+155pp. 5⅜ x 8½. 42829-X

INTRODUCTION TO QUANTUM MECHANICS WITH APPLICATIONS TO CHEMISTRY, Linus Pauling & E. Bright Wilson, Jr. Classic undergraduate text by Nobel Prize winner applies quantum mechanics to chemical and physical problems. Numerous tables and figures enhance the text. Chapter bibliographies. Appendices. Index. 468pp. 5⅜ x 8½. 64871-0

METHODS OF THERMODYNAMICS, Howard Reiss. Outstanding text focuses on physical technique of thermodynamics, typical problem areas of understanding, and significance and use of thermodynamic potential. 1965 edition. 238pp. 5⅜ x 8½. 69445-3

TENSOR ANALYSIS FOR PHYSICISTS, J. A. Schouten. Concise exposition of the mathematical basis of tensor analysis, integrated with well-chosen physical examples of the theory. Exercises. Index. Bibliography. 289pp. 5⅜ x 8½. 65582-2

THE ELECTROMAGNETIC FIELD, Albert Shadowitz. Comprehensive undergraduate text covers basics of electric and magnetic fields, builds up to electromagnetic theory. Also related topics, including relativity. Over 900 problems. 768pp. 5⅝ x 8¼. 65660-8

GREAT EXPERIMENTS IN PHYSICS: Firsthand Accounts from Galileo to Einstein, Morris H. Shamos (ed.). 25 crucial discoveries: Newton's laws of motion, Chadwick's study of the neutron, Hertz on electromagnetic waves, more. Original accounts clearly annotated. 370pp. 5⅜ x 8½. 25346-5

RELATIVITY, THERMODYNAMICS AND COSMOLOGY, Richard C. Tolman. Landmark study extends thermodynamics to special, general relativity; also applications of relativistic mechanics, thermodynamics to cosmological models. 501pp. 5⅜ x 8½. 65383-8

STATISTICAL PHYSICS, Gregory H. Wannier. Classic text combines thermodynamics, statistical mechanics, and kinetic theory in one unified presentation of thermal physics. Problems with solutions. Bibliography. 532pp. 5⅜ x 8½. 65401-X